Radiation Injury Prevention and Mitigation in Humans

Radiation Injury Prevention and Mitigation in Humans

Kedar N. Prasad, Ph.D.

CRC Press
Taylor & Francis Group
Boca Raton London New York

CRC Press is an imprint of the
Taylor & Francis Group, an **informa** business

CRC Press
Taylor & Francis Group
6000 Broken Sound Parkway NW, Suite 300
Boca Raton, FL 33487-2742

First issued in paperback 2018

© 2012 by Taylor & Francis Group, LLC
CRC Press is an imprint of Taylor & Francis Group, an Informa business

No claim to original U.S. Government works

ISBN-13: 978-1-4398-7424-0 (hbk)
ISBN-13: 978-1-138-37460-7 (pbk)

Visit the Taylor & Francis Web site at
http://www.taylorandfrancis.com

and the CRC Press Web site at
http://www.crcpress.com

Contents

Preface

There are two types of radiation: ionizing radiation, such as x-rays and gamma rays, and nonionizing radiation, such as ultraviolet radiation and electromagnetic radiation. Although ionizing radiation–induced injuries and their prevention and mitigation have been studied for decades primarily on animal models, recent events related to this form of radiation have alarmed radiation scientists and civilians alike. For example, after the tragic events of 9/11, the threat of an explosion of a radiological weapon (primarily a dirty bomb) by terrorist groups exists in the United States and certain regions of the world. Such an explosion can cause a few casualties, but can create a mass panic, and can contaminate water, air, food, and food sources with radioactive isotopes, some of which can remain active for decades. The possibility of unintentional nuclear conflict remains a possibility as long as nations have nuclear arsenals. Such an event can cause mass casualties as well as contaminate water, air, and food sources with radioactive isotopes, some of which can continue to emit radiation for decades. The survivors of radiation exposure may have increased risk of developing neoplastic and nonneoplastic diseases several years after exposure. In addition, rapid growth of nuclear power plants around the world to meet the demand for energy raises the possibility of a nuclear accident. Indeed, a major nuclear power plant accident occurred at the Chernobyl nuclear power plant in Russia (now Ukraine) in April 1986; and more recently in Fukushima, Japan. Such major accidents can cause a few deaths, but they can expose populations to low levels of radiation that can increase the risk of cancer and heritable mutations that can increase the risk of diseases in future generations. The National Aeronautics and Space Administration (NASA) plans to establish a human base camp on the lunar surface during the next 8–10 years in order to explore the lunar landscape and then voyage to Mars. During lunar exploration, the astronauts may receive high, even lethal, doses of proton radiation from solar flares. With our current technology, warning of a solar flare can be relayed only 30 minutes before its actual occurrence.

In civilians as well as the military, growing use of x-ray-based diagnostic procedures has raised concerns about potential long-term health risks, especially of cancer. It has been estimated that about 5 billion imaging examinations are performed worldwide each year, and 2 out of 3 involve ionizing radiation. In 2006, the estimated medical radiation exposure dose in the United States reached 3.2 mSv, which is more than sixfold higher than that estimated in 2004. In 2006, about 20 million nuclear medicine examinations were performed, and in 2008, over 60 million computed tomography (CT) scans were performed in the United States. These estimates did not include other diagnostic procedures such as chest x-rays, dental x-rays, fluoroscopic imaging, positron emission tomography (PET), and other nuclear medicine scans. Therefore, it is likely that many more patients were exposed to diagnostic doses of radiation than the current estimates. Potential risks of diagnostic doses of radiation include cancer as well as nonneoplastic diseases. In addition, they also include somatic and heritable mutations. The increased risk of thyroid cancer after repeated dental x-rays has been reported, but this study has been disputed by the British Dental Association. Military and civilian pilots and flight attendants are exposed to cosmic ionizing radiation, potential chemical carcinogens (fuel and jet engine exhaust), and electromagnetic fields from cockpit instruments. Several epidemiologic studies have shown an increased risk of cancer in these populations due to exposures to cosmic radiation. At present, there is no biological protection strategy against the adverse health effects of low doses of radiation, including diagnostic radiation procedures. The current use of physical radiation protection is not adequate to reduce the health risks of low doses of radiation.

Because of growing concerns about radiation injury, prevention and mitigation of radiation damage in humans have become an urgent issue for reducing the risk of diseases before and after radiation exposures. At present, there are no strategies to provide biological protection before radiation exposure and inadequate resources to provide biological protection after irradiation. The available physical protection strategies, which include lead shielding of the unexposed area, increased distance between the radiation source and individuals, reduced time of exposure, and recommendations of as low as reasonably achievable (ALARA) can reduce the dose levels only if implemented before radiation exposure. These physical protection strategies are not always applicable before irradiation, and they are totally ineffective after irradiation. Therefore, it is essential to identify nontoxic and cost-effective agents that can reduce radiation damage when administered orally before and/or after irradiation in order to prevent and mitigate acute as well as late adverse health effects of radiation. Such agents would serve as a bioshield against radiation damage. In recent years, several laboratory studies have identified many radioprotective agents (for prevention) and radiation-mitigating agents (for treatment) primarily in animal models. Among various radioprotective agents and radiation mitigating agents, antioxidants appear to act as an ideal bioshield against radiation injury because they satisfy all three criteria of an ideal bioshield. These criteria include extensive laboratory and limited human studies supporting the value of antioxidants as radioprotective and radiation-mitigating agents, and lack of toxicity when administered orally.

The major purpose of this book is to briefly describe the following:

1. Physics of ionizing radiation and radiological weapons, principles of nuclear reactors, and the types of radiological weapons and consequences of their explosions
2. Acute and late health effects of high and low doses of radiation
3. The efficacy of Food and Drug Administration (FDA)-approved and unapproved radioprotective and radiation mitigating agents
4. The efficacy of radioprotective and radiation mitigating agents not requiring FDA) approval (antioxidants and herbs)
5. Scientific data and rationale in support of using micronutrient preparations containing dietary and endogenous antioxidants for prevention of acute radiation sickness (ARS) and in combination with standard therapy for mitigation of ARS
6. Scientific data and rationale in support of using a micronutrient preparation containing dietary and endogenous antioxidants for prevention and mitigation of late adverse health effects among survivors of high and low doses of radiation

This book also discusses implementation plans for physical and biological protection strategies for the first responders, radiation workers, astronauts, and civilians who might be exposed to high or low doses of radiation.

Exposure to nonionizing radiation, such as ultraviolet (UV) light and electromagnetic radiation emitted from cell phone use has raised health concerns after prolonged exposure to these forms of nonionizing radiation. This book discusses recent advances in research on the effects of these types of nonionizing radiation on cellular and genetic levels, and proposes an implementation plan of physical and biological protection strategies for those individuals who are exposed to UV light and electromagnetic radiation in excessive amounts, as well as those planning to be exposed.

This book contains up-to-date references for each topic.

About the Author

Dr. Kedar N. Prasad obtained a master's degree in zoology from the University of Bihar, Ranchi, India, and a PhD in radiation biology from the University of Iowa, Iowa City, in 1963. He received postdoctoral training at the Brookhaven National Laboratory, Long Island, New York, and joined the Department of Radiology at the University of Colorado Health Sciences Center, where he became a professor and director for the Center for Vitamins and Cancer Research. He has published over 200 articles in peer-reviewed journals, and authored and edited 15 books in the areas of radiation biology, nutrition and cancer, and nutrition and neurological diseases, particularly Alzheimer's disease and Parkinson's disease. These articles were published in highly prestigious journals such as *Science*, *Nature*, and *Proceedings of the National Academy of Sciences* in the United States. Dr. Prasad has received many honors, including an invitation by the Nobel Prize Committee to nominate a candidate for the Nobel Prize in Medicine for 1982; the 1999 Harold Harper Lecture at the meeting of the American College of Advancement in Medicine; and an award for the best review of 1998–1999 on antioxidants and cancer and 1999–2000 on antioxidants and Parkinson's disease by the American College of Nutrition. He is a Fellow of the American College of Nutrition and served as president of the International Society of Nutrition and Cancer, 1992–2000. He belongs to several professional societies, such as the American Association for Cancer Research and the Radiation Research Society. Currently, he is chief scientific officer of the Premier Micronutrient Corporation.

1 Growing Health Concerns with Respect to Low Doses of Ionizing Radiation
Can We Prevent and/or Mitigate Them?

INTRODUCTION

All living organisms and plants, including humans, have been exposed to ionizing and nonionizing radiation during the entire period of evolutionary processes. Ionizing radiation is known to induce mutations, some of which may have been harmful, and eliminated them, while others may have been beneficial to the organisms that were participating in the struggle for survival during the period of evolution when the Earth's atmosphere was rapidly changing from the anaerobic to aerobic condition. It is likely that radiation-induced mutations may be one of the factors that could have accelerated the rate of evolution of the species. At present, humans are exposed daily to background radiation. Individuals living at higher altitudes receive higher doses of background radiation than those residing at sea level. It is generally assumed that the background radiation is totally safe; however, this assumption may be incorrect. It is more likely that the background radiation may be causing mutations that do not appear during your lifetime, or that it may be causing mutations in inactive parts of the genome without having significant impact on health, or it may be contributing to chronic diseases such as cancer. In addition to ionizing radiation, the atmosphere has varying levels of toxic chemicals, depending upon the region, which can also induce mutations similar to those produced by ionizing radiation. The combination of the two can markedly increase the risk of diseases. At this time, we do not know to what extent the prevalence of human diseases are associated with mutations induced by background radiation alone. Most human diseases should be considered as a result of the interactions among background radiation, environmental chemicals, and dietary and lifestyle-related factors.

Since the discovery of radioactivity, nuclear fission, and nuclear fusion reactions, it has been possible to build nuclear weapons, such as atom bombs and hydrogen

1

bombs, as well as nuclear power plants to generate energy. Since the terrorist attack of 9/11, the threat of an explosion of a radiological bomb by the terrorists exists in the United States and certain other regions of the world. The possibility of unintentional nuclear conflicts remains as long as nations have nuclear arsenals. These events, if they occur, can cause mass casualties as well as vast devastation of infrastructures, and radioactive contamination of water, soil, and growing fruits and vegetables over large areas from the point of explosion. In contrast to the explosion of a nuclear bomb, an explosion of a dirty bomb that is made of radioactive materials may cause blast damage but not mass casualties. The radioactive dust may contaminate soil, water, and fruits and vegetables. The survivors of radiation damage may have increased risk of diseases, including cancers, and they can transmit the radiation-induced genetic defects to future generations. These mutations may remain recessive or may appear as a dominant trait in the first generation after exposure.

Despite the recent nuclear power plant accident in Fukushima, Japan, rapid growth of nuclear power plants around the world to meet the demand for energy could occur. This would increase the possibility of a nuclear accident due to a human error or natural disaster, such as an earthquake or tsunami. Nuclear power plant accidents have occurred many times in the past, in the United States and abroad. The most serious nuclear accident occurred at the Chernobyl nuclear plant in Russia (now Ukraine) in April 1986. In the United States, a well-publicized nuclear accident occurred at the Three Mile Island nuclear plant near Middletown, Pennsylvania, on March 28, 1979. More recently, a serious nuclear power plant accident occurred in Fukushima, Japan, because of a serious earthquake followed by a tsunami in the area. The nuclear accident can expose workers at the plant to high doses of gamma rays, which can cause lethality in some. In addition, radioactive materials that are released into the atmosphere can travel farther from the site of the accident and contaminate water, soil, and fruits and vegetables, and can increase the health risk of individuals residing in these areas for a long period of time. Such an accident may increase the levels of background radiation in the affected region.

The National Aeronautics and Space Administration (NASA) plans to establish a human base camp at the lunar surface during next 8–10 years in order to explore the lunar landscape and then voyage to Mars. During lunar exploration, the astronauts may receive high, even lethal doses of proton radiation from solar flares. With our current technology, the warning of solar flares can be relayed only 30 minutes before its actual occurrence. If the astronauts are exploring the lunar surface away from the shielded base camp, they may not have sufficient time to return to the base camp where an appropriate shielding against radiation exposure can be available. Consequently, they may be exposed to high doses of proton radiation, which is more damaging than x-rays or gamma rays.

In the civilian sector as well as in the military, growing use of x-ray-based diagnostic procedures have raised concerns about potential long-term health risks, especially cancer. It has been estimated that about 5 billion imaging examinations are performed worldwide each year, and 2 out of 3 involve ionizing radiation. In 2006, the estimated medical radiation exposure dose in the United States reached 3.2 mSv, which is more than sixfold higher than that estimated in 2004. In 2006, about 20 million nuclear medicine examinations were performed, and in 2008, over

60 million computed tomography (CT) scans were performed in the United States. These estimates did not include other diagnostic procedures such as chest x-rays, dental x-rays, fluoroscopic imaging, positron emission tomography (PET), and other nuclear medicine scans. Therefore, it is likely that many more patients were exposed to diagnostic doses of radiation than current estimates. Potential risks of diagnostic doses of radiation include cancer as well as nonneoplastic diseases. In addition, they include somatic and heritable mutations. The increased risk of thyroid cancer after repeated dental x-rays has been reported, but this study has been disputed by the British Dental Association.

Military and civilian pilots and flight attendants are exposed to cosmic ionizing radiation, potential chemical carcinogens (fuel, jet engine exhaust), and electromagnetic fields from cockpit instruments. Several epidemiologic studies have shown increased risk of cancer in these populations due to exposures to cosmic radiation during long and frequent flights. In addition, frequent flyers are exposed to higher doses of radiation than those who fly infrequently. More recently, x-ray based whole-body scanners have been placed at the security gates of most airports in the United States. Pilots and the public who fly frequently have raised health concerns over the long-term consequences of radiation exposure from these scanners.

Exposure to nonionizing radiation, such as UV light and electromagnetic radiation emitted during cell phone use, has raised health concerns after prolonged exposure to these forms of nonionizing radiation. The increased risk of skin cancer, especially melanoma, occurs after exposure to UV radiation, and increased risk of brain tumors has been linked with the excessive use of cell phones over a long period of time.

PREVENTION OF RADIATION INJURIES

Prevention generally refers to devices, procedures, or chemicals that can reduce radiation damage when administered before irradiation. This can be accomplished by reducing the radiation dose or tissue damage. At present, prevention of radiation damage is primarily based on three physical principles: increasing the distance from the radiation source, reducing the exposure time, and shielding. In addition, adoption of the principle of as low as reasonably achievable (ALARA) has been recommended in order to reduce radiation doses. Adopting these principles, no doubt, will reduce radiation doses to individuals; however, they are not applicable under all conditions of radiation exposure. In addition, these physical strategies of prevention of radiation injury do not protect tissue damage during irradiation. In order to develop a biological strategy for protecting against tissue damage during irradiation, it is essential to identify a few critical biochemical events that initiate and promote radiation damage. It is now established that excessive amounts of free radicals generated during irradiation initiate radiation damage, whereas long-lived free radicals and products of acute inflammation, such as reactive oxygen species, pro-inflammatory cytokines, prostaglandins, adhesion molecules, and complement proteins, contribute to the progression of damage after irradiation. Since the atomic bombing of Hiroshima and Nagasaki during the World War II, many agents that can neutralize free radicals or reduce inflammation have been shown to reduce damage when administered before irradiation. These agents are called *radioprotective agents*. However, most agents at radioprotective doses were found to be toxic in humans. A group of dietary and endogenous antioxidants, which are nontoxic in humans, protects against radiation injury by neutralizing free radicals and reducing inflammation. Although antioxidants are readily available, they are not being given serious consideration for use in reducing radiation damage in humans at this time. This book will provide scientific rationale and data with a recommendation of adding antioxidants for tissue protection strategy that should complement the physical protection strategy for dose reduction in order to provide a maximal protection against radiation injury.

MITIGATION OF RADIATION INJURIES

Mitigation generally refers to chemicals (synthetic or natural) that can reduce the progression of radiation damage, allowing increased survival rate or survival time when administered after irradiation. At present, replacement therapy, which includes administration of antibiotics, blood, and electrolyte infusions when indicated, can be used to mitigate some of the symptoms of acute radiation sickness (ARS), including mortality to a certain degree. In addition, certain growth factors and bone marrow transplants have been used to reduce the symptoms of ARS. These mitigation strategies are not adequate. In addition, these agents cannot be used on a long-term basis to reduce the risk of late adverse health effects of ionizing radiation. During the last five years, some promising radiation mitigating agents have been identified primarily on rodent models; however, their efficacy and safety in humans remain unknown. In the past, most radioprotective or radiomitigating agents that showed great promise on rodent models were found to be toxic to humans at radioprotective doses. This book will provide scientific rationale and data with a recommendation of adding antioxidants to the mitigation strategy used that would complement the existing radiation mitigating agents, such as replacement therapy, in order to increase the survival rate maximally in lethally irradiated humans.

THE CONCEPT OF A BIOSHIELD AGAINST RADIATION DAMAGE

In contrast to physical protection, which may include lead shielding, the *bioshield* concept refers to chemicals or procedures that can provide tissue protection against low as well as high doses of ionizing radiation when administered before and/or after irradiation. The bioshield includes both radioprotective and radiation mitigating agents. At present, there is no effective bioshield against radiation injury in humans. Ideally, an effective bioshield may include agents that can reduce radiation damage when administered before and/or after irradiation, and that can be administered orally for a long period of time without any toxicity in order to reduce the late adverse health effects of radiation. Among various agents that have been identified, a radioprotective or radiation mitigating agent group of dietary and endogenous antioxidants may be considered as an ideal bioshield. This book will provide scientific rationale and data with a recommendation of adopting an effective bioshield that could enhance the present strategies of prevention and mitigation of radiation injury after irradiation with low as well as high doses of radiation.

SUMMARY

The threat of an explosion of a radiological bomb by terrorists exists. The possibility of unintentional nuclear conflicts remains as long as nations have nuclear arsenals. If these events occur, they can cause mass casualties as well as vast devastation of infrastructures and radioactive contamination of water, soil, and fruits and vegetables. The explosion of a dirty bomb may cause blast damage, but not mass casualties. The radioactive dust from the dirty bomb may contaminate soil, water, and fruits and vegetables. Rapid growth of nuclear power plants around the world to meet the demand for energy is occurring. This increases the possibility of nuclear accidents that have occurred many times in the past, both here and abroad. The nuclear accident can expose workers at the plant to high doses of gamma rays, which can be lethal. In addition, radioactive materials can contaminate water, soil, and fruits and vegetables over large areas. During lunar exploration, the astronauts may receive high, even lethal doses of proton radiation from solar flares.

In the civilian sector as well as in the military, growing use of x-ray-based diagnostic procedures has raised concerns about potential long-term health risks, especially cancer. Military and civilian pilots, flight attendants, and frequent flyers are exposed to cosmic ionizing radiation, potential chemical carcinogens (fuel and jet engine exhausts), and electromagnetic fields from cockpit instruments. More recently, x-ray-based whole-body scanners have been placed at the security gates of most airports in the United States. Health concerns are being raised for the long-term consequences of radiation exposure from these scanners.

Exposure to nonionizing radiation, such as UV light and electromagnetic radiation emitted during cell phone use, has raised health concerns after prolonged exposure to these forms of nonionizing radiation.

At present, there are no effective tissue radiation protection (prevention or mitigation) strategies except for the physical protection strategies that reduce the level of radiation dose. Although several radioprotective and radiation mitigating agents have been identified in rodents, only a group of dietary and endogenous antioxidants are effective and safe for use in humans. Therefore, they can be considered as an ideal bioshield for providing tissue protection against low as well as high doses of radiation. The efforts toward developing a bioshield that can include both radiation preventing and mitigating agents are in progress. Since free radicals and inflammation contribute to the initiation and progression of radiation injury, agents that can attenuate these two biological processes would be good candidates for a bioshield. This book will provide scientific rationale and evidence for the use of such agents, which can reduce acute as well as late radiation injuries.

2 Physics of Radiological Weapons and Nuclear Reactors

INTRODUCTION

Radiological weapons include nuclear bombs and radiological dispersion devices or dirty bombs. In order to understand the differences between nuclear bombs and dirty bombs, it is necessary to understand the fundamental principles of atomic and nuclear physics. Therefore, this chapter briefly describes only those concepts of physics that are essential for an understanding of this difference. These principles are discussed in a broad and conceptual form that can be understood by radiation biologists and medical professionals. Some important references on the subject are listed at the end of this chapter.[1–17]

STRUCTURE OF AN ATOM

In 1913, Dr. Neils Bohr proposed the structure of an atom. According to his concept, an atom consists of a central mass called the *nucleus*, which is surrounded by the orbital *electrons*, which are negatively charged (Figure 2.1). The nucleus of the atom is composed of two particles, a *proton*, which has a positive charge, and *neutron*, which has no charge. Later, Bohr's concept of the atom, with sharply defined electron orbits, was replaced by the electron cloud theory, which treats electrons as a kind of three-dimensional cloud spread around the nucleus as electron waves. However, in order to conceptually visualize the structure of the atom, Bohr's concept has been used in Figure 2.1.

The atom has a diameter of about 10^{-8} cm, whereas the nucleus has a diameter of about 10^{-12} cm. In the stable state, also called the *neutral state*, the negative charges of electrons are balanced against the positive charges of the nucleus. Once this balance in atomic charge is disrupted, either by the loss or gain of electric charge, the atom becomes ionized.

Electron: Electrons orbit the atomic nucleus in a precisely defined path, each path being characterized by its unique energy level. Electrons are positioned in shells or energy levels that surround the nucleus. The innermost shell (identified as the *K shell*) holds only 2 electrons, but the *L shells* may contain up to 8 electrons, and the third or *M shell* may hold up to 18 electrons. The outermost electron shell of an atom never contains more than 8 electrons, which are referred to as *valence electrons*, which

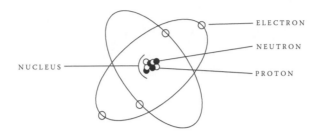

FIGURE 2.1   Diagrammatic structure of an atom.[1]

determine, to a large degree, the chemical properties of an atom. An atom with an outer shell filled with electrons seldom reacts chemically. These atoms constitute elements known as the *inert gases*, such as helium, neon, argon, krypton, and xenon.

Electrons are held in their shell by a combination of *centrifugal force* (which tends to propel them away from the nucleus, and *electrostatic force* (which binds them with the nucleus). Since electrons are bound to the nucleus by electrostatic force, a certain amount of energy is required to move an electron from its shell to another shell or completely outside the atom. The energy required to completely remove an electron from an atom is called the *binding energy* (*EB*). Binding energies are negative because they represent amounts of energy that must be supplied to remove electrons from atoms. Electron shells often are described in terms of the binding energy of the electrons occupying the shells. Electrons lose or gain energy only when they jump from one orbit to another. No change in energy occurs as long as the electrons remain in a specified orbit. Vacancies or holes exist in electron shells from which electrons have been removed. The vacancies are filled promptly by electrons moving farther from the nucleus. As the vacancies are filled, energy in the form of electromagnetic radiation is released. During the transition of a particular electron, the energy released equals the difference in binding energy between the original and the final energy level for the electron. In most cases, energy is released as photons. Occasionally, the energy may be used to eject a second electron, usually from the same shell as the cascading electrons. The ejected electron is termed an *Auger electron*. Electromagnetic radiation released during electron transition is termed *characteristic radiation*, because the photon energies are characteristic of differences in the binding energy of electrons in a specified atom.

Nucleus: The nucleus of an atom is composed of two particles, protons and neutrons, referred to collectively as *nucleons*. Each proton has a positive charge of 1.6×10^{-19} coulomb (c), equal in magnitude but opposite in sign to the charge of an electron. The mass of a proton is 1.6724×10^{-27} Kg. The neutron has no charge but has the same mass as a proton. The number of protons in the nucleus is the atomic number of the atom, whereas the total number of protons and neutrons in the nucleus together constitute the *mass number*. Isotopes of an element can have the same atomic number but a different mass number. Nuclei with an even number of protons and neutrons are generally stable compared to nuclei with an odd number of protons or neutrons. The nuclei with a low atomic number are stable because they contain equal numbers of protons and neutrons. As the atomic number of nuclei increases, the number of neutrons increases more rapidly than the number of protons. If the ratio of neutrons to protons is much greater than one, nuclei may become unstable. The unstable nuclei emit radiation of different energy.

Electrostatic repulsive forces exist between particles of similar charge. Because the distance between protons is less than the diameter of the nucleus, the protons remain together due to the existence of a nuclear force. This force is stronger than the electrostatic repulsive force and binds neutrons and protons together with their nucleus. The mass of a nucleus is less than the sum of masses of nucleons (proton and neutrons) in the nucleus. The mass difference is called the *mass defect* and represents an amount of energy that must be supplied to separate the nucleus into individual nucleons. This amount of energy is the binding energy of the nucleus. The relationship between mass and energy is described by Einstein's formula: $E = mc^2$ (E represents the amount of energy equivalent to a mass [m in kilograms], and c represents the speed of light in vacuum [3×10^8 meters/second]).

ISOTOPES

Atoms that have the same atomic number but different mass number are known as *isotopes*. The isotope could be stable or could be radioactive and emit gamma radiation. The terms *isotope* and *nuclide* are not the same. The term *isotope* does not always refer to a distinct species of atom, whereas the term *nuclide* refers to a species of atom characterized by its nuclear properties, protons, neutrons, and the energy contents. Thus, different isotopes of an element are composed of nuclides having the same atomic number but different mass number.

RADIOACTIVITY

Atoms with unstable nuclei are radioactive. Elements with an atomic number above 83 are naturally occurring radioactive isotopes; however, this property can be induced in other elements. In order to reach a more stable state, a radioactive element may undergo decay releasing radiation in the form of particle radiation, such as alpha particles and beta particles, or electromagnetic radiation, such as gamma radiation and x-radiation. Gamma-radiation is emitted directly from the nucleus, whereas x-ray radiation is produced by radiative and collisional interactions of electrons outside the nucleus. Certain naturally occurring radioactive isotopes, such as uranium-235 (^{235}U), undergo fission after absorbing a neutron.

Alpha particles: In 1902, Dr. Ernest Rutherford at the University of Cambridge in the United Kingdom discovered alpha particles that were initially referred to by him as alpha rays emitting from ^{235}U. Certain radioactive elements, such as radium (^{226}Ra) decays to a stable state by emitting alpha particles at a high speed. An alpha particle has two protons and two neutrons; however, when it annexes two electrons, it becomes stable helium. During nuclear disintegration or decay, an alpha particle can travel at speeds of 9,000 to 20,000 miles per second, but it slows down rapidly as it passes through matter, and eventually becomes stable helium.

$$_{88}Ra\text{-}226 \rightarrow {}_{86}Rn\text{-}222 + {}_2H\text{-}4 + \text{gamma radiation}$$

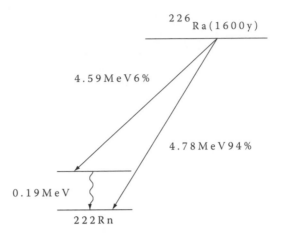

FIGURE 2.2   Decay scheme of Ra-226.[1]

The total energy during the radioactive decay of a nucleus is termed the *transition energy*. During decay, energy is released as kinetic energy of alpha particles and as gamma radiation with their specific energy. The decay scheme of ^{226}Ra is illustrated in Figure 2.2.

During decay of 94% of ^{226}Ra, alpha particles with energy of 4.78 MeV are released, and during the remaining 6% of transition, alpha particles with energy of 4.59 MeV are accompanied by gamma rays of energy of 0.19 MeV. The transition energy is the same for all transitions of ^{226}Ra (4.59 + 0.19 = 4.78 MeV).

Beta-particles: In 1896, Dr. Henri Becquerel discovered the emission of energetic electrons from uranium salts. A beta particle in the form of a negatively charged electron (*negatron*) or positively charged electron (*positron*) is ejected from the nucleus, probably resulting from spontaneous conversion of a neutron into a proton. The ratio of neutrons to protons (n/p) in a negatron-emitting nuclide is greater than that required for maximum stability of the nucleus. Negatron decay results in an increase in atomic number by one and a constant mass number. During this transition, one neutron is converted into one proton. In addition, neutrinos and gamma rays are released. *Neutrinos* are uncharged particles with undetectable mass. There are two forms of neutrinos. One form describes particles ejected during positron decay. The second form is called an *antineutrino*, which is released during negatron decay. Negatron transition is illustrated here.

$$_{55}\text{Cs-137} \rightarrow {}_{56}\text{Ba-137} + {}_{-1}\beta\text{-0} + \text{neutrino} + \text{gamma-rays}$$

During the decay of a particular nucleus, the transition energy already exceeds the sum of the energy as gamma radiation and the kinetic energy of the ejected electrons. The energy unaccounted for during each transition is possessed by the second particles, termed the neutrinos.

$$\text{Energy of neutrino} = \text{E max} - \text{kinetic energy of electron}$$

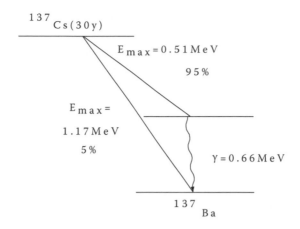

FIGURE 2.3 Decay scheme of Cs-137.

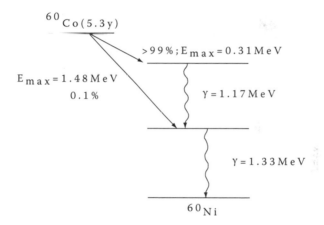

FIGURE 2.4 Decay scheme of Co-60.[1]

Decay schemes of ^{137}Cs and ^{60}Co are presented in Figures 2.3 and 2.4.

Positron-emitting radioisotopes are generated by a cyclotron. They include ^{11}C, ^{15}O, and ^{18}F. Positron-emitting nuclides possess n/p ratios lower than those required for maximum stability. Positron decay is accompanied by a decrease in the atomic number by one and a constant mass number. One proton is transformed into one neutron. A representative positron transition is illustrated here.

$$_{15}P\text{-}30 \rightarrow {}_{14}Si\text{-}30 + {}_{+1}\beta + \text{neutrino}$$

The positron is unstable. On a close approach to a negatron, each will annihilate the other, and their mass appears as electromagnetic radiation —usually two 0.5 MeV photons moving in opposite directions. The nuclei that cannot furnish at least 1.02 MeV for transition do not decay by positron emission. These nuclei increase their n/p ratio by *electron capture*. Most electrons are captured from the K shell, although electrons may be captured from other shells. This creates a hole in the K shell. This vacancy is filled by an electron cascading from an energy level farther from the nucleus. Energy released during this transition appears as x-ray radiation or as the kinetic energy of an Auger electron. During electron capture, one proton captures an electron and becomes a neutron. Nuclei with transition energy greater than 1.02 MeV may decay both by positron emission and electron capture.

Unit of radioactivity: The unit of radioactivity is the Curie (Ci), named after Dr. Marie Curie, who discovered radioactive material.

$$1 \text{ Ci} = 3.7 \times 10^{10} \text{ disintegrations per second (dps)}$$

$$1 \text{ mCi} = 3.7 \times 10^{7} \text{ dps}$$

$$1 \text{ } \mu\text{Ci} = 3.7 \times 10^{4} \text{ dps}$$

Physical half-lives of the radioactive isotopes: The physical half-life is defined as the time required to decay half of its original activity. The half-lives of radioactive isotopes found in nature or produced by cyclotron markedly differ, varying from a few milliseconds to a few billion years. Some examples are provided here.

^{238}U: 4.5 billion years
^{239}Pu: 24,000 years
^{14}C: 5,730 years
^{131}I: 8 days
^{3}H: 12 years
^{11}C: 20.4 minutes
^{13}N: 9.7 minutes
^{15}O: 2.04 minutes

Biological half-lives of radioactive isotopes: The biological half-life is defined as the time required to reduce the body or organ radioactivity to half of its original activity. The biological half-life varies from one nuclide to another, depending upon the turnover of the nuclide in the organ. The physical half-life of ^{45}Ca is 163 days, but its biological half-life in the bone is 18,000 days. The physical half-life of ^{3}H is 12 years; however, the biological half-life depends upon the form of ^{3}H-labeled compounds. Average biological half-life of ^{3}H is 10–12 days, which can be reduced by a factor of 2–3 by ingesting excessive amounts of water. However, if ^{3}H is ingested as a ^{3}H-thymidine (a precursor of DNA), the biological half-life of ^{3}H in the cells could be months.

INTERACTIONS OF RADIATION WITH MATTER

Particulate radiation: An interaction is considered elastic if the sum of the kinetic energy of the interacting entities is unchanged by the interaction. If the sum of the kinetic energy is changed, then the interaction is considered inelastic. Particle radiation, such as protons, deuterons, alpha particles, and other heavy particles lose most of their energy when they interact inelastically with electrons of the absorbing materials. Part of the energy lost by incident particles is used to raise electrons in the absorber to an energy level farther from the nucleus. This process is called *excitation*. Sometimes electrons are ejected from their atoms—a process known as *ionization*. Electrons ejected from atoms by incident radiation are referred to as *primary electrons*. Some primary electrons have enough kinetic energy to produce an additional *ion pair* as they migrate from their site of release. Electrons ejected during the interaction of primary electrons are termed *secondary electrons*. An ejected electron and the residual positive ion constitute an ion pair.

The *specific ionization* is the number of primary and secondary ion pairs produced per unit length of path of the incident radiation. The linear energy transfer (LET) is the average loss in energy per unit length of path of the incident energy and is expressed as KeV/μm. LET depends upon the mass, charge, and velocity of the particles. A particle with greater mass and charge, but with lower velocity, will have a higher LET. Heavy particles, such as alpha particles and protons, are considered as high LET radiation, whereas x-rays and gamma rays are considered as low LET radiation.

Alpha particles: The specific ionization and LET are not constant along the entire path of monoenergetic charged particles traversing a homogeneous medium. As the particles, such as alpha particles, slow down, the specific ionization increases because nearby atoms are influenced for a longer period of time. The region of the increased specific ionization is called the *Bragg peak*. Finally, the particles capture two electrons, become neutral, and can no longer ionize; therefore, the rate of ionization falls abruptly.

The range of particle radiation depends upon the energy, mass, charge, and atomic number of the medium through which the particles are passing. For the same energy, alpha particles have less range than deuterons, protons, or electrons. Protons and neutrons have similar mass; however, for the same energy, neutrons have greater range than protons because they have no charge.

Neutrons: In 1932, Dr. James Chadwick discovered the neutron. The availability of the neutron made possible the production of radioactive isotopes of biological and medical interest. Slow (0 to 0.1 KeV), intermediate (0.1 to 20 KeV), fast (20 to 10 MeV), and high-energy neutrons are produced by the nuclear reactor. Neutrons with various kinetic energy are also emitted by ^{252}Cf, which fissions spontaneously. Neutrons lose energy either by elastic or inelastic collision with the nuclei of absorbing materials. The probability of elastic collision is greatest with nuclei of similar mass. Therefore, in biological tissue during elastic collision, neutrons transfer most of their energy to hydrogen nuclei, protons, which in turn cause ionization like any other heavy charged particles. Certain elements, such as lithium, boron, cadmium, and uranium, exhibit a high cross section for the capture of slow neutrons. Energetic positively charged particles may be ejected by certain nuclei, such as boron ($_5$B-10) after capturing a neutron. For example:

$$_5\text{B-10} + \text{neutron} \rightarrow {_3}\text{Li-7} + \text{alpha particles}$$

For neutrons with kinetic energy above 10 MeV, inelastic collision also contributes to the energy lost in the tissues. For example, inelastic interactions account for about 30% of the energy deposited in tissues by 14.1 MeV neutrons. Most inelastic collisions occur with nuclei other than hydrogen. Energetic charged particles (e.g., protons or alpha particles) often are ejected from nuclei excited by inelastic interactions.

Neutron capture: Certain elements (e.g., boron, lithium, and uranium) exhibit a high cross section for the capture of slow neutrons. During this process, positively charged particles may be ejected by certain nuclei. For example:

$$^{10}B + neutron \rightarrow {}^7Li + alpha\ particles$$

Other nuclei, such as ^{235}U and ^{239}Pu, undergo fission spontaneously after absorbing a neutron.

Electrons: Interaction of negative and positive electrons may be divided into three categories:

1. Scattering electrons
2. Elastic scattering by nuclei
3. Inelastic scattering by nuclei

Negative and positive electrons traversing an absorbing medium transfer energy to electrons of the medium. Incident electrons lose energy and are deflected at some angle with respect to their original direction. An electron receiving energy may be raised to an electron shell farther from the nucleus or may be ejected from the atom. The kinetic energy of an ejected electron equals the energy received minus the binding energy of the electron. The probability of electron–electron scattering increases with the atomic number of the absorber and decreases rapidly with increasing kinetic energy of the incident electrons. Incident electrons are deflected with reduced energy during elastic interaction with nuclei of an absorbing medium. The probability of elastic scattering varies with the square of the atomic number of the absorber and approximately with $1/Ek^2$, where Ek represents the kinetic energy of the incident electrons. The probability for elastic scattering by nuclei is slightly less for positrons than for negatrons with the same kinetic energy. Backscattering of negatrons and positrons in a radioactive sample is due primarily to elastic scattering by nuclei. A negative or positive electron passing near a nucleus may be deflected with reduced velocity. The interaction is inelastic if energy is released as electromagnetic radiation during the encounter. A sudden deceleration or braking of electrons gives rise to *bremsstrahlung* (the German term for braking radiation). This type of radiation was later termed *x-ray*. The probability of x-ray production varies with the square of the atomic number of the absorbing medium.

X-rays and gamma rays: X-rays or gamma rays are identical except for their origin and method of production. When a negatively charged electron approaches the positively charged nucleus, it may be deflected from its original direction by the attractive forces of the nucleus. The change in direction causes deceleration of the electrons or a loss of some of its kinetic energy. The energy lost by the electron is emitted as an x-ray photon. The radiation produced by this type of interaction is called bremsstrahlung. The energy of the x-ray depends upon the original kinetic energy of the electron, how close the electron comes to the nucleus, and the charge of the nucleus. It appears that in this type of interaction, only a portion of the kinetic energy of the electron is lost; therefore, it may have one or more similar interactions with other atoms before expending all of its energy. This results in the production of x-rays with a wide range of energies. On the other hand, gamma rays are emitted during decay of radioactive nuclei. X-ray and gamma ray photons interact with materials in a similar manner. They are attenuated (absorbed or scattered) as they traverse the medium. Of the various alternating processes, *photoelectric* and *Compton interactions* are the most important for radiation-induced damage in biological tissues. Interactions of less importance from a radiobiology point of view include coherent scattering, pair production, and photodisintegration. The rate at which x-ray or gamma ray photons are attenuated is a function of photon energy as well as the physical property of the absorbing materials.

The photoelectric effect: When an x-ray or gamma ray photon collides with an atom, it may transfer all of its energy to an orbital electron, which is ejected out with kinetic energy. The process of energy absorption is called the *photoelectric effect*, and the ejected electrons are referred to as *photoelectrons*. The kinetic energy of the ejected electrons equals the energy of the incident photon minus the binding energy of the electron. Thus, the photoelectric effect involves bound electrons whose ejection probability is highest if the photon has just enough energy to knock the electron from its shell. The photoelectric absorption is dominant up to photon energy of 50 KeV. A diagrammatic presentation of the photoelectric effect is shown in Figure 2.5.

The Compton effect: When an x-ray or gamma ray photon collides with an electron, the electron is ejected with a kinetic energy and a scattered photon with a reduced energy. If the released photon has enough energy, it may repeat the above process in another atom. This process is called the *Compton effect*. The Compton effect is independent of atomic number and decreases with increasing energy. It is a dominant form of interaction with photon energy between 200 KeV and 2 MeV. A diagrammatic presentation of the Compton effect is shown in Figure 2.6.

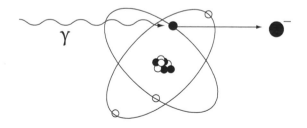

Energy of photoelectron =

energy of photon −

binding energy (BE) of electron

FIGURE 2.5   Photoelectric effect of ionizing radiation.[1]

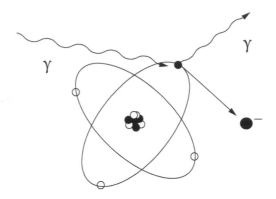

Energy of recoil electron =

energy of photon −

(B.E. of electron + energy
of scattered photon)

FIGURE 2.6   Compton effect of ionizing radiation.[1]

Pair production: In this process, an x-ray or gamma ray photon passes near the nucleus and is subjected to the strong field of the nucleus. During this event, the photon suddenly disappears and becomes a positive and negative electron pair. This is a good example of conversion of energy into mass. The process of pair production must be considered as a collision between the photon and the nucleus; in this collision, the nucleus recoils with some momentum. It also has a little energy, but the energy involved is too small in comparison with energies given to the positron and electron, and therefore can be neglected. The energy equivalent of one electron mass is 0.511 MeV. Since two particles (positron and electron) of equal mass are formed, the minimum energy of the incident photon required to produce pair production must be $2 \times 0.511 = 1.022$ MeV. A diagrammatic presentation of pair production is shown in Figure 2.7.

> *Summary of Absorption X-rays or Gamma Rays in Soft Tissue*
>> Up to 50 KeV: The photoelectric effect is important.
>> 60 to 90 KeV: The photoelectric effect and the Compton effect are equally important.
>> 200 to 2 MeV: The Compton effect is dominant.
>> 5 to 10 MeV: Pair production begins to be important.
>> 50 to 100 MeV: Pair production is the most important type of absorption.

Coherent scattering: Photons are scattered with negligible loss of energy during coherent scattering.

Photodisintegration: Photodisintegration occurs with threshold x-ray or gamma-ray energy of 1.65 MeV. A beryllium foil emits neutrons after irradiation by photons of energy in excess of 1.65 MeV. A silver foil adjacent to the beryllium is activated by the neutrons and emits gamma rays.

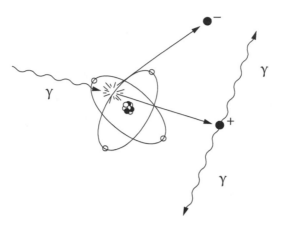

FIGURE 2.7   Pair production by ionizing radiation.

Units of ionizing radiation: The first effect of ionizing radiation on the skin appeared as an acute radiation dermatitis or so-called radiation burn among investigators working with radium. This skin reaction, which was referred to as the *threshold erythema dose* (TED), was used as a unit of radiation dose. In 1962, the International Commission on Radiological Units and Measurements defined the unit of ionizing radiation more explicitly. To honor Dr. Wilhelm Roentgen who discovered x-rays, the unit of exposure dose in air was called an *R*. Later, this unit was replaced by *rad* (radiation absorbed dose). One rad = 100 ergs of energy absorbed per gram of any material. To honor the contribution of Dr. Louis Harold Gray, a new unit, the Gy (Gray) was proposed to measure radiation doses; 1 Gy = 100 rads. This unit is currently used in radiobiology studies. In order to account for the differences between the biological effects of radiation of high LET and low LET, a new unit, rem (relative equivalence man) was established. Rem is calculated as follows:

$$\text{rem} = \text{rad} \times \text{RBE (relative biological equivalence)}$$

RBE = dose to produce an effect with x-rays or gamma rays/dose to produce the same effect with a new radiation

To honor Dr. Rolf Sievert, a Swedish physicist, a new unit Sv (Sievert) was established; 1 Sv = 100 rems.

NUCLEAR FISSION

Nuclear fission is a process in which a nucleus with high mass numbers separates, or fissions, into two parts, each with an average binding energy per nucleon greater than that of the original nucleus (Figure 2.8). Certain nuclei with high mass numbers, such as uranium-235 (^{235}U), pulutonium-239 (^{239}Pu), and uranium-233 (^{233}U), fission spontaneously after absorbing a slowly moving neutron. An example of a nuclear fission of ^{235}U is described here.

Uranium-235 Fission

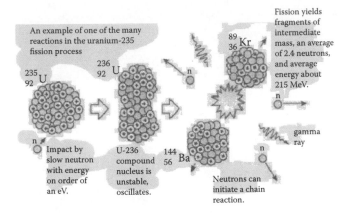

FIGURE 2.8   Fission products of Uranium-235.

$$_{92}U\text{-}235 + neutron \rightarrow {}_{92}U\text{-}236 \rightarrow {}_{36}Kr\text{-}92 + {}_{56}Ba\text{-}141 + 3 \text{ neutrons} + Q$$

The energy released is designated as Q and averages about 200 MeV/fission.

Another example of a fission reaction that was used in the first atomic bomb and is still used in nuclear reactors is shown here.

$$^{235}U + neutron \rightarrow {}^{236}U \rightarrow {}^{134}Xe + {}^{100}Sr + 2 \text{ neutrons} + Energy$$

Fission reactions can produce any combination of lighter nuclei as long as the total number of protons and neutrons are the same as the initial fissioning nucleus. Fission occurs because of the strong electrostatic repulsion created by the large number of positively charged protons in a nucleus. Once the larger nucleus overcomes the strong nuclear binding energy that holds it together, it undergoes a fission reaction. Figure 2.8 is a diagrammatic representation of the fission of ^{235}U.

Some fission products of ^{235}U are listed here with their half-life.

^{141}Ba (barium): 18.27 seconds
$^{222}Radon$: 4 days
^{131}I (iodine): 8 days
^{85}Kr (krypton): 10 years
^{3}H (Tritium): 12 years
^{90}Sr (strontium): 29 years
^{137}Cs (cesium): 30 years
^{241}Am (Americium): 433 years
^{226}Ra (radium): 1,622 years
^{239}Pu (plutonium): 24,000 years
^{134}Xe (Xenon, gas): Stable

When a neutron released from the fission of ^{235}U is absorbed by another ^{235}U nucleus, the chain reaction is sustained.

Energy from Uranium-235 Fission

Form of Energy Released	Amount of Energy Released in MeV
Kinetic energy of two fission fragments	168
Immediate gamma rays	7
Delayed gamma rays	3–12
Fission neutrons	5
Energy of Decay Products of Fission Fragments	
Gamma rays	7
Beta particles	8
Neutrons	12
Average total energy release	215

Plutonium-239 formation: In 1940, Dr. Glenn Seaborg of the University of California at Berkeley discovered Pu-239, but its publication was withdrawn because it was found that this element could undergo nuclear fission in a way that might be useful in an atomic bomb. Plutonium-238 was produced by deuteron bombardment of uranium in a cyclotron. All 15 plutonium isotopes are radioactive and decay emitting particles and some gamma radiation. All plutonium isotopes are fissionable with fast neutrons; therefore, they are significant in a fast neutron reactor. However, only Pu-239 has a major role in a conventional light water power reactor.

There are two different kinds of plutonium: reactor grade and weapon grade. The reactor-grade plutonium can be recovered as a by-product of uranium fuel used in a nuclear reactor, after the fuel has been irradiated for about 3 years. The weapon-grade plutonium is recovered from uranium fuel that has been irradiated for only about 2–3 months in a plutonium production reactor. Reactor-grade plutonium is defined as that which contains 19% or more of Pu-240, whereas weapon-grade plutonium contains only 8% of Pu-240. It has been estimated that 1 kilogram of Pu-239 can generate about 10 million kilowatt-hours of electricity. About 10 kilograms of nearly pure Pu-239 is needed to make a bomb.

The main isotopes of plutonium and their decay schemes are listed below.

^{238}Pu: Alpha decay, physical half-life 88 years
^{239}Pu: Fissile, alpha decay, physical half-life 24,000 years
^{240}Pu: Alpha decay, physical half-life 6,560 years
^{241}Pu: Fissile, beta decay, physical half-life 14.4 years
^{242}Pu: Alpha decay, physical half-life 374,000 years

NUCLEAR FUSION

Fusion is a nuclear reaction in which two light nuclei combine to form a single heavier nucleus with an average binding energy per nucleon greater than that of either of the original nuclei. The nuclear fusion is accompanied by the release of large amounts of energy. This is due to the fact that when two light nuclei fuse, the sum of the masses of the product nuclei is less than the sum of the masses of the initial fusing nuclei. The difference in the mass is converted into energy. For example, fusion of two hydrogen isotopes forms a helium nucleus (^{4}H) as follows.

$$^2H + 3H \rightarrow {}^4H + \text{neutron} + \text{Energy (18 MeV)}$$

Since both nuclei of the hydrogen isotopes are positively charged, there is a large electrostatic repulsive force that has to be overcome before the fusion of nuclei can occur. Nuclei moving at very high velocities possess enough momentum to overcome the repulsive force of the nuclei. The high velocity needed can be achieved by heating a sample containing low-atomic-number nuclei to a temperature greater than 12×10^6 K, roughly equivalent to the temperature in the inner region of the sun. Temperatures this high are attained on Earth only in the center of a fission explosion. Fusion reaction is the basis of a hydrogen or thermonuclear bomb.

In a star, hydrogen isotopes fuse to form a helium nucleus. Helium nuclei can fuse and form heavier nuclei. These fusion reactions continue until the nuclei reach the mass of iron, around 60, which has the most nuclear binding energy. When a star has converted a large fraction of its core's mass to iron, it has almost reached the end of its life.

Radiological weapons: Radiological weapons can be divided into two categories: dirty bombs, also referred to as radiological dispersion devices (RDDs), and nuclear weapons, which include atom bombs, hydrogen bombs, and neutron bombs. In order to understand the potential risk of detonation of radiological weapons, it is essential to understand the differences between dirty bombs and nuclear bombs, which may provide guidance for developing an effective countermeasure against their acute and long-term health hazards.

Dirty bomb: A dirty bomb is made of radiation-emitting radioactive isotopes and is designed to detonate using a conventional explosive. Radioactive isotopes are released following the explosion. These radioactive materials can travel away from the explosion site, depending on the direction and speed of the wind. They can also contaminate water and food, and food sources, for a long time to come, depending upon the half-lives of the dispersed radioactive isotopes. Types of radiation released may include gamma rays, beta radiation, or alpha particles of various energies, depending upon the type of radioactive isotopes. Some radioactive isotopes, such as ^{60}Co, ^{137}Cs, and ^{192}Ir emit primarily gamma radiation; ^{3}H, ^{131}I, ^{14}C, and ^{32}P emit primarily beta particles, and ^{252}Cf, ^{226}Ra, ^{235}U, and ^{219}Po emit primarily alpha particles. It appears that dirty bombs can be constructed with only those radioactive isotopes that have longer half-lives. Dirty bombs remain a radiological weapon of aspiration for terrorist groups because many of the radioactive isotopes that are used in biological, medical, and industrial research are commercially available. They can be purchased from commercial sources and transported from one place to another with less difficulty than a nuclear bomb. Only a conventional explosive is needed to detonate a dirty bomb.

It's unlikely that the explosion of a dirty bomb would cause mass casualties, although a few people present in the vicinity of the explosion may die because of exposure to lethal doses of gamma rays. However, the long-term adverse health risk of the population residing in the contaminated area persists for generations. Populations far away from the site of the explosion can also be adversely affected because the dust containing radioactive materials can travel through the air. The ultimate goal of the explosion of a dirty bomb would be to create panic, psychological stress, and a serious adverse financial impact—casualties are a secondary consideration. Therefore, a dirty bomb should be considered as a weapon of mass disruption rather than as a weapon of mass destruction.

Nuclear weapons: Nuclear weapons are the most destructive technology ever discovered by a group of scientists. A single small nuclear weapon is capable of destroying an entire city within a few hours. Based on the nuclear reactions, nuclear weapons can be divided into two classes: fission bombs and fusion bombs. A fission bomb is also referred to as an atom bomb, atomic bomb, or A bomb. Fusion bombs are referred to as hydrogen bombs, H bombs, or thermonuclear bombs. A nuclear weapon generates its enormous destructive force from a nuclear reaction through fission or fusion. There are other types of nuclear bombs, such as neutron bombs.

Atom bomb: On December 2, 1942, Dr. Enrico Fermi and his associates at the University of Chicago accomplished a chain reaction from the fission of ^{235}U in a pile of graphite blocks. This remarkable discovery became the basis for manufacturing the atom bomb and the nuclear reactor. The isotopes responsible for the large-scale fission reaction include ^{235}U, ^{239}Pu, and ^{233}U. The binding energy of these isotopes is so low that when a neutron is captured by one of these isotopes, the energy released by nuclear rearrangement exceeds it. The nucleus, after capturing a neutron, becomes unstable and splits into two pieces. Since the fission occurs regardless of the kinetic energy of the neutrons, this type of fission is called *slow fission*. However, when an isotope ^{238}U captures a neutron, its binding energy has a deficit of 1 MeV after internal nuclear rearrangement. If ^{238}U captures a neutron with kinetic energy exceeding 1 MeV, fission can occur. This process of fission is called *fast fission*, because it requires a fast neutron with a large kinetic energy. Generally, the slow-fissionable isotopes more easily undergo fission than fast-fissionable isotopes for neutrons of all energies.

An atom bomb requires the production of explosive energy through a nuclear fission reaction (breaking up of nuclei of atoms) alone. The fissionable materials needed to make atom bombs include enriched radioactive uranium (U-235) or plutonium (Pu-239). When the nucleus of uranium-235 captures a neutron, it is split into two different products (Kr and Ba), and releases 2–3 neutrons and enormous amounts of energy. One of the neutrons is then captured by the nucleus of another uranium-235 and repeats the above reaction again and again. This creates a chain reaction generating enormous amounts of energy. This chain reaction can be controlled, such as in a nuclear power plant, or uncontrolled, which is required for a nuclear weapon. A mass of uranium-235 is assembled into a supercritical mass needed to start a chain reaction—the release of huge, uncontrolled amounts of gamma radiation with energy as measured in MeV (million electron volts). The supercritical mass is achieved by bringing together very rapidly two subcritical masses of uranium-235 by means of a chemical explosive. At this point, neutrons are injected into the supercritical mass of uranium-235 to initiate a chain reaction.

The amounts of energy released by atom bombs can range from less than a ton of trinitrotoluene (TNT) to around 500 kilotons. The explosion of an atom bomb can deliver high doses of neutron radiation, gamma radiation, and shock waves instantly to humans, animals, and plants, and can affect a large surface area from the center of denotation. The water, soil, and plants around the explosion site become contaminated with radioactive materials released from the bomb. Some radioactive materials are released into the upper atmosphere as dust or gas and may subsequently be deposited far away from the site of the explosion.

Hydrogen bomb: This type of bomb releases energy through nuclear fusion, the combining of two nuclei together, rather than through fission (the breaking up of a nucleus as in an atom bomb). This type of bomb can deliver more than a thousand times more gamma radiation with energy as measured in MeV than the atom bomb.

The fusion reaction requires that an atom bomb be detonated in a specially manufactured compartment adjacent to a fusion fuel (tritium, deuterium, or lithium deuteride). This is due to the fact that a high temperature, equivalent to the temperature in the inner region of the sun, is needed for the fusion of two nuclei. The gamma and x-rays emitted by this explosion compress and heat a capsule of fusion fuel, which starts a fusion reaction. Neutrons emitted by this fusion reaction can induce a final fission stage in depleted uranium that surrounds the fusion fuel. This process increases the yield of gamma radiation considerably, as well as the amount of radioactive fallout. The largest hydrogen bomb that was detonated by the USSR in 1955 released energy equivalent to more than 50 million tons of TNT. Other types include a boosted fission bomb, which increases its explosive yield through a small number of fusion reactions, but it is not a hydrogen bomb. The fusion reaction between deuterium and tritium produces extensive amounts of energy. The most efficient fusion reaction is illustrated here.

^{2}H (deuterium-2) + ^{3}H (tritium-3) $\rightarrow$ ^{4}He (Helium-4) + neutron + energy (18 MeV)

Other fusion reactions include:

^{2}H (deuterium-2) + ^{2}H (deuterium-2) $\rightarrow$ ^{3}He (Helium-3) + neutron + energy (3.268 MeV)

^{2}H (deuterium-2) + ^{2}H (deuterium-2) $\rightarrow$ ^{3}H (tritium-3) + proton + energy (4.03 MeV)

Neutron bomb: Neutron bombs, also referred to as *enhanced radiation (ER) warheads*, are small thermonuclear weapons in which the burst of energy generated by the fusion reaction is not allowed to be absorbed inside the weapon, but is allowed to escape. This intense burst of high-energy neutrons is responsible for destructive effects. Neutrons are more penetrating than other forms of radiation. This is due to the fact that they have no charge. Shielding that is effective against gamma radiation is not as effective against neutron radiation. A neutron bomb yields a relatively small explosion but releases large amounts of neutron and gamma radiation—radiation that can cause massive casualties. In addition, a neutron bomb explosion can make many elements—such as boron in the soil or water—radioactive, which then become sources of increased radiation levels in the environment for generations to come. In addition, some elements in building materials that have a high cross section to capture slow neutrons may become radioactive for a long time, depending upon the half-lives of the newly formed radioactive materials. Neutron bombs have never been used in any warfare. All neutron bombs in the United States have been retired and dismantled. Therefore, the likelihood of acquiring this form of nuclear weapon by any terrorist group is virtually zero.

Energy released after a nuclear explosion: A nuclear explosion releases several distinct forms of energy that can cause mass destruction, including the physical destruction of buildings and infrastructures, and massive human and animal casualties. These forms of energy include blast, thermal radiation, prompt radiation (ionizing radiation), radioactive fallout, and electromagnetic pulse (EMP). The extent of damage caused by a nuclear explosion depends on various factors, including the size of the nuclear weapon, the height at which it is detonated, and the geography of the target.

Blast: The major portion (about 60%) of energy from an explosion of a nuclear bomb is released in the form of blast and shock waves. The rapid release of energy from an explosion creates a shock wave of overpressure. It has been estimated that the overpressure near the center of a nuclear explosion is equivalent to several thousand pounds per square inch (psi). The overpressure can crush objects. Human lungs are crushed at about 30 psi overpressure. Brick buildings or houses are destroyed at about 10–15 psi overpressure. The blast can also generate hurricane-force winds of several hundred miles per hour, which can turn humans or objects into missiles. At about 10–15 psi overpressure, the winds can lift and accelerate a person at several hundred kilometers (1.6 kilometers = 1 mile) per hour. The blast from a nuclear bomb equivalent in yield to the atom bomb dropped on Hiroshima would totally flatten all wooden or unreinforced masonry structures within one mile from the point of explosion (ground zero). The Hiroshima bomb was detonated at about 2,000 feet elevation to maximize the blast damage.

Thermal radiation: It has been estimated that about 35% of the energy released from an explosion of a nuclear bomb is in the form of thermal radiation, which includes ultraviolet light, visible light, and infrared, which can be seen from hundreds of miles away from the center of the explosion. The light is so intense that it can explode sand, blind people many miles away, and can ignite flammable materials located many miles away.

Immediately after a thermonuclear explosion, the temperature at ground zero may exceed 100 million degrees centigrade (°C). This is about 10 times the temperature of the surface of the sun. At these temperatures, matter cannot exist in its solid, liquid, or gaseous state; it is vaporized. Thermal radiation also creates a fireball that rapidly expands outward, consuming oxygen and combining with the blast to cause total destruction miles away from ground zero. In case of a nuclear bomb the size of the Hiroshima bomb, the firestorm generated after the explosion could incinerate everything within about 12 miles (1.9 kilometers).

Prompt radiation (ionizing radiation): It has been estimated that about 5% of the energy released from a nuclear explosion is in the form of prompt radiation, which includes neutrons, gamma rays, x-rays, and alpha and beta particles. Gamma rays, x-rays, and neutrons are highly penetrating types of radiation, whereas alpha and beta particles have very short penetrating range. For example, alpha particles cannot penetrate human skin, but if ingested, can cause more damage than gamma radiation of the same total dose. The prompt radiation from a Hiroshima-sized nuclear bomb explosion can deliver 5 Sv to individuals within about 1.3 miles (2 kilometers) of ground zero.

Radioactive fallout: It has been estimated that about 10% of the energy released from a nuclear explosion is in the form of radioactive dust particles, which are pushed high into the atmosphere by the force of the explosion. These radioactive dust particles also contain nonradioactive particles from the Earth's surface and buildings. Some of them will fall back to the Earth's surface within a few minutes and others will take weeks. The rising and descending debris generated by the nuclear explosion form the so-called mushroom cloud that is seen at the site of a nuclear explosion. The radioactive dust particles can travel through the air hundreds of miles away from the site of the explosion, and eventually return to the Earth's surface, causing radioactive contamination of water and soil covering thousands of square miles. The intensity and duration of contamination from fallout vary with the size of the nuclear bomb, with how close it exploded to the ground, and with the half-lives of radioactive isotopes. Nuclear weapons detonated at or close to ground level produce the most fallout. It has been estimated that for about 2 days following an explosion of a Hiroshima-sized nuclear bomb, individuals within about 1.7 miles (2.75 kilometers) of ground zero would receive a 5 Sv radiation dose from fallout. Radioactive substances in the contaminated areas would not decay to safe levels for several years. The time required to reach safe levels depends upon the physical half-life of the contaminating nuclides. Generally, it requires the decay of original activity by 10 half-lives of the particular radioactive element present in the contaminated area.

Electromagnetic pulse: The ionizing radiation from a nuclear explosion creates a strong EMP. This is due to the fact that radiation causes ionization of air particles that act as an electric charge, which interacts with the magnetic field of the Earth, producing a surge of electromagnetic pulse. This effect is similar to the electric surge produced by a lightning bolt. A large nuclear explosion can generate electric surges of 25,000 to 50,000 volts per meter, which can destroy all electronic devices, such as unprotected electronic equipment, medical equipment, microchips found in newer cars, shut down electric grids and communications networks, and erase a computer's memory. If a nuclear weapon is detonated on or close to the ground, the effect of EMP is produced only over a relatively small area; however, if it is detonated at high altitude, the effect of EMP can be felt over a radius of hundreds of miles. It is very unlikely that terrorists would have the technical capability to detonate a nuclear weapon at high altitude.

A nuclear bomb can be delivered by missiles or dropped from an airplane. The fission materials needed for an atom bomb, uranium-235 or plutonium-239, cannot easily be obtained or assembled, although these materials can be transported by virtue of their small weight. However, this type of bomb is easily detected by radiation detectors. Because detonating an atom bomb requires complex steps that include assembly of uranium-235 into a critical mass and injection of neutrons, it is not a weapon of choice for terrorists. In contrast, a dirty bomb is a weapon of choice for terrorists because the ingredients needed to make a dirty bomb are relatively easy to obtain, easy to assemble, and easy to detonate. Therefore, fear that terrorists might explode a nuclear bomb anywhere in the world may not be justified.

It is believed by some that Russia and the United States have a few dozen suitcase nuclear weapons (a miniature atom bomb), but neither country has admitted to having them. It is also believed by some that these suitcase nuclear weapons can be detonated by a remote control device. If this is true, the detonation of such weapons can cause mass casualties similar to that produced by a small atom bomb.

Explosion of an atom bomb: The first testing of a nuclear bomb, named The Gadget, occurred near the Jemez Mountains in northern New Mexico at 5:29:45 a.m. when the sky was still dark. The brilliant light from the explosion turned orange as a fireball began shooting upward at the speed of 360 feet per second, reddening and pulsing as it cooled. The characteristic mushroom cloud of radioactive vapor formed at the height of 30,000 feet. Except for fragments of jade green radioactive glass created by the intensity of the heat, nothing remained below the mushroom cloud. The flashes of intense light as bright as the sun were seen miles from the site of the explosion. Since then, several countries have tested nuclear bombs on the ground and underground.

In human history, the nuclear bomb was used twice, during World War II. On August 6, 1945, at 8:15 a.m, a uranium bomb nicknamed Little Boy was dropped from a plane on the city of Hiroshima, Japan. It has been estimated that 66,000 people were killed instantly from burns, the blast, and lethal doses of radiation, and another 69,000 people were injured. An area one-half mile in diameter was totally vaporized and the total destruction covered one mile in diameter. Severe damage due to the atomic explosion was seen in the area covering over two miles in diameter. Everything in an area covering 2.5 miles was totally burned. Serious structural damage and blazes were seen in the areas covering over three miles from the site of the explosion.

A few days later on August 9, 1945, a plutonium bomb nicknamed Fat Boy was dropped from a plane on the city of Nagasaki, Japan. The explosion totally destroyed about half of the city in a split second. About 39,000 people were instantly killed and another 25,000 were injured.

In addition to causing instant destruction of humans, animals, and plants, an atomic explosion creates radioactive fallout that can expose individuals on the ground to lethal doses of radiation. Many survivors of the Hiroshima and Nagasaki bombs died later from the radiation from the radioactive fallout.

Because of the technological complexity of manufacturing and detonation of a nuclear bomb, it is very unlikely that terrorists can obtain this form of bomb. An unintentional nuclear conflict in which one nation drops a conventional powerful bomb on the nuclear power plants of another country, or a conflict in which two nations exchange nuclear attacks—although a remote possibility—remains a viable scenario in the future.

PHYSICS OF NUCLEAR REACTORS

A nuclear reactor is a device to initiate and control a sustained nuclear chain reaction generated by fission of ^{235}U. The concept of a nuclear chain reaction was first proposed by Hungarian scientist Dr. Leo Szilard in 1933. In 1942, Dr. Enrico Fermi and his associates at the University of Chicago succeeded in constructing the first nuclear reactor, referred to as Chicago Pile-1. The nuclear chain reaction is controlled by neutron poisons, which absorb excess neutrons, and neutron moderators, which reduce the velocity of fast neutrons to thermal neutrons, which are easily captured by other nuclei. Commonly used neutron moderators include light water, heavy water, and solid graphite. Most reactors utilize light water as a neutron moderator. Reactors are commonly used to generate electric power for civilian, military, and industrial use.

Conventional power plants generate electricity by converting burning fossils to thermal energy. In contrast, nuclear reactors convert released nuclear energy to thermal energy. For example, the kinetic energy of fission products is converted to thermal energy. In addition, gamma rays produced during fission are also converted to heat by the reactor. Thermal energy is also produced by the radioactive decay of fission products and materials that have been made radioactive by neutron capture. Thermal energy is used to generate pressurized steam for the turbines that generate electricity.

All commercial power reactors are based on nuclear fission. They utilize uranium and its product plutonium as nuclear fuel. There are two classes of fission reactors: thermal reactors that use slow or thermal neutrons to initiate fission reaction, and fast neutron reactors that utilize fast neutrons to cause fission. Almost all current reactors belong to the thermal reactor class.

In a nuclear reactor, when the neutron population remains steady, the fission reaction is considered self-sustaining, and the reactor condition is called *critical*; when the neutron production exceeds neutron losses, it is considered *supercritical*; and when the neutron losses exceed neutron production, it is referred to as *subcritical*. Most nuclear reactors utilize a mixture of an alpha particle emitter, such as ^{241}Am (Americium-241) with a lightweight isotope ^{9}Be (beryllium-9) as a "starter" neutron source to initiate the chain reaction in ^{235}U. In operation reactors, a combination of antimony and beryllium is used as a secondary source of neutrons. Antimony becomes activated in the reactor, producing high-energy gamma radiation that produces neutrons from beryllium. The primary neutron starter source can be removed from the core to prevent damage from the high neutron influx in the operating core. The secondary neutron sources are allowed to remain in the reactor in order to provide a background reference level for control of criticality.

Any element that strongly absorbs neutrons is called a *reactor poison* because it can shut down the fission chain reaction. Some reactor poisons, such as boron or cadmium, are used in control rods in fission reactor cores to control the fission reaction. Many short-lived and long-lived reactor poisons are produced by the fission process, and increased accumulation of these neutron absorbers can reduce the chain reaction. The lifetime of a nuclear fuel is dependent upon the amounts of long-lived neutron absorbers accumulated. Therefore, it becomes essential to reprocess the nuclear fuel by removing the fission products chemically; the reprocessed nuclear fuel can then be used again.

One of the examples of a short-lived neutron poison produced as a fission product is ^{135}Xe (xenon-135) with a half-life of about 9 hours. In a normal operating nuclear reactor, each nucleus of ^{135}Xe is destroyed by neutron capture as soon as it is created, thus preventing its buildup in the core. When a reactor shuts down, the level of ^{135}Xe builds up in the core for about 9 hours before beginning to decay. Thus, it becomes impossible to start the chain reaction after reaction has been shut down for about

6–8 hours. The chain reaction can be achieved after the decay of ^{135}Xe. It is believed that ^{135}Xe played a large part in the Chernobyl nuclear accident. After about 8 hours of a scheduled maintenance shutdown, attempts were made to bring the reactor to a zero power critical condition to test the control circuit. However, since the core was already loaded with ^{135}Xe, the chain reaction grew rapidly and uncontrollably, leading to steam explosion in the core, fire, and total destruction of the facility, releasing radiation into the atmosphere.

NUCLEAR ACCIDENTS

Since the establishment of many nuclear plants for generating electricity or manufacturing nuclear bombs, many accidents have occurred. They have caused fatalities and increased risk of adverse health effects such as cancer, and radioactive contamination of the atmosphere, water, and soil. Most accidents were due to human errors, negligence, poor training, and inadequate attention to the safety procedures. A few examples of nuclear accidents at nuclear power plants are described below.

Chernobyl nuclear power plant accident: The most serious accident in the history of the nuclear industry occurred in the Chernobyl nuclear plant on April 26, 1986. This plant is located in Ukraine, and at the time of the accident had four working reactors. The fire lasted for 10 days. In addition, huge amounts of radioactive materials were released into the atmosphere and a cloud of radioactive dust spread over much of Europe. The greatest contamination to the environment, water, and soil occurred around the areas of the reactor that are now part of Belarus, Russia, and Ukraine. Since the accident, about 600,000 first responders have been involved in cleaning and recovery operations and all of them have received higher doses of radiation; however, only a few of them were exposed to lethal doses of radiation. Approximately 1,000 first responders who were onsite during the first day of the accident received the highest doses of radiation. After the accident, civilians were exposed to radiation both directly from the radioactive cloud, radioactive materials deposited on the ground, and through ingesting radioactive contaminated food, water, or through breathing contaminated air. The children were exposed to radiation from drinking milk contaminated with radioactive iodine. At present, about 100,000 people living in contaminated areas still receive higher doses of radiation than those living further away from the area of the accident. Following the accident, 28 first responders died from acute radiation syndrome and 15 died from thyroid cancer. It is estimated that about 4,000 additional cancers may develop among the 600,000 people who were exposed to higher doses of radiation. The risk of cancer among civilians who were exposed to low doses of radiation after the accident has not been fully evaluated.

Tokaimura nuclear plant accident: In 1999, a serious accident occurred at the Tokaimura nuclear plant in Japan. A total of 119 people received doses of radiation higher than 1 mSv from the accident, and two people died from high doses of radiation. Long-term adverse health effects of those exposed to higher doses of radiation are unknown.

Mayak nuclear plant accident: In 1952, a serious accident occurred at the Mayak nuclear plant near the city of Kyshtym in the Soviet Union (now Russia). The cooling system in a tank containing 70 tons of radioactive waste failed; consequently, a rise in temperature caused a nonnuclear explosion of the tank, releasing radioactive materials into the atmosphere. About 10,000 people were evacuated and 200 people died of cancer.

Boris Kidrich Institute nuclear plant accident: In 1958, a serious nuclear accident occurred at the Boris Kidrich Institute nuclear plant in the city of Vinca, Yugoslavia. Six scientists received radiation doses between 2 to 4 Sv; five of them died from acute radiation syndrome.

Tomsk-7 Siberian Chemical Enterprise Plutonium Reprocessing Facility accident: In 1993, a serious accident occurred at the Siberian Chemical Enterprise Plutonium Reprocessing Facility in Tomsk, Russia. This accident exposed 160 onsite workers and about 2,000 first responders to doses up to 50 mSv.

Three Mile Island nuclear plant accident: In 1979, a nuclear accident occurred at Three Mile Island near Harrisburg, Pennsylvania. The reactor was totally destroyed. Some radioactive gas was released a couple of days after the accident, but did not affect the level of background radiation. This accident aroused a lot of public concern regarding the safety of the nuclear plant. Fortunately, there was no reported injury or long-term adverse health effects from this accident.

Nuclear plant accident in Hamm-Uentrop, Germany: In 1986, a nuclear accident occurred at the nuclear plant in Hamm-Uentrop, Germany (then West Germany). This accident released radiation that could be detected up to two kilometers from the reactor.

Nuclear plant accident in the United Kingdom: In 2005, a nuclear accident occurred at the nuclear fuel processing plant in Sellafield, England. This accident involved leaking of twenty metric tons of uranium and 160 kilograms of plutonium dissolved in 83,000 liters of nitric acid from a storage tank over several months. The levels of contamination in the plant or surrounding areas have not been reported.

Erwin Nuclear Service Plant accident: In 2006, an accident occurred at the Erwin Nuclear Service Plant, in Erwin, Tennessee. This accident involved leaking of 35 liters of a highly enriched solution of radioactive materials during transfer into a laboratory. This incident caused shutdown of the plant for seven months. There were no reports of injuries or contamination.

Wood River Junction nuclear plant accident: In 1964, a nuclear accident occurred at the Wood River Junction nuclear plant in Charlestown, Rhode Island. In this accident, an operator accidently dropped a concentrated uranium solution into an agitated tank containing sodium carbonate, causing a critical nuclear reaction that exposed the operator to a lethal dose of radiation (100 Gy). Ninety minutes later, a plant manager returned to the building and turned off the agitator, exposing himself and another administrator to doses of up to 1 Gy without immediate adverse health effects.

Japanese Uranium Reprocessing Facility accident: In 1999, an accident occurred at the Japanese Uranium Reprocessing Facility in Tokai, Ibaraki. In this accident, workers at the plant added a mixture of uranyl nitrate solution into a precipitation tank that was not designed to dissolve this type of solution, causing a critical mass to form. This accident caused the death of two workers from the lethal doses of radiation.

Nuclear accidents such as those described suggest that in order to reduce the risk of an accident, it is essential that workers are properly trained, safety regulations are strictly implemented, and structural components of the plant are frequently monitored to detect any potential wear and tear. In spite of all precautions, some accidents may occur due to human error or a natural disaster, such as an earthquake or tsunami. This risk of an accident is increasing because of the proliferation of nuclear power plants around the world to meet energy needs.

Fukushima nuclear power plant accident in Japan: Following a major earthquake of magnitude 7.0, and subsequent tsunami on March 11, 2011, the nuclear power plants in Fukushima were damaged and the radioactive elements ^{131}I, ^{137}Cs, ^{239}Pu, and ^{90}Sr were released into the ocean and the atmosphere. The levels of radiation released are now considered similar to those released during the Chernobyl nuclear plant accident in 1986. The levels of radiation in the atmosphere varied depending upon the distance from the nuclear plants. The value ranged 1000 mSv per hour inside the plant and then dropped off sharply further away from the plant.

According to the International Atomic Energy Agency (IAEA), the situation at the nuclear power plant remained serious as of June 2, 2011. The Tokyo Electric Power Company (TEPCO) is working to improve the conditions in the plants. Although some people residing near the plants may have been initially exposed to higher levels of ^{131}I (physical half-life of 8 days), ^{137}Cs (physical half-life of 29 years), and ^{90}Sr (physical half-life of 30 years), the levels of radiation from these radioisotopes are decreasing. However, the decrease in the levels of gamma rays has slowed down, since the short-lived radionuclides, such as ^{131}I, have decayed away. Food and water sources and marine water and marine life contained higher levels of radiation than those allowed by the federal government. The Ministry of Health, Labor, and Welfare of Japan has maintained continued restrictions of raw unprocessed milk, turnips, bamboo shoots, spinach, and ostrich fern from certain areas of Fukushima. It is too early to estimate the long-term effects of this accident on the human population residing near the power plant. One can get some idea of the possible long-term health risks from the data obtained from individuals exposed after the Chernobyl nuclear power plant accident and the testing of nuclear bombs in Nevada. It is estimated that among the 600,000 people who were exposed to higher doses of radiation after the Chernobyl accident, about 4,000 additional cancers may develop. The risk of cancer among civilians who were exposed to low doses of radiation after the accident has not been fully evaluated. The US National Cancer Institute estimated that fallout from the Nevada nuclear testing sites may cause 49,000 excess cases of thyroid cancer (http://books.nap.edu/catalog.php?record id=6283). Therefore, it is expected that people who were exposed to radiation from radioactive iodine, cesium, and strontium may have an increased risk of thyroid cancer and other neoplastic diseases, such as leukemia. Some of them may suffer from heritable mutations that could increase the risk of diseases in future generations.

SUMMARY

In order to understand the differences between nuclear bombs and the dirty bombs, it is necessary to understand the fundamental principles of atomic and nuclear physics. An atom consists of a central mass called the *nucleus*, which is surrounded by the orbital *electrons*, which are negatively charged. The nucleus of the atom is composed of two particles: a proton, which has a positive charge, and a neutron, which has no charge. Atoms having the same atomic number but having different mass numbers are known as *isotopes*. The isotope could be stable or could be radioactive, if the nuclei are unstable. In order to reach a more stable state, radioactive elements may undergo decay and release radiation in the form of particle radiation, such as alpha particles and beta particles, or electromagnetic radiation such as gamma radiation and x-radiation. Gamma radiation is emitted directly from the nucleus, whereas x-radiation is produced by radiative and collisional interactions of electrons outside the nucleus. Certain naturally occurring radioactive isotopes such as uranium-235 (^{235}U) undergo fission after absorbing a neutron. The physical half-life of a radioactive isotope is defined as the time required for decaying of half of its original activity. The half-lives of radioactive isotopes found in nature or produced by cyclotron markedly differ, varying from a few milliseconds to a few billion years. The biological half-life of a radioisotope is defined as the time required to reduce the body or the organ radioactivity to half of its original activity. The biological half-life varies from one nuclide to another, depending upon turnover of the nuclide in the organ. The physical half-life of ^{45}Ca is 163 days, but its biological half-life in the bone is 18,000 days.

Particle radiation, such as protons, deuterons, alpha particles, and other heavy particles lose most of their energy when they interact inelastically with electrons of the absorbing materials. Part of the energy lost by incident particles is used to raise electrons in the absorber to an energy level farther from the nucleus. This process is called *excitation*. Sometimes electrons are ejected from their atoms—a process known as *ionization*. The heavy particles have a higher LET (linear energy transfer) than x-rays or gamma rays. A neutron interacts with matter through protons. A slow neutron can be captured by an element such as boron-10 that becomes radioactive and emits alpha particles. X-rays or gamma rays interact with matter most commonly through photoelectric effect and/or Compton effect, depending upon the energy of the photons.

Nuclear fission is a process in which a nucleus with a high mass number separates, or fissions, into two parts, each with an average binding energy per nucleon greater than that of the original nucleus. Certain nuclei with high mass numbers, such as uranium-235 (^{235}U), plutonium-239 (^{239}Pu), and uranium-233 (^{233}U), fission spontaneously after absorbing a slowly moving neutron. This reaction is the basis of nuclear bombs and nuclear reactors. Fusion is a nuclear reaction in which two light nuclei combine to form a single heavier nucleus with an average binding energy per nucleon greater than that of either of the original nuclei. The nuclear fusion is accompanied by the release of large amounts of energy. This is due to the fact that when two light nuclei fuse, the sum of the masses of the product nuclei is less than the sum of the masses of the initial fusing nuclei. The difference in the mass is converted into energy.

Radiological weapons can be divided into two categories: dirty bombs, also referred to as RDDs, and nuclear weapons, which include the atom bomb, hydrogen bomb, and neutron bomb. A dirty bomb is made of radiation-emitting radioactive isotopes and is designed to detonate using a conventional explosive. Radioactive isotopes are released following the explosion of a dirty bomb. These radioactive materials can travel away from the explosion site, depending on the direction and speed of the wind. They can also contaminate water, soil, and food, and food sources for a long time to come, depending upon the half-lives of the dispersed radioactive isotopes. Types of radiation released may include gamma rays, beta radiation, and/or alpha particles of various energies, depending upon the type of radioactive isotope. It appears that dirty bombs can be constructed with only those radioactive isotopes that have longer half-lives. Dirty bombs remain a radiological weapon of choice for terrorists because many radioactive isotopes that are used in biological, medical, and industrial research are commercially available. They can be purchased from commercial sources, and can be transported from one place to another with less difficulty than a nuclear bomb. Only a conventional explosive is needed to detonate a dirty bomb.

Based on the nuclear reactions, nuclear weapons can be divided into two classes: fission bombs and fusion bombs. A fission bomb is also referred to as an atom bomb, atomic bomb, or A bomb, whereas a fusion bomb is referred to as a hydrogen bomb, H bomb, or thermonuclear bomb. A nuclear weapon generates its enormous destructive force from a nuclear reaction through fission or fusion. There are other types of nuclear bombs, such as neutron bombs. A nuclear explosion releases several distinct forms of energy that can cause mass destruction, including the physical destruction of buildings and infrastructure, and massive human and animal casualties. These forms of energy include blast, thermal radiation, prompt radiation (ionizing radiation), radioactive fallout, and EMP. The extent of damage caused by a nuclear explosion depends on various factors, including the size of the nuclear weapon, the height at which it is detonated, and the geography of the target.

In human history, the nuclear bomb was used twice, during World War II. In 1945, a uranium bomb nicknamed Little Boy was dropped from a plane on the city of Hiroshima, Japan. It has been estimated that 66,000 people were killed instantly from burns, the blast, and lethal doses of radiation; 69,000 people were injured. An area of one-half mile in diameter was totally vaporized, and there was total destruction of an area covering one mile in diameter. Severe damage due to the atomic explosion was seen in an area covering over two miles in diameter. Everything in a 2.5-mile area was totally burned. Serious structural damage and blazes were seen in areas covering over three miles from the site of explosion. A few days later, a plutonium bomb nicknamed Fat Boy was dropped from a plane on the city of Nagasaki, Japan. The explosion totally destroyed about half of the city in a split second. About 39,000 people were instantly killed and 25,000 were injured. In addition to causing instant destruction of humans, animals, and plants, atomic explosions create radioactive fallout that can expose individuals on the ground to lethal doses of radiation. Many survivors of Hiroshima and Nagasaki later died from the radiation emitted from the radioactive fallout.

Because of the technological complexity of manufacturing and detonating a nuclear bomb, it is very unlikely that terrorists can obtain this form of bomb. An unintentional nuclear conflict in which one nation drops a conventional powerful bomb on the nuclear power plants of another country or two nations exchange nuclear attacks, although a remote possibility, remains a viable scenario in the future.

A nuclear reactor is a device to initiate and control a sustained nuclear chain reaction generated by fission of ^{235}U. The concept of a nuclear chain reaction was first proposed by Hungarian scientist Dr. Leo Szilard in 1933. In 1942, Dr. Enrico Fermi and his associates at the University of Chicago succeeded in constructing the first nuclear reactor, referred to as Chicago Pile-1. The nuclear chain reaction is controlled by neutron poisons, which absorb excess neutrons, and neutron moderators, which reduce the velocity of fast neutrons to thermal neutrons, which are easily captured by other nuclei. Commonly used neutron moderators include light water, heavy water, and solid graphite. Most reactors utilize light water as a neutron moderator. Reactors are commonly used to generate electric power for civilian, military, and industrial use.

Since the establishment of many nuclear plants for generating electricity or manufacturing nuclear bombs, many accidents have occurred. They have caused fatalities and increased risk of adverse health effects such as cancer, and radioactive contamination of the atmosphere, water, soil, and food sources. Most accidents were due to human errors, negligence, poor training, and inadequate attention to safety procedures. This risk of accident is increasing because of the proliferation of nuclear power plants around the world to meet energy needs. In order to reduce the risk of an accident, it is essential that workers are properly trained, safety regulations are strictly implemented, and structural components of the plant are frequently monitored to detect any potential wear and tear. In spite of all precautions, some accidents may occur due to human error or natural disasters, such as earthquakes and tsunami.

REFERENCES

1. Prasad, K. N., *Handbook of Radiobiology*, 2nd ed. CRC Press, Boca Raton, FL, 1995.
2. Bellis, M., History of the atomic bomb and the Mahattan Project, http://inventors.about.com/od/astartinventions./a/atomic_bomb_2.htm, 2010.
3. Ware, A., Nuclear weapons: The basics, http://www.nuclearfiles.org/menu/key-issues/nuclear-weapons/basics/weapons-basics.htm, 2010.
4. Anonymous, Introduction to nuclear weapon physics and design, http://nuclearweaponarchive.org/Nwfaq/Nfaq2.html, 1999.
5. Anonymous, Nuclear terrorism, http://www.nti.org/h_learnmore/nuctutorial/chapter02_06.html, 2010.
6. US Department of Energy, *Fundamentals Handbook: Nuclear Physics and Reactor Theory*, http://www.hss.energy.gov/NuclearSafety/techstds/standard/hdbk1019/h1019v2.pdf, 1993.
7. Tiwari, J. G., and Gray, C. J., US nuclear weapons accidents, http://www.cdi.org/issues/nukeaccidents/accidents.htm, n.d.
8. Wikipedia, List of civilian nuclear accidents, http://en.wikipedia.org/wiki/List_of_civilian_nuclear_accidents, 2010.
9. McCarthy, J., Technical Supplement on Nuclear Energy, http://www.formal.stanford.edu/jmc/progress/nuclear-supplement.html, 1995.
10. Dobrzynski, L., Blinowski, K., *Neutrons and Solid State Physics*, Ellis Horwood Limited, New York, 1994.
11. Mulligan, J. F., *Practical Physics: The Production and Conservation of Energy*, McGraw Hill, New York, 1980.
12. Anonymous, Georgia State University, http://hyperphysics.phy-astr.gsu.edu/hbase/nucene/u235chn.html, n.d.
13. Wick, O. J., *Plutonium Handbook: A Guide to the Technology*, Vol. 1 and II. American Nuclear Society, La Grange, IL, 1980.

14. A. Medical, *Fundamentals of X-Ray Physics*, Integrated Publishing, Port Richey, FL, 2007.
15. Nuclear Regulatory Commission, Dirty bombs, http://www.nrc.gov/reading-rm/doc-collections/fact-sheets/dirty-bombs.html.
16. Johnson, W. R., Database of radiological accidents and related events, http://www.johnstonsarchive.net/nuclear/radevents/index.html, April 2, 2011.
17. Nero, A. V., *A Guide Book to Nuclear Reactors*, University of California Press, Berkeley, 1979.

3 Acute Radiation Damage by High Doses of Ionizing Radiation in Humans

INTRODUCTION

Ionizing radiation doses of more than 0.50 Gy delivered to the whole body in a single dose can be defined as a high dose. The effects of high doses have been studied extensively in animal models; however, data on the effects of high doses of radiation in humans primarily come from populations exposed to the atomic explosions in Hiroshima and Nagasaki during World War II. These studies have been described in detail in books.[1–3] Before discussing acute radiation damage in humans, it is important to understand how radiation initiates damage, and why some cells are more radiosensitive than others (concept of *radiosensitivity*).

Ionizing radiation can initiate damage by generating free radicals in the cells and by causing ionization of molecules. Low linear energy transfer (LET) radiation such as x-rays or gamma rays initiates about two-thirds of the injury primarily by generating excessive amounts of inorganic and organic free radicals, and the remaining injury by producing ionization. On the other hand, high LET radiation such as proton and alpha particles initiate radiation damage primarily by causing ionization in molecules. Free radicals can damage every component of the cells, such as ribonucleic acid (RNA), deoxyribose nucleic acid (DNA), proteins, lipids, membranes, and carbohydrates. Ionized molecules become inactive. The energy released per ionization is about 33 eV, which is sufficient to break a chemical bond. For example, the energy associated with a C = C bond is only about 4.9 eV; therefore, an ionization can break this bond.[1]

Since high LET radiation is more damaging than low LET, the efficacy of particular radiation in causing specific damage is expressed as the relative biological effectiveness (RBE).

RBE = Dose to produce an effect with x-rays/dose to produce the same effect with a new radiation.

For x-rays or gamma rays, the RBE value is one; however, for heavy particles such as alpha particles, this value can be as high as 10, depending upon the energy of the alpha particles and the criteria of radiation damage.

RADIATION-INDUCED FREE RADICALS

The radicals referred to as free radicals are atoms, molecules, or ions with unpaired electrons. These unpaired electrons are highly reactive and play an important role in several biochemical reactions and gene expressions. Since about 80% of the cell contents represent water, the effects of x-rays on the generation of free radicals were investigated in water. When x-rays interact with the water, two types of free radicals are formed:

$$\text{X-rays} + H_2O \rightarrow H^{\bullet} \text{ (hydrogen free radical)} + OH^{\bullet} \text{ (hydroxyl free radical)}$$

The recombination of these free radicals yields the following:

$$H^{\bullet} + H^{\bullet} \rightarrow H_2 \text{ (hydrogen)}$$

$$OH^{\bullet} + OH^{\bullet} \rightarrow H_2O_2 \text{ (hydrogen peroxide)}$$

The presence of an excess of free oxygen during irradiation of cells allows the formation of additional types of free radicals:

$$H^{\bullet} + O_2 \rightarrow HO_2^{\bullet} \text{ (hydroperoxy free radical)}$$

$$HO_2^{\bullet} + HO_2^{\bullet} \rightarrow H_2O_2 + O_2 \text{ (oxygen)}$$

When an organic molecule, designated as RH, combines with a hydroxyl free radical, an organic free radical is formed:

$$RH + OH^{\bullet} \rightarrow R^{\bullet} \text{ (organic free radical)} + H_2O$$

The organic free radical then combines with O_2 to form peroxy free radicals:

$$R^{\bullet} + O_2 \rightarrow RO_2^{\bullet} \text{ (peroxy free radicals)}$$

It should be noted that the presence of excess oxygen allows the formation of two additional types of free radicals—hydroperoxy and peroxy free radicals—which, in part, may account for the increased radiation damage in the presence of excess oxygen.

Most of the free radicals are very short-lived and either readily combine with each other or attack biological molecules in cells. The lifetime of free radicals is generally less than 10^{-10} seconds. However, there are a few radicals derived from complex organic substances that are stable and do not readily combine with each other. Initially, it was thought that free radicals were generated in the cells only during irradiation, but it has been demonstrated that long-lived free radicals exist after irradiation. Furthermore, free radicals are also released during inflammatory reactions that occur immediately after radiation injury to cells.

CONCEPT OF RADIOSENSITIVITY OF CELLS

In order to appreciate radiation damage, it is essential to understand the concept of radiosensitivity of cells. In comparing the radiosensitivity of different types of cells, tissue, organs, or whole organisms, one must define the criteria for radiation damage. A statement about the radiosensitivity of cells without reference to the criterion for radiation injury is meaningless. Some cells may be considered radioresistant as judged by one criterion of damage, but highly radiosensitive by another.

Law of Bergonie and Tribondeau: One of the most important concepts in radiobiology is the Law of Bergonié and Tribondeau. In 1906, these two French investigators, while working on the effects of radiation of the rat testis, formulated a concept of the radiosensitivity of cells that has been remarkably valid in general terms. According to this concept, dividing cells are more radiosensitive than non-dividing cells on the criterion of reproductive death or cell death, and embryonic, undifferentiated dividing cells are more radiosensitive than adult dividing cells. Indeed, using the criterion of cell death, organs containing rapidly dividing cells, such as bone marrow, small intestine, and gonads are more radiosensitive than those organs containing nondividing cells, such as liver, kidney, and brain. The cells that exhibit an exception to this law are peripheral lymphocytes and oocytes, which do not divide and are highly differentiated cells. The Law of Bergonié and Tribondeau is not always applicable to tumor cells in vivo, especially tumor cells that recur after completion of radiation therapy.

On the cellular level, the nucleus is considered more radiosensitive than the cytoplasm, and DNA is more sensitive to radiation than RNA or proteins on the criterion of reproductive death. Initially, it was thought that cells undergo different phases of mitosis. Later, using mammalian cells in culture, it was discovered that a cell undergoes four different phases to complete one cell cycle. The time needed to complete one cell cycle is called *generation time*, and the time required for a cell population to double in number is called *doubling time*. The doubling time differs from one cell type to another. The phases of the cell cycle are: mitosis (M), pre-DNA synthesis (G_1), DNA synthesis (S), and post-DNA synthesis (G_2). The radiosensitivity of cells in each phase varies, depending upon the criterion of damage (Table 3.1).

TABLE 3.1
Radiosensitivity of Cells in Various Phases of the Cell Cycle

Criteria of Radiosensitivity	Most Radiosensitive Phase of the Cell Cycle
Reproductive death	Mitosis (M)
Chromosomal damage	Post-DNA synthesis phase (G2)
Division delay	Post-DNA synthesis phase (G2)
DNA synthesis	Early Pre-DNA synthesis (G1)

CELLULAR RADIOBIOLOGY

Dose–response (survival) curves: Most important radiobiology concepts have been developed on mammalian cells growing in culture. Therefore, it is important to understand something about the nature of cell cultures, and various parameters of a mammalian radiation survival curve. At present, tumor cells, immortalized cells, and normal cells in culture are used for radiobiology studies. Previously, most of our understanding of cellular radiobiology was developed on tumor cells in culture. Two types of dose–response curves on the criterion of survival after irradiation were established—nonlinear and linear. The nonlinear dose–response curve shows a threshold dose with a shoulder, indicating doses that cause sublethal damage, which can repair fully provided sufficient time is allowed between two doses. The survival after the threshold dose decreases exponentially as a function of dose. The linear dose–response curve has no threshold and no shoulder, and the survival decreases exponentially as a function of dose.

Effect of radiation factors on the dose–response curve: From the studies on cellular radiobiology, the following important principles have been established:

1. The higher the dose rate the greater the radiation damage.
2. Fractionation of a radiation dose reduced radiation injury compared to a single dose, with total doses being the same.
3. Protraction of a dose (delivered at a very slow dose rate) reduced radiation injury compared to a single dose, with total dose being the same.
4. High LET radiation is more damaging than low LET radiation.
5. The shape of the dose–response (survival) curve after irradiation with high LET radiation is linear without any shoulder. It is sigmoid with a shoulder after irradiation with low LET radiation. However, the dose–response curve after irradiation with low LET radiation at a very high dose rate becomes linear.
6. The presence of free oxygen during irradiation increases radiation damage on most criteria. The hypoxic cells (reduced amounts of available oxygen) become radioresistant compared to oxygenated cells.

Bystander effect: The central radiobiological concept has been that damaging effects of radiation are due to the direct effect on cell structures by ionization or free radicals; however, since the early 1990s, the bystander effect of radiation has been observed repeatedly. According to the concept of the *bystander effect*, irradiated cells release agents that can reduce survival, and induce cytogenetic alterations, apoptosis, and other biochemical changes in neighboring or distant nonirradiated cells. Several mechanisms for the bystander effect have been proposed. They include gap-junction-mediated intercellular and intracellular communications, secreted soluble factors, oxidative metabolism, plasma membrane–bound lipid rafts, and calcium flux.[4] The bystander cells can exhibit a number of abnormalities similar to those observed in irradiated cells. They include genetic and epigenetic changes, alterations in gene expression profiles, activation of signal transduction pathways, and late adverse effects in their progeny.

The investigation of changes in global gene transcript levels in human erythro-leukemic cells induced by direct irradiation (4 Gy) and by bystander effects revealed that alterations in the levels of transcripts were similar in 72% of the transcripts, whereas only 0.6% of the transcripts showed an opposite response[5] 36 hours after irradiation. It has been reported that unirradiated cells growing in the presence of irradiated cells exhibited increased proliferation; however, unirradiated cells must be in close proximity to irradiated cells.[6] This study suggests that direct cell-to-cell contact is essential for transmitting proliferative signals from irradiated cells to unir-radiated cells. In order to understand further the mechanisms of the bystander effect, the time course kinetics of intracellular distribution of protein kinase C (PKC) iso-forms (PKC-betaII, PKC-alpha/beta, PKC-theta) in irradiated and bystander human lung fibroblasts in culture was investigated.[7] The results showed that bystander cells had higher activation of PKC isoforms compared to irradiated and sham-irradiated cells. Proteins of PKC isoforms accumulated more in the nuclear fraction than in the cytosolic fraction.

It has been reported that the frequency of micronuclei formation in unirradiated bystander human normal fibroblasts increased from about 6.5% to about 9–13% from the irradiated cells receiving 0.1–10 Gy.[8] In addition, the levels of p21Waf protein and foci of gamma-H2AX (the phosphorylated form of the histone variant H2AX) increased in bystander cells, which were independent of doses received by the irradi-ated cells, whereas the survival of bystander cells decreased. The levels of reactive oxygen species (ROS) increased in both irradiated and bystander cells. Addition of Cu-Zn superoxide dismutase (SOD) and catalase to the medium decreased the for-mation of micronuclei and induction of p21Waf protein and foci of gamma-H2AX, but did not affect the survival of bystander cells. This suggests that irradiated cells release other toxic agents, in addition to ROS, and that free radicals released from the irradiated cells may be involved in increasing the frequency of micronuclei forma-tion and induction of p21Waf protein and foci of gamma-H2AX.

More recently, it has been demonstrated that p53-dependent release of cyto-chrome c from mitochondria of irradiation hepatoma cells was responsible for pro-ducing the bystander effect on normal Chang liver cells, which were co-cultures with irradiated hepatoma cells.[9]

The effects of irradiated glioblastoma cells on bystander cells are mediated through the release of cytokines IL-6 and IL-8.[10] It has been demonstrated that early activation of NF-kappaB-dependent gene expression levels (IL-6, IL-8, IL-33, TNF) first occurred in directly irradiated cells and then in bystander cells. An inhibition of IL-33 transmitting functions with an anti-IL-33 monoclonal antibody decreased the activation of NF-kappaB in both directly irradiated cells and bystander human fibro-blasts.[11] This suggested that IL-33, one of the agents released from the irradiated cells, was acting as a signal transmission to the changes observed in the bystander cells.

Mitochondrial DNA: Mitochondrial DNA (mtDNA) contains 13 genes that encode proteins of the oxidative phosphorylation complex, which are involved in adenosine triphosphate (ATP) generation. Radiation-induced damage to mtDNA may increase the radiosensitivity of the cell.[12]

Proteasome: Ionizing radiation causes damage by activating certain signal trans-duction pathways that are regulated by posttranscriptional as well as transcriptional mechanisms. One of the most important posttranscriptional pathways is the ATP- and ubiquitin-dependent degradation of proteins by the 26S proteasome. Irradiation with high (8–20 Gy) or low (0.25 Gy) doses of radiation immediately inhibited 20S and 26S proteasomes in vitro or in the intact mammalian cells.[13] This inhibitory effect of radiation on proteasomes is independent of the availability of the known endogenous proteasome inhibitor heat shock protein 70 (hsp90).

Repair of cellular radiation damage: X-irradiation of the cells simultaneously sets in motion the processes of repair and damage. Therefore, whether the cells live or die depends upon the relative dominance of the repair or damage process. Although it had been known for a long time that irradiated cells repair damage, Elkind and Sutton-Gilbert were first to quantitatively define the repair processes in mammalian cells in culture.[2] Cellular radiation damage can be divided into three categories: (1) sublethal damage, (2) potential damage, and (3) lethal damage. Damage that can be completely repaired by the cells, provided that a sufficient time interval after irradiation is allowed, is called sublethal damage. In the case of potential damage, the cells can repair radiation injury provided that an appropriate treatment is given after irradiation. Lethal damage cannot be repaired. This definition of lethal damage may not be totally accurate because lethal damage that can lead to apoptosis can be prevented by an appropriate treatment. All irradiated cells qualitatively sustain the same damage; the difference between sublethal, potentially lethal, and lethal damage is one of quantity.[1]

The repair of sublethal damage is independent of DNA, RNA, and protein synthesis, but it is reduced at very low temperature. Reduced oxygen interferes with the repair of sublethal damage. Although repair of sublethal damage occurs in all phases of the cell cycle, it is most efficient in the S phase. The repair of potential damage is also reduced at low temperature and when RNA synthesis is decreased.

BIOCHEMICAL CHANGES

Effects of ionizing radiation on biochemical changes that include DNA, RNA, proteins, carbohydrates, and lipids have been investigated. These effects have been described in detail elsewhere.[1]

DNA: The effects of radiation on DNA are dependent upon the dose, time, and phase of the cell cycle. X-irradiation of DNA solution with a dose of 10 Gy produces the following types of damage: (1) breakage of hydrogen bonds, (2) chain breaks, (3) cross linkage, (4) disruption of the sugar-phosphate backbone of DNA, (5) impairment of transforming ability of DNA, (6) DNA base damage, and (7) inability of the DNA to act as a template for the synthesis of a new DNA strand. When DNA is irradiated in vitro, single-stranded DNA is more radiosensitive than double-stranded DNA, on the criterion of base damage.[14] Several properties of DNA, such as DNA content per cell, deoxyribose level, viscosity, sedimentation, and priming ability, do not show any significant change when DNA is isolated immediately after irradiation of cells with radiation exposures of 750–1000R (about 7.5–10 Gy); however, when more time after irradiation is allowed, some changes of DNA properties can be demonstrated after high doses of radiation.[15]

The initiation of DNA synthesis in the slowly dividing human amnion cells in culture is much more radiosensitive than that in rapidly dividing cells.[16] In HeLa cells in culture, DNA synthesis is suppressed by about 50% after irradiation with 500–750 R (about 5–7.5 Gy) in the S phase of the cell cycle. These radiation exposures also prolong the total period of the S phase.[17] An exposure of 1500 R (about 15 Gy) to the liver during any period of regeneration completely inhibits DNA synthesis 24 hours after irradiation.[18] After irradiation of mouse leukemic cells, a new phenomenon of unscheduled DNA synthesis, also called DNA repair synthesis, was observed. The unscheduled DNA synthesis differs from S-phase DNA synthesis in the following aspects: (1) it occurs in all phases of the cell cycle, and (2) it is not inhibited by an inhibitor of normal DNA synthesis. The mechanisms of radiation-induced inhibition of DNA synthesis include a decrease in the DNA polymerase activity and the pool size of DNA precursors.

RNA: Generally, the total RNA synthesis is less radiosensitive than the DNA synthesis. In regenerating rat liver, an exposure of 1500 R (about 15 Gy) causes marked inhibition of DNA synthesis, but even an exposure of 3000 R (about 30 Gy) has no effect on the total RNA synthesis 24 hours after irradiation.

Proteins: Generally, proteins are very radiosensitive when they are being synthesized. The fully developed proteins are very radioresistant on the criterion of structural changes. However, the activities of certain enzymes can show increase, no change, or decrease after irradiation. For example, the activity of monoamine oxidase (MAO) in mouse liver decreases by about 71% after irradiation with 20 Gy; however, the enzyme activity does not change in the kidney, heart, and brain.[19] A temporary dose-dependent increase in MAO activity was observed in the liver, kidney, heart, brain, and retina of fish immediately after irradiation, but no such effect was observed in any organs of the mice.

Lipid and carbohydrate: The synthesis of lipids is enhanced during the first day after an exposure to 300–5000 R (about 3–50 Gy). The following factors influenced the rate of the synthesis of lipids: (1) dietary state, (2) type of lipid precursor utilized, (3) species, (4) age of the animal, (5) radiation dose, and (6) post-irradiation time of assay. The oxidative phosphorylation in radiosensitive cells is very sensitive to radiation, but it is radioresistant in cells that are resistant to radiation.

Other biochemical indicators: The deoxycytidine level in human urine markedly increases after radiation therapy treatment.[20] In rat urine, the level of deoxycytidine increased linearly as a function of whole-body exposures of 200–650 R (about 2–6.5 Gy). Beta-aminoisobutyric acid, a metabolite of thymidine, increased in the urine of irradiated individuals.[21] In addition, the urine levels of several amino acids, cysteic acid, valine, leucine, hydroxyproline, phenylalanine, arginine, aspartic acid, proline, threonine, creatine, and tryptophan were markedly elevated in irradiated individuals with exposures of 37–410 R (about 0.37–4.1 Gy).[21,22]

RADIATION SYNDROMES

Most studies with high doses of radiation were performed in rodents (mice and rats) as well as in cell culture models of normal and cancer cells derived from both rodents and humans. The sources of radiation have been primarily x-rays from x-ray machines of varying energy, delivering radiation dose at varying dose rates, and gamma radiation from radioactive isotopes, such as ^{60}Co and ^{137}Cs of varying energy, delivering radiation dose at varying dose rates. Occasionally, a linear accelerator has been used to deliver radiation at a very high dose rate.

The extent of radiation damage depends upon the dose, dose rate, mode of delivery (single vs. fractionated dose), surface area irradiated (whole body vs. partial body) and radiosensitivity of target organs. The most radiosensitive organs on the criterion of cell death include bone marrow, small intestine, hair follicles, and gonads. The responses to high-dose radiation differ among mammals, and are generally measured in terms of mortality rate and survival time after irradiation. The mortality rate can be measured as LD_{50} (a dose that produces 50% lethality) or LD_{100} (a dose that produces 100% lethality).

The signs and symptoms produced by delivering high doses of radiation in a single dose to the whole body are referred to as *radiation syndrome*. Radiation syndromes have been divided into three major categories: (1) bone marrow syndrome (BM), (2) gastrointestinal (GI) syndrome, and (3) central nervous system (CNS) syndrome. Each radiation syndrome is characterized by specific doses, survival, survival time, and signs and symptoms. Acute radiation sickness (ARS) includes both BM syndrome and GI syndrome. Most studies on radiation syndromes have been obtained from animals, primarily rodents who received whole-body radiation delivered in a single dose; however, very limited data on humans have been obtained from the atomic bombing of Hiroshima and Nagasaki and accidents at nuclear plants. These studies reveal that on the criteria of bone marrow syndrome, GI syndrome, and CNS syndrome, humans are more sensitive than rodents with respect to dose requirement and survival time. Each radiation syndrome is briefly described in subsequent sections. The *Handbook of Radiobiology* discusses radiation syndromes in great detail.[1]

Bone marrow syndrome: In order to produce BM syndrome in humans, low LET radiation doses delivered to the whole body in a single dose can vary from 2–4 Gy. The BM syndrome dose is often expressed as the LD_{50} dose. The LD_{50} to produce BM syndrome varies from one species to another (Table 3.2). The LD_{50} dose for rodents is about threefold higher than for humans. The BM syndrome in most species is often expressed in terms of $^{30}LD_{50}$ (50% lethality within 30 days), because all deaths occur within 30 days of irradiation; however, in humans, 50% of irradiated individuals die within 60 days of irradiation. Therefore, for humans, the dose of BM syndrome is expressed as $^{60}LD_{50}$.

Clinical signs and symptoms in humans: The signs and symptoms of BM syndrome are caused by radiation damage to the bone marrow. The extensive damage to the lymphatic system and to other radiosensitive organs plays a minor role in causing death. Nausea, vomiting, and fatigue are commonly observed in individuals exposed to doses that produce BM syndrome. These symptoms are sometimes associated with diarrhea. Infection and bleeding are prominent. The severity of symptoms depends upon the radiation dose. Loss of hair may occur 2 to 6 weeks after irradiation. The BM syndrome in humans has been described in great detail in two books.[23,24]

Laboratory findings: The pattern of changes in peripheral blood varies depending upon the doses of radiation and time after irradiation. Initially, granulocytosis is observed within the first 2–4 days after irradiation, followed by leukopenia during

TABLE 3.2
Variations in Bone Marrow Syndrome Doses
in Different Species

Species	LD_{50} Value in Gy
Human	2.7–3.0
Monkey	3.98
Rodents (rat and mice)	8.5–9.0
Desert mice	11.0–12.0
Hamster	9.0
Rabbit	8.4
Dog	2.65
Sheep	1.55
Swine	1.95
Guinea pig	2.55
Marmoset	2.0

Source: Data summarized from K. Prasad, *Handbook of Radiobiology*, 2nd edition (Boca Raton, FL: CRC Press, 1995) and V. P. Bond, "Radiation Mortality in Different Species," in *Comarative Cellular and Species Radiosensitivity*, ed. V. P. Bond and Tsutomu Sugahara, Tokyo: Igaku Shoin, 1969), p. 5.

weeks 4 and 5. Recovery of leukocytes can be observed at 36 days after irradiation. Changes in the lymphocyte counts are less variable. The number of lymphocytes decreases maximally (about 800–900/mm³ within 3–4 days of irradiation), and remained at the same level for at least 5 weeks. The number of lymphocytes increases slightly thereafter, but does not return to normal values for weeks or months. The severe decline in platelets is not seen until 28 days after irradiation. After 24 days, the number of platelets dropped below 50,000/mm³. A transient mild anemia associated with reticulocytosis may be found in most irradiated individuals. A *late critical phase* occurs during weeks 4 and 5. This phase is characterized by a severe granulocytopenia and thrombocytopenia. The *late critical phase* includes nausea, vomiting, fever, and diarrhea. About 50% of the Hiroshima and Nagasaki population exposed to BM syndrome doses died during this late critical phase. The survival of irradiated individuals depends upon recovery of the bone marrow. The major cause of death is infection. This is supported by the fact that controlling infection alone can increase the survival rate of irradiated individuals.

Animal studies have revealed [25,26] that age has an impact on the value of $^{30}LD_{50}$. For example, a group of 3-month-old animals require 850 R (about 8.5 Gy) to cause $^{30}LD_{50}$, whereas a group of 7-month- old animals needed 900 R (about 9 Gy) to produce the same. This suggested that young animals were more sensitive to radiation on the criterion of BM syndrome. A higher dose rate also increases the value of $^{30}LD_{50}$.

Radiosensitivity of bone marrow–derived stem cells: The recovery of bone marrow after irradiation depends upon the repopulation of resident stem cells, which have capacity for both self-renewal and differentiation. These cells remain in a non-proliferative state that protects them from radiation-induced cell cycle–specific killing as well as from cytokine and free radical–mediated cell killing.[27] Irradiation of brain with 10 Gy or less causes minimal histopathological changes, but it can induce variable degrees of cognitive dysfunction that are due to loss of neural stem cells. Irradiation of neural stem cells in culture with a dose of 5 Gy caused increased apoptosis and G2M checkpoint delay. Increased levels of oxidative and nitrosylative stress was found in irradiated human neural stem cells.[28]

Gastrointestinal (GI) syndrome: The signs and symptoms associated with the GI syndrome may be due to the failure of the intestinal mucosa and the bone marrow, but the damage to the intestinal crypt cells plays a dominant role in causing death of irradiated individuals. The GI syndrome in humans is not as well defined as it is in animals in terms of the dose–effect relationship. The GI syndrome in humans has been described in great detail in two books by Bond et al. and Rubin and Casarett.[24,29]

The dose requirement to produce GI syndrome varies between rodents and humans. In rodents, it is 12–40.0 Gy, and in humans it is 5–40.0 Gy (delivered to the whole body in a single dose). Thus, in contrast to the range of doses that produce BM syndromes, the range of doses that produce GI syndrome is wide. The doses that produce GI syndrome cause 100% mortality in 3.5 days in rodents, and 14 days in humans. Again, humans require about a twofold lower radiation dose than rodents to produce GI syndrome. The survival time in humans is also longer than that in rodents.

Clinical signs and symptoms in humans: The initial symptoms resemble airsickness or seasickness and may be influenced by psychological factors and individual susceptibility. The common signs and symptoms include abrupt (within 2 hours) loss of appetite, gastric complaint, and apathy, soon followed by nausea and vomiting, which increases to a maximal intensity at about 8 hours after irradiation. These symptoms subside as rapidly as they develop. On the second day, the general health of the irradiated individuals appears much improved, although nausea and occasional vomiting persist. The symptoms subside on the third day. The clinical condition on the third day may be deceptively encouraging for a day or two before abrupt onset of the symptoms of GI syndrome. They include malaise, anorexia, nausea, vomiting, high fever, persistent diarrhea (may be bloody in some cases), electrolyte imbalance, and abdominal distension leading to the clinical picture of severe paralytic ileus. During the second week after irradiation, severe dehydration, hemoconcentration, and circulatory collapse are observed, eventually leading to death. The major cause of death is the denudation of villi of the small intestine; however, other factors, such as infection, hemorrhage, and electrolyte imbalance, contribute to the rate of progression of injury and seriousness of illness.

Laboratory findings: A marked decrease in granulocyte, lymphocytes, and platelet counts can be seen. The severity of these changes is dose dependent. The crypt of the small intestine becomes devoid of cells, and thus, the elongated structure of villus collapses with a few cells lining its wall. It has been reported that radiation-induced acute intestinal inflammation differs following whole-body vs. abdominopelvic irradiation with gamma rays in ferrets. The animals were exposed to whole-body or abdominopelvic irradiation with a dose of 5 Gy and various parameters of inflammation were measured. The results showed that the activity of myeloperoxidase increased in the ileum 2 days after abdominopelvic irradiation, but it was reduced after whole-body irradiation.[30] The levels of prostaglandin E2 (PGE2) in the ileum increased after abdominopelvic or whole-body irradiation. Whole-body irradiation of rats with a dose of 10 Gy gamma radiation increased the activity of inducible nitric oxide synthase (iNOS) in the ileum, but not in the colon.[31] The result suggests that iNOS-derived NO may participate in radiation-induced ileal dysfunction, independent of the development of an inflammatory response.

Central nervous system (CNS) syndrome: The CNS syndrome in humans is poorly defined; therefore, animal data are primarily discussed. The doses of 50 Gy or more delivered to the whole body in a single dose can produce CNS syndrome in both humans and rodents. The survival time is dose dependent. The larger the dose, the shorter the survival time. All humans or animals die within 24 hours of irradiation. The CNS syndrome in humans is described in two books.[24,29]

Clinical signs and symptoms in humans: This syndrome is characterized by periods of agitation, marked apathy, followed by signs of disorientation, balance problems, ataxia, diarrhea, vomiting, opisthotonus, convulsion, coma, and death within 24 hours. The major causes of death are increased inflammatory reactions damaging blood vessels, neurons, and enhanced intracranial pressure. The death occurs so soon that no treatment is possible. The time of onset and progression of subsequent signs and symptoms are dose dependent. The histopathology of the brain is characterized by acute inflammatory reactions throughout the brain, and death is due to the damage to neuronal cells and increased intracranial pressure.

ACUTE RADIATION SICKNESS (ARS) IN ACCIDENTLY EXPOSED INDIVIDUALS

ARS includes both BM syndrome and GI syndrome. In the Chernobyl accidents, based on the clinical symptoms of nausea, vomiting, and diarrhea, 237 individuals initially were considered to have been exposed to radiation (beta/gamma radiation) doses that can produce ARS. Later, the diagnosis of ARS was confirmed in only 134 cases. Of the 28 individuals who received whole-body doses in excess of 6.5 Gy, 95% died within a short period of time, possibly from GI syndrome.[32] Those who died during the first two months exhibited bone marrow syndrome, and some of them who may have received doses of 6 Gy or more may have died of GI syndrome. It was further revealed that among 134 individuals who exhibited ARS, 54 patients suffered from cutaneous radiation syndrome (CRS), and 16 of 28 irradiated dead individuals actually died of CRS rather than BM syndrome. Fifteen survivors of the accident who received localized radiation exposure from radioactive fallout were followed up between the years 1991 and 2000. All patients showed varying degrees of CRS. Some of the symptoms of CRS included xerosis, cutaneous telangiectasias and subungual splinter hemorrhage, hemangiomas and lymphangiomas, epidermal atrophy, disseminated keratoses, extensive dermal and subcutaneous fibrosis with partial ulcerations, and pigment changes.[33]

DAMAGE TO ORGAN SYSTEMS

The effects of ionizing radiation on organ systems have been extensively studied and have been described in detail in two books.[1,24]

Skin: Extensive works have been published on the effects of ionizing radiation on the skin. As a matter of fact, the first effect of ionizing radiation appeared as acute radiation dermatitis, also called *radium burn*, among investigators who were working with radium. Prior to the introduction of the unit R as a measure of radiation exposure in air, radiation therapists utilized a skin reaction referred to as the *threshold erythema dose* (TED) to express radiation dose. The degree of skin reactions after irradiation depends on several factors: (1) total radiation dose, (2) dose rate, (3) mode of delivery of radiation dose (single vs. fractionate dose), (4) LET, (5) size of the field, and (6) anatomical location. The radiosensitivity of the skin of the same anatomical location markedly differs from one individual to another. Furthermore, the various structures of the skin display marked differences in radiosensitivity. The extent and rate of recovery depend upon the degree of radiation injuries. The hair follicles of the skin are the most radiosensitive. The hair follicles of children are more sensitive to radiation than those of adults for the same anatomical location. The radiosensitivity of the sebaceous glands is similar to that of the hair follicles, but the damage persists longer than that in the skin. Sweat glands are not as radiosensitive as the hair follicles or the sebaceous glands. Degenerative changes in the sebaceous glands are primarily responsible for the dryness and the tendency toward scaling of the irradiated skin.

Radiation responses of the skin under condition of radiation therapy (delivered to the skin either 2 Gy per day, 5 days a week, for a total dose of 40 Gy, or 2.5 Gy per day, 5 days a week, for a total of 50 Gy) have been described in detail by Rubin and Casarett.[24] The responses that are seen within 6 months include the following.

Initial erythema: The initial erythema is generally seen within a few hours to a few days after irradiation and lasts only a day or so. This type of reaction is primarily due to the dilation of the capillaries caused by the release of histamine-like substances.

Dry desquamation: This condition gradually develops after initial erythema. The dry desquamation is characterized by atrophy of epidermal papillae, epidermal hypoplasia, and vascular changes, and is generally accompanied by temporary depilation and incomplete sloughing of the epidermis.

Erythema proper: The erythema proper generally develops in the third or fourth week after irradiation. The skin becomes red, warm, edematous, and tender, and it exhibits a burning sensation. The erythema proper appears to be associated with obstructive changes in arterioles.

Moist desquamation (exudative radiation dermatitis): If the skin reaction during the period of erythema proper is severe, acute radiation dermatitis may develop into moist desquamation in the fourth week after irradiation. Some of the pathological changes include the following: (1) blister formation in the epidermis, (2) dermal hypoplasia, (3) edema, (4) inflammatory cell infiltration, (5) damage of vascular and connective tissues, and (6) permanent depilation.

Recovery of skin: The recovery of the skin depends primarily on the extent of damage to the vascular and connective tissues of the dermis. If the damage is not severe, reepithelialization begins with the island of cells that generally coalesces in 6–8 weeks. The new skin is thin and pink, but it returns to normal appearance in about 2–3 months.

Development of necrosis: Necrosis of the skin after irradiation seldom develops; however, infection may cause necrosis of the irradiated area. Even without infection, if the blood vessels and connective tissues are damaged due to high radiation doses or preexisting diseases, a necrotic ulcer may develop in about 2 months.

Hyperpigmentation: Following erythema proper, an increase in pigmentation, which is primarily due to an increase in the synthesis of melanin, may occur. The degree of pigmentation varies from one region of the skin to another in the same individual, and from one individual to another.

Mucous membrane: The response of mucous membrane to radiation is called *radiation mucositis*. The radiation-induced changes in the mucous membranes are similar to those observed in the skin. However, they appear somewhat earlier in the mucosa than in the skin. The rate of recovery in the mucous membranes may be more rapid than in the skin.

At the end of the second week after irradiation, patients complain of soreness, dysphagia, and pain when swallowing. Dryness of the mouth may be observed, and coarse and seasoned foods are poorly tolerated. Taste may be altered and appetite may decrease. Mastication of dry, solid food is more difficult. Erythema, prominence of papillae, and patchy radiation mucositis limited to the palate and uvula may appear. There may be hypersecretion by the mucous glands.

At the end of the third week after irradiation, the throat may become congested and the tongue becomes swollen. The saliva appears thick and tenacious. The patchy mucositis becomes more confluent and extends onto the tonsil pillars.

At the end of the fourth week, the degree of damage increases. The mucositis extends onto the buccal mucosa, and the epithelial surface is denuded.

At the end of the fifth week, the damage to mucosa is at a maximal level, and maintenance of nutrition becomes very difficult. The mucositis at this time appears as a whitish to yellowish pseudomembrane, which is confluent and limited to the irradiated area. It is difficult to remove by scrapping. If this pseudomembrane is removed, a superficial ulceration that bleeds can be observed. Regeneration of mucosa can begin after irradiation, and may be completed in about 2–3 months after irradiation, if the damage is not severe. After irradiation, varying degrees of squamous-cell metaplasia and hyperplasia may be observed in the mucous glands of the oral mucosa.

Nervous system: It is generally assumed that adult nervous tissue is highly radio-resistant. This may be true on the criterion of cell death that may appear within a few months. Functional changes in neuron or glial cells occur at comparatively low doses. The extent of radiation damage to the brain depends on those radiation and biological factors that have been described in the section of the skin.

During development of the nervous system, some cells lining the inner layer of the neural tube differentiate into neuroblasts, which do not divide. Each neuroblast matures into a neuron. Some cells of the neural tube differentiate into spongioblasts, which give rise to supporting elements of the nervous system. The neuroblasts are most radiosensitive on the criterion of cell death; even a few rads of ionizing radiation can kill neuroblasts. Significant loss of neuroblasts can lead to severe brain abnormalities.

Neuron and its associated structures: Adult neurons are very radioresistant on the criteria of cell death. A dose of 100 Gy or more may be needed to kill neurons. Dendrites and axons are also very radioresistant. In contrast, activity of synapse is sensitive to radiation. A dose of 2 Gy can alter synaptic activity.

Glial cells: In the adult CNS, glial cells normally divide infrequently, but may proliferate rapidly following a proper stimulus such as tissue damage. Different types of glial cells in the adult brain are less radioresistant than the adult neurons. Damage to oligodendrocytes can cause transient demyelination.

White matter vs. gray matter: The white matter of the CNS appears to be more sensitive to radiation than the gray matter on the criterion of necrosis, which occurs due to the damage of the vascular system.

The signs and symptoms of acute brain damage may include sudden increase in headache, lethargy, vomiting, and papilloedema. If irradiation is continued in spite of these symptoms, convulsive seizures may occur.

Spinal cord: Radiation responses of the spinal cord are called *radiation myelitis* because of the presence of products of acute inflammation. Serious damage to the spinal cord may lead to invalidism and even death. The spinal cord lesions after irradiation include hyaline thickening of the blood vessels with obliterated lumen and dissolution of white matter. The neurological signs include lower-limb motor weakness, which rapidly leads to paralysis with involvement of the sphincter.

Peripheral nerves: Peripheral nerves are also very radioresistant. They are more resistant than the spinal cord. This is shown by the fact that doses of ionizing radiation that induce necrosis in the spinal cord have no effect on the peripheral nerves.

Reproductive Tract: The gonads are extremely radiosensitive. Although there are sufficient data available on the effect of ionizing radiation on the reproductive tracts of animals, data on human gonads, especially testis, are limited. This is due to the fact that testis are rarely irradiated in clinical settings. Most of the information on testis is based on a few accidental radiation exposures and some clinical observations.

Testis: The processes of forming sperm in the testis is referred to as *spermatogenesis*. The highly proliferating spermatogonia lining the seminiferous tubules give rise to primary and secondary spermatocytes, which then form spermatids. The spermatids do not divide. Each spermatid matures into one spermatozoon. On the criterion of cell death, spermatogonia are most radiosensitive, whereas spermatids and spermatozoa are most radioresistant. However, on the criterion of genetic damage, all cells during spermatogenesis are equally radiosensitive. The interstitial cells, which produce male hormone, are very radioresistant. This is supported by the fact that men exposed to a sterilizing radiation dose of 500–600 rads (5–6 Gy) retain their fertility and produce seminal and prostate fluid for a while. The extent of damage depends upon the same factors that were described in the section on skin. Radiation doses that can cause sterility in men are provided in Table 3.3.

Ovary: Ovary is very radiosensitive. The hormone-producing cells are moderately radiosensitive. Consequently, radiation-induced sterilization may be associated with the development of menopause. Effects of radiation on sterility in humans are shown in Table 3.3. Age is a very important factor in producing radiation-induced sterility.

TABLE 3.3

Effect of Ionizing Radiation on Sterility in Humans

Sex	Radiation Exposure (R)	Sterility Status
Male	500–600	Permanent
Male	250	Temporary for 12 months
Female	320–625	Permanent
Female	125–150	Amenorrhea in 50% of women
Female	170	Temporary sterility for 1–3 years

Source: Data summarized from K. Prasad, *Handbook of Radiobiology*, 2nd edition (Boca Raton, FL: CRC Press, 1995).

Younger females (less than 40 years) require larger doses than those over the age of 40 years in order to induce menopause. Doses greater than 150 R (about 1.5 Gy) may sterilize 90% of the older women, whereas only 50% of the younger women may be affected by these doses. Exposure to doses in excess of 200 R (about 2 Gy) causes cessation of the menstrual function, which is temporary in young women but permanent in those over 40. Radiation doses of 550–650 R (about 5.5–6.5 Gy) substantially reduced urinary estrogen excretion in women with normal menstruation.[34]

Cardiovascular system: The radiation response of the heart is called *radiation carditis* because of the presence of the products of acute inflammation. The extent of radiation damage to the heart depends on those radiation and biological factors that have been described in the section on the skin. The myocardial cells are very resistant to ionizing radiation on the criterion of morphological changes; however, the pericardial cells and fine vasculatures are moderately sensitive to radiation. Radiation-induced alterations include swelling of the fiber, loss of longitudinal and cross-striation, homogenization of the sarcoplasm, complete disappearance of the protoplasm, and persistence of hollow sarcolemma. Nuclei may show pyknosis, fragmentation, and lysis. The intima of small arterioles may become thick. Myocardial infarction rarely occurs after irradiation; however, there have been frequent reports of pericarditis, pericardial effusions, pericardial adhesion, and pericardial fibrosis. Hemorrhage is the main manifestation of radiation damage to the major vessels. Doses in the range of 5000 R (about 50 Gy) or more have produced rupture of the aorta. The damage of blood vessels may cause thickening of intima and formation of thrombi.

Respiratory system: The radiation response of the respiratory system is called *radiation pneumonitis* because of the presence of products of acute inflammatory reactions after irradiation. The richly ramified vascular system and lymphatic tissues are radiosensitive, but cartilage of air passages and the pleura are radioresistant. The symptoms of acute inflammatory reactions may subside or may lead to chronic lung fibrosis. The lung injury is characterized by the accumulation of fibrin-rich exudates within alveoli, the thickening of alveolar septa by fibrillar materials, and cellular proliferation. The fibrin condenses at the alveolar walls to produce the so-called *hyaline membrane.*

The radiation and biological factors mentioned in the section on the skin are also applicable to the lung. The size of the surface area irradiated is very important for the extent of lung injury. If less than 25% of the lung is irradiated with fractionated total doses of 3000 R (about 30 Gy), no clinical effects of lung injury are seen. If larger areas of the lung are irradiated, the degree of radiation pneumonitis is severe. If both lungs are irradiated with the same total doses, it is generally fatal. The dose rate also affects the severity of radiation pneumonitis. The presence of chronic pulmonary diseases, such as interstitial fibrosis, emphysema, and pneumoconiosis, probably increases the sensitivity of the lung to radiation. It has been suggested that monocyte/macrophage and lymphocyte recruitment and activation are essential for radiation-induced pulmonary fibrosis.[35]

Urinary tract: The radiation response of the kidney is called *radiation nephritis* and occurs 6 months to a year after irradiation. The kidney is considered moderately radiosensitive. The major complaints are swelling of the legs, shortness of breath, headache, and vomiting. The main features of radiation nephritis are an early hyperemia and increased capillary permeability, which result in interstitial edema. Endothelial cells of the fine vasculature show degeneration and necrotic changes, followed by a rapid proliferation, which eventually may block the lumen of the vessels. The kidney cortex becomes pale, probably due to the blockage of the afferent glomerular arterioles.

The occurrence of hypertension depends upon radiation doses. With larger doses, the hypertension may develop earlier and to a greater degree. Hypertension causes degeneration and thickening of the small arterioles and the larger vessels of the kidney. Previously irradiated renal vessels are more susceptible to hypertension-induced necrosis than are renal vessels that are not previously irradiated. Three young children died of renal failure and hypertension within 2 months after irradiation with therapeutic doses of 5200–6850 R (about 52–68.5 Gy) over the renal area. Five adults who received radiation doses of 2300 R (23 Gy) to the kidney died of acute radiation nephritis 506 months after irradiation.

Liver: The radiation response of the liver is called *radiation hepatitis*. The liver is a highly radioresistant organ on the criterion of morphological changes. A radiation therapeutic dose of 4000 rads (40 Gy) can produce radiation hepatitis in about 75% of patients. The characteristic features of radiation hepatitis include severe sinusoidal congestion, hyperemia or hemorrhage, some atrophy of the central hepatic cells, and mild dilation of central veins. Children are more sensitive than the adults on the criterion of radiation hepatitis.

The regenerating liver is highly radiosensitive, like any proliferating system; however, the liver maintains the capacity to regenerate even after high doses of radiation. Even after a dose of 20,000 R (about 200 Gy) delivered to the liver immediately after partial hepatectomy, regeneration occurred and the liver attained 90% of the original weight.

Pancreas and bile ducts: The radiation response of the pancreas is called *radiation pancreatitis*. Very little data is available for radiation pancreatitis in humans; therefore, animal studies are presented in this section. The pancreas is very radioresistant. The insular epithelium is more radioresistant than the acinous epithelium. Pyknosis and necrosis of the islet beta cells were observed in monkeys who died within 8 days after whole-body irradiation with doses of 10,000 to 50,000 R (100–500 Gy). The alpha cells are relatively more radiosensitive than the beta cells on criterion of cell death. The LD_{50} value for the alpha cells is 5000 R (about 50 Gy), whereas the LD_{50} value for the beta cells is 20,000 R (about 200 Gy).

It has been suggested that, in humans, epithelial cells of the biliary ducts—particularly the small and medium ducts—appear to be more radiosensitive than the liver cells. The secretion of bile salt in dogs was unaffected even after high radiation doses.

Glands: Glands in general are very radioresistant on the criterion of morphological changes. The effects of irradiation on glands are described below.

Salivary glands: The salivary glands are radioresistant on the criterion of cell death. The symptoms of radiation damage include dryness of the mouth within 4–6 hours after irradiation. Enlargement of the parotid and submaxillary glands usually start at this time, and the gland reaches an enormous size. The saliva becomes viscous and frothy because of increased mucous secretion. This results in cessation of secretion and drying of the oropharyngeal mucosa.

Adrenal gland: Adrenal glands are highly radioresistant on the criterion of cell death. The ability of the adrenal glands to respond to stress was markedly reduced.

Thyroid and parathyroid glands: The thyroid and parathyroid glands are very radioresistant on the criterion of morphological changes. The radiosensitivity of the thyroid is influenced by its metabolic activity. The more active thyroid gland exhibits higher radiosensitivity than the normal active thyroid gland. The damage to the thyroid is primarily due to damage of the vascular system. High doses of radiation may produce edema, inflammation, and hemorrhage in the thyroid. The colloid contents of damaged follicles become reduced after irradiation. The follicular basement membrane may rupture following irradiation. Children's thyroid glands are more radiosensitive than those of adults.

Muscle and bones: Striated muscle is highly radioresistant on the criterion of morphological changes; however, delayed necrosis of muscle has been observed after therapeutic doses of radiation.

Growing bones and cartilage are relatively more radiosensitive. The cervical spine can be altered after irradiation. Osteophytic formation and diminution of disc interspaces are also observed after irradiation. Mandibular growth impairment can also be seen after irradiation in some cases, causing facial asymmetry.

The early changes in growing cartilage following irradiation include: (a) reduction in mitotic index of chondroblasts, (b) degeneration and necrosis of chondroblasts, (c) reduction in the number of chondroblasts, and (d) enlargement of lacunae.

The osteoblasts are much more resistant to radiation than the proliferating chondroblasts, and consequently osteoid deposition and osteogenesis may continue after cartilage growth has stopped. Mature cartilage and bone are very radioresistant. One of the most important delayed effects of radiation damage to bone is increased susceptibility to fracture from strain or traumatic injury.

Teeth: Teeth are very radioresistant on the criterion of morphological changes. However, high doses of radiation can induce hypersensitivity of the teeth to heat, cold, and sweet food. Decay of the neck of the tooth may be seen after irradiation. Irradiation of children's teeth causes a marked retardation in tooth and bone development.

Eye: The eyes consist of complex structures that show variation in radiosensitivity. The lens is the most sensitive structure; the conjunctiva and cornea are moderately radiosensitive, and the retina and optic nerve are radioresistant. In the lens, radiation-induced inhibition of mitosis and death of epithelial cells interfere with the progression of differentiation and deposition of lens fibers. The degree and duration of this process of lens fiber disorganization determines the degree of opacity or cataract. After irradiation with therapeutic doses, conjunctivitis may appear and photophobia becomes pronounced. In addition, the secretion of the lachrymal gland is suppressed, and the eye is bathed in thicker secretions. The eyes are more readily irritated and become susceptible to infection. The normal mucous membrane of the conjunctiva is converted into a roughened keratinized surface. The retina of the young is more radiosensitive than the retina of older individuals. However, in the mature retina, necrosis may develop secondary to the damage of the fine vasculature. Massive doses of radiation can cause complete blindness.

Ear: The radiation response of the middle ear is called *otitis media*, whereas the radiation response of the inner ear is referred to as labyrinthitis. The radiation damage to the middle and inner layers is primarily due to damage of capillaries and fine vasculature. These changes may cause disturbances in the cochlear function, which can lead to an abnormal sensitivity to loud. The damage to the cochlear function can lead to hearing loss.

GENETIC BASIS OF RADIATION RESPONSE IN VITRO AND IN VIVO

Genetic changes following high doses of radiation are very complex and depend upon the levels of radiation doses, types of cells, genes of interest, and time after irradiation when assay of changes in genetic activity are measured.

Oncogenes in radiosensitivity: Increased expressions of certain oncogenes, such as c-myc, H-ras, and raf appear to be associated with enhanced radioresistance.[36–38] The combination of high expression of v-myc and H-ras produces a synergistic effect on radioresistance.[39] The mechanisms of oncogene-induced radioresistance are very complex. Some of them are described briefly here. The membrane localization of ras-encoded p21 protein is critical for the maintenance of the radioresistance phenotype of mammalian cells in culture.[40] In addition, both RalA and RalB, which are downstream effectors of ras function contribute to K-ras-dependent radioresistance of the pancreatic cancer–derived cell line.[41] This is supported by the fact that suppression of Ral expression increased radiosensitivity of these cells due in part to decreased efficiency of repair that requires both Ral1A and Ral1B, and increased susceptibility to apoptosis, which needs only Ral1B. Increased expression of ras enhanced the level of the Ku80 protein, which is one of the key enzymes that participate in the repair of DNA double-strand breaks. The role of Ku80 in mediating radioresistant effects of ras is further supported by the fact that overexpression of Ku80 in mouse fibroblasts increased radioresistance of the cell.[42] The radioresistance of overexpression of ras is also mediated through phosphoinositide 3 (PI3)-induced phosphorylation Akt (protein kinase B). This is supported by the observation in which selective inhibition of PI3 kinase reduced phosphorylation of Akt and survival of irradiated cells.[43] It has been demonstrated that an inhibitor of Akt increased radiosensitivity of glioma cells in culture.[44] It appears that Akt activation protects against radiation-induced apoptosis through the regulation of phosphorylation and expression of acinus. Thus the Akt/Nf-KappaB/acinus pathway acts as one of the important regulatory mechanisms for modulating radiosensitivity of mammalian cells.[45] The c-myc oncogene-induced radioresistance appears to be mediated through two proteins, MLH1 and MLH2, which play a role in DNA mismatch repair. It has been reported that downregulation of myc inhibited the levels of MLH1 and MLH2, and increased radiosensitivity of melanoma cells in culture. This effect is independent of p53.[46] The role of raf in radioresistance is further supported by the fact that inhibition of raf-1 expression by antisense oligonucleotides increased the radiosensitivity of hepatocarcinoma cells and decreased the capacity for the repair of potentially lethal damage.[47] It is interesting to note that increase in the expression of raf-1 was found to be associated with an increase in the expression of superoxide dismutase (SOD), which may in part account for raf-induced radioresistance in mammalian cells.[48] It has been reported that the expression of the immediate early response gene c-fos increased after irradiation. It has been suggested that c-fos enhanced both p53-dependent and radiation- and p53-independent steroid-induced apoptosis in T-lymphocytes.[49]

X-irradiation increases the expression of proto-oncogene, c-jun in mammalian cells.[50] This rise in c-jun expression can be prevented by pretreatment of the cells with an antioxidant, n-acetylcysteine or an inhibitor of protein kinase, H7.[51] These studies suggest that protein kinase C and free radicals may be involved in radiation-induced enhancement of expression of c-jun. It has been demonstrated that doses of 10–20 Gy markedly enhanced the expression of c-jun and c-fos at the time of onset of DNA fragmentation and apoptosis; however, further studies revealed that increased expressions of c-jun and c-fos are not a general requirement for radiation-induced apoptosis in Jurkat T cells in culture.[52] The role of PKC in radiosensitivity of mammalian cells is further confirmed by the fact that an inhibitor of PKC increased the sensitivity of cells to radiation and repressed c-fos transcription in normal LM cells (fibroblasts).[53] Bax accelerates apoptosis in normal cells, but it is weakly expressed in tumor cells. Overexpression of Bax (Bcl2-associated X protein) increased radiosensitivity of breast cancer cells in culture.[54] It has been demonstrated that increased expression of p210 bcr/abl increased radiosensitivity of hematopoietic progenitor cells in culture. This effect was considered due to amplification of radiation damage to DNA by inhibiting the DNA repair process.[55]

Increased expression of melanoma differentiation-associated 7 (MDA-7), also called *interleukin-24*, inhibited the growth of a variety of tumor cells in culture, but has no effect on normal cells. It enhances the radiosensitivity of glioma cells in culture.[56] It has been demonstrated that irradiation causes an increase in the expression of the double-stranded RNA-activated protein kinase R (PKR) in a dose- and time-dependent manner that leads to an increase in radioresistance of mouse embryo fibroblasts in culture. This effect of PKR is in part mediated through the protective effects of NF-kappaB and Akt activation.[57] Overexpression of Bcl-2 increases the radioresistance of cells. It has been shown that inhibition of Bcl-2 increases radiosensitivity of human prostate cancer cells in culture.[58] The anti-apoptotic B-cell lymphoma-2 (Bcl-2) family of proteins, such as Bcl-2 and Bcl-x, increase radioresistance of tumor cells in culture. Downregulation of the myeloid leukemia cell differentiation gene (Mcl-1), another member of the Bcl-2 family, increases the radiosensitivity of melanoma cells in culture.[59] Radiation-induced apoptosis is associated with an increase in expression of the pro-apoptotic gene Bax. The insertion of

TABLE 3.4

Role of Wild Type p53 in BM Syndrome and GI Syndrome in Animals

Deletion of p53	Protect animals from developing BM syndrome[65]
Overexpression of p53	Promote BM syndrome in animals[65]
Deletion of Bax and Bak1	Protect animals from developing BM syndrome[66]
Deletion of p53	Promoted GI syndrome in mice[76]
Overexpression of p53	Protected mice from developing GI syndrome[76]
Deletion of Bax and Bak1	Did not protect mice from GI syndrome[76]

the Bax gene into radioresistant esophageal cancer cells enhanced the radiosensitivity in association with increased apoptotic cell death.[60]

Notch-1, a transmembrane protein, plays a key role in the development of many tissues and organs, and is present in tumor tissue. It has been demonstrated that human prostate cancer cells contain high levels of Notch-1, which promotes metastasis of human prostate cancer cells.[61] Silencing the activity of Notch-1 prevents invasion of human prostate cancer. Inhibition of the Notch-1 pathway increases the radiosensitivity of glioma cells. Indeed, overexpression of Notch-1 or Notch-2 protects glioma cells against radiation damage.[62] It has been demonstrated that Notch regulates expression of the epidermal growth factor receptor (EGFR). Notch inhibition reduced EGFR mRNA and EGFR protein in glioma and other cancer cell lines. This effect of Notch-1 was mediated via p53 in glioma cells.[63] Valproic acid, an inhibitor of histone deacetylase, activated the Notch-1 pathway in small-cell lung carcinoma.[64] Based on these observations, one can predict that pretreatment of cells with valproic acid would prevent radiation injury by increasing the level of the Notch pathway.

Role of P53 in BM syndrome and GI syndrome: The role of p53, a tumor suppressor protein and a regulator of the cell cycle, in regulating BM syndrome is totally different from that in GI syndrome. It has been reported that the deficiency of PUMA (p53 upregulated mediator of apoptosis) in mice increases radioresistance of hematopoietic cells.[65] Even the loss of one allele of PUMA is sufficient to induce radioresistance in hematopoietic cells. Null mutation in PUMA protects both primitive and differentiated hematopoietic cells from damage produced by low doses of radiation, but selectively protects hematopoietic stem cells and hematopoietic progenitor cells. These results suggest that PUMA is required for radiation-induced apoptosis in hematopoietic cells. Therefore, the elevated levels of p53 may promote bone marrow syndrome. This is further supported by the fact that deletion of PUMA protects mice against BM syndrome.[66] This would suggest that drugs that can block the intrinsic pathway of apoptosis may be useful in protecting against BM syndrome. Indeed, an inhibitor of p53, pifithrin-alpha, protects mice against radiation-induced BM syndrome.[67] In a recent study, common binding sites of 56 transcriptional factors that might be involved in the early radiation response of hematopoietic cells in vivo have been identified.[68]

Ionizing radiation activates p53 in GI, the epithelium cell that induces PUMA to kill cells by the intrinsic pathway of apoptosis.[69–71] Deletion of either p53 or PUMA blocks radiation-induced apoptosis of most epithelial cells of the intestine. Therefore, p53-mediated apoptosis has been implicated in the GI syndrome.[71] Others have reported that deletion of p53 does not alter the GI syndrome, and proposed that radiation-induced apoptosis of endothelial cells plays an important role in the GI syndrome.[72] Although radiation-induced apoptosis in endothelial cells can proceed in the absence of p53 or PUMA,[71,73] the intrinsic pathway of apoptosis may still be involved because transgenic mice with deleted Bak1 and Bax showed reduced apoptosis in the endothelial cells after irradiation.[74,75] Others have suggested that p53 may protect mice from radiation-induced GI syndrome.[67] It has been reported that selective deletion of pro-apoptotic genes Bak1 and Bax from the GI epithelium or from endothelial cells did not protect mice from radiation-induced GI syndrome.[76] Selective deletion of p53 or PUMA from the GI epithelium, but not from the endothelial cells, sensitizes irradiated mice to develop GI syndrome, whereas transgenic mice with overexpressed wild-type p53 in all tissues protected the mice from developing GI syndrome.[76] An inhibitor of p53, pifithrin-alpha, did not protect mice against radiation-induced GI syndrome.[67]

Role of p53 in cellular radiation response: The p53 gene, a tumor suppressor gene, plays an important role in genomic stability, cell cycle, apoptosis, and radiation response of normal and tumor cells. It has been demonstrated that exposure to ionizing radiation causes a G1 cell cycle delay (also, referred to as a G1 check point) due to p53-mediated inhibition of G1 cyclin kinase and retinoblastoma (Rb) protein function in normal fibroblasts. Radiation does not induce the G1 check point in tumor cells, which contain mutated p53.[77]

It has been demonstrated that p53 mutant mouse embryonic fibroblasts are more radiosensitive than those with wild-type p53.[78] A dose of 6 Gy increases p53 by 2–3 times in wild-type p53 cells, but it remained unchanged in cells with mutant p53. The same dose of radiation produced G2/M arrest in all cell lines, indicating that p53 is probably not involved in radiation-induced G2/M arrest. On the other hand, no p53 mutant cells exhibited G1/S delay after irradiation, and no apoptosis was observed in these cells. These studies suggest that the mutant of p53 affects radiosensitivity of cells independent of apoptotic pathway.

Wild-type p53 plays a role in the repair of radiation damage to DNA, whereas mutated p53 increases resistance of cells to ionizing radiation.[79] Irradiation of mice with a whole-body dose of 7 Gy caused an increase in the levels of proteins of p53 and PUMA in the granulocytes.[80] However, irradiation of mice did not change the levels of mitogen-activated protein kinase (MAPK) p38 in the nucleus and the cytoplasm of the granulocytes, but it induced phosphorylation of p38 and its accumulation in the cytoplasm during apoptosis signaling.

In response to damage to DNA, p53 stimulates the CDKN1A gene, which codes for p21, which leads to inhibition of cyclin-dependent kinase activity and cell cycle (G1/S) arrest. Irradiation with x-rays or high LET radiation decreased CD kinase (CDKN1A) activity within 24 hours, whereas it increased the level of p21, which peaked at 3–6 hours and remained at the high level after 24 hours.[81] The localization of p21 after irradiation depends upon the LET. It has been demonstrated that in normal human fibroblasts in culture, p21 is localized in the nucleus after irradiation with high LET radiation, where it is diffused within the cells after x-irradiation.[82] Irradiation of primary human fetal osteoblasts induced phosphorylation of p53, c-Jun, and p38 and expression of p21, as well as cell-cycle arrest and apoptosis.[83] Irradiation also induced phosphorylation of NF-kappaBp65 and NF-kappaB activation in human fetal osteoblast cells in culture. An inhibition of NF-kappaB expression by siRNA upregulated p21 expression and induced apoptosis in human fetal osteoblast cells in culture. This suggests that NF-kappaB plays an important role in protecting osteoblasts against radiation damage.

Changes in expression of gene profiles: Investigation of expression of human gene profiles following irradiation reveled the following: (a) changes in gene profiles after irradiation with low LET radiation were similar to those observed after irradiation with high LET radiation, suggesting the genetic pathway damage may be independent of LET; and (b) about one-third of genes differently expressed in cells irradiated at a high dose rate (1 Gy/minute) compared to a low dose rate (1 Gy/22 hours), total dose being the same.[84] It is established that a higher dose rate causes greater damage compared to a low dose rate. The previously cited study showed that the pathway of damage caused by a high dose rate may in part be different from that of a low dose rate.

Microarray analysis of gene expression profiles revealed that the predominant gene expression response in neonatal rat cardiac myocytes in culture was upregulated, whereas it was downregulated in rat neonatal fibroblasts.[85] Genomic instability is characterized by the increased rate of accumulation of changes in mammalian DNA. Ionizing radiation induced genetic instability by increasing oxidative stress.[86,87] Genomic stability is closely linked with the development of cancer.

Repair of DNA double-strand breaks: It is now established that DNA double-strand breaks (DSBs) are critical radiation-induced cellular lesions, which if left unrepaired can lead to mutations or chromosomal aberrations, and cell death. In addition, they promote the onset of chronic diseases associated with genomic instability, such as cancer. One of the most important cellular responses to DSBs is the accumulation and localization of a plethora of DNA damage–signaling and repair proteins in the vicinity of the lesion, initiated by ATM-mediated phosphorylation of histone H2AX (gamma-H2AX), resulting in the formation of microscopically visible nuclear domains referred to as radiation-induced foci (RIF). As a matter of fact, DNA damage sensor p53-binding protein (53BP1) is localized to the sites of DSBs within seconds to minutes after irradiation.[88–90] It has been demonstrated that H2AX was phosphorylated in the liver and kidney of ATM gene knockout mice, suggesting that ATM kinase is not essential for phosphorylation of H2AX in these organs after x-irradiation in vivo.[91]

Whole-body irradiation of rats with doses ranging from a $^{30}LD_{50}$ to $^{30}LD_{100}$ increased the concentration of serum proteins associated with acute responses in a dose-dependent manner. These proteins include alpha2-macroglobulin, haptoglobin, fibrinogen, and cysteine protease inhibitor, which increased by 2 to 5 times compared to unirradiated cells, whereas alpha1-acid glycoprotein was increased by 6 times.[92]

Increased expression of EGFR is associated with increased radioresistance of tumor cells.[93] It has been proposed that EGFR signaling pathways may be associated with DNA repair mechanisms, especially nonhomologous end joining repair of DNA double-strand breaks.[94] It has been demonstrated that radiation-induced stabilization and activation of src kinase, which initiates caveolin-1-driven EGFR internalization and nuclear transport. This effect of radiation can be mimicked by the addition of hydroxyl-nonenal (HNE), one of the major products of lipid peroxide.[95] Inhibition of internalization and nuclear transport of EGFR- blocked radiation-induced phosphorylation of DNA-dependent protein kinase (DNA-PK) and interfered with repair of radiation-induced DNA double-strand breaks.[96]

RAD 51 is a human gene that is a highly conserved gene from yeast to humans.[97,98] The RDA51 group of proteins includes RDA 51 paralogs, RAD52, RAD54, and RAD55. They assist in the repair of radiation-induced DNA double-strand breaks and play a major role in homologous recombination of DNA during the repair of double-strand breaks. RDA51 protein binds to the DNA at the site of a break and encases it in a protein sheath, which is a necessary first step in the repair of damage to DNA. The RDA51 protein in the nuclei of many cells may interact with other proteins, including breast cancer genes (BRCA1 and BRCA2) to repair damaged DNA. The BRCA2 protein regulates RDA51 protein activity by transporting it to the sites of DNA damage in the nucleus. BRCA1 protein also appears to activate the RDA51 protein in response to DNA damage. It has been demonstrated that irradiation of cells causes accumulation of the RDA51 group of proteins at the sites of DNA damage as a foci. In the yeast, absence of RAD52 prevents foci formation by RAD51 and RAD54, whereas the foci of RAD51 are formed in the absence of RAD55 and RAD51 paralog. In contrast to yeast, RAD52 is not needed for foci formation of RAD51 and RAD54 in mammalian cells.[99] In addition, radiation-induced foci formation of RAD51 and RAD54 is impaired in all RAD51 paralog and BRCA2 mutant cell lines, whereas RAD52 foci formation is not affected by a mutation in any of these proteins. These data suggest that despite their evolutionary conservation from yeast to mammals, RAD52 appears to contribute to the DNA responses to damage in a different way.

It has been demonstrated that BCCIP, an interacting protein with BRCA2 protein and P21[100,101] is critical for BRCA2- and RAD51-dependent responses to DNA damage and homologous recombination repair.[102] It has been suggested that BCCIP-dependent homologous recombination repair is an early RAD51 response to ionizing radiation–induced DNA damage, and that RAD52-dependent homologous recombination repair occurs later.[103]

MUTATIONS AND CHROMOSOMAL DAMAGE

In 1927, Dr. H. J. Muller of Columbia University in New York, first demonstrated that x-irradiation of the sperm of the fruit fly (*Drosophila melanogaster*) induced mutation in the offspring. Most mutations in key genes are considered harmful for the organisms; however, mutations have been the basis for evolution of animal species. Mutation may result from a change in nucleotides, which can alter the normal coding for amino acids in proteins. This is called *gene mutation*. Mutation may also result from *damage to chromosomes*. The chromosomal damage may result in loss, duplication, and rearrangement of the chromosomes, and may be caused by a single break or a double break of the chromosome. Most (up to 90%, depending upon the experimental system) radiation-induced mutations are deletion type (loss of a part of chromosome) and are recessive in nature. Radiation-induced genetic changes are described in detail elsewhere.[104]

The frequency of gene mutation increases linearly with dose, and is independent of dose rate (at least in the fruit fly), fractionation, LET, and temperature. The dose–response curve for single-break chromosomal damage is linear, and it resembles the dose–response curve of gene mutation. However, the dose–response curve for double-break chromosomal damage is of the sigmoid type, requiring a threshold dose for an effect.

Ionizing radiation often causes chromosomal breakage that can result in loss of genetic materials. Examples of such damage are *terminal deletion*, which requires one break, and *interstitial deletion*, which requires two breaks in a chromosome.

Ionizing radiation can cause *duplication* of a chromosome, which requires one break. The free piece of chromosome may be added onto another chromosome, so that cells after mitosis contain a duplication of certain genes. *Translocation* chromosomal damage requires two breaks and occurs when an exchange of segments between two chromosomes occur without any loss of genetic materials. *Inversion* chromosomal damage requires two breaks and occurs when there are two breaks on the same chromosome and the resulting piece is inverted and reinserted into the same chromosome. *Dicentric ring* chromosomal damage occurs when the rearrangement of a chromosome can lead to the configuration of a dicentric ring. *Nondisjunction* chromosomal damage may produce trisomy (can cause Down's syndrome) and monosomy (can cause Turner's syndrome).

Unlike in *Drosophila melanogaster*, the frequency of mutation in mice is dependent upon the dose rate.[105] Irradiation of spermatogonia at a lower dose rate of 0.009 R/min (about 0.008 cGy) markedly reduced the frequency of mutations in comparison to those irradiated with the same total dose at a higher dose rate of 90 R/min (about 0.9 Gy). In spermatogonia, the frequency of mutations from a high dose rate to a low dose rate is reduced by a factor of 3. There is no evidence of a threshold dose rate for inducing mutations in male mice.

In female mice, the effect of dose rate on the frequency of mutations is different from that found in males. For example, the mutation frequency in oocytes at a high dose rate of 90 R/min (about 0.9 Gy) is greater and at a low dose rate of 0.009 R/min (about 0.009 cGy) less than that of spermatogonia. In females, the frequency of mutation did not change when radiation exposure of 100 R (about 1 Gy) was delivered in two fractions 24 hours apart; however, in males, the frequency of mutations markedly enhanced under the same radiation conditions. The incidence of radiation-induced dominant lethal mutations in spermatozoa after irradiation with a dose of about 4 Gy is about 50% higher in rats than in mice.

CELLULAR IMMUNE RESPONSE TO IONIZING RADIATION

Ionizing radiation suppresses the immune response of mammals, including humans. It does so by inhibiting innate immunity, which includes destruction of mechanical barrier, cellular defense mechanisms, clearance mechanisms, bactericidal efficiency of the blood, and the properdin system, or by inhibiting the formation of antibodies. These issues in detail have been discussed previously.[1,106]

TABLE 3.5
Effects of Ionizing Radiation on the Lethality of Rat Embryos

Gestational Age (day)	LD_{50} Dose (Gy) at the Term
0–5	0.30–1.20
9	1.20
10	1.6
11	2.1
12	2.4
Adult	60

Source: Summarized from R. L. Brent, "The Effect of Radiation on Mammalian Fetus," *Clin Obstet Gynecol* 3 (1960): 928–935.

Mechanical barriers: It has been shown that irradiation with high doses can impair the barrier function of certain tissues that normally prevent the penetration of infective agents into the blood stream. Examples of such tissues are the skin, lung, and intestinal mucosa.

Cellular defense mechanisms: Radiation damages phagocytic mechanisms. For example, intestinal microorganisms are found in the spleen and liver cultures of mice 60 hours after whole-body irradiation with 12 Gy. This showed that phagocytic cells in the spleen and liver were unable to engulf and destroy the microorganisms. The macrophages are markedly radioresistant on the criterion of morphological changes.

Clearance mechanisms: The natural resistance of organisms to certain infective agents depends on their ability to eliminate them from their bodies within a short time. Ionizing radiation is known to impair this ability.

Bactericidal efficiency of the blood: It has also been found that ionizing radiation impairs the normal bactericidal activity of rabbit serum.

The properdin system: In some animals, the production of properdin, a serum protein, takes place without any antigenic stimulation. The serum protein participates in the destruction of bacteria and viruses. The properdin level in dogs irradiated with high doses of radiation decreases. However, the level of properdin increases among those who survive after irradiation. Administration of properdin before irradiation partially protected rats irradiated with the $^{30}LD_{50}$ dose.

Antibody formation: The effect of radiation on antibody formation depends on whether the organisms are in contact with antigens for the first time (primary response) or whether they have previously been in contact with antigens (secondary response).

Primary response: Primary response is generally more radiosensitive than the secondary response. Ionization radiation reduces the formation of antibodies. The extent of the radiation-induced decrease in antibody formation depends upon the dose, post-irradiation time of antigenic stimulus, and nature of the antigen. A dose of 1.75 Gy given 3 days before the antigen administration caused a greater depression of the hemolysin response than the dose of 6 Gy given 2 days after the antigenic stimulus. The relationship between the effect of irradiation on hemolysin formation and the time of antigen administration is summarized as follows:

1. The maximum reduction of antibody occurs when irradiation was given 12–24 hours before the administration of antigen.
2. When the antigen is administered just after irradiation, the number of antibodies produced is almost equal to normal values, but the production rate is less than the normal rate.
3. When the antigen is administered shortly before irradiation, the amount of antibodies produced is almost equal to normal values, but the production rate is higher than the normal rate.

Similar results were obtained by other investigators using different antigens and different species of animals. Certain quantitative differences appear to be due to the differences in species, in the half-life of antigens, and in different rates of metabolism of antibodies in the body.

The formation of antibodies may involve three specific periods: (1) the preinduction period, (2) the induction period, and (3) the production period. The radiosensitivity of these periods differs.

Pre-induction period: This period covers a short time (1–4 hours) during which antibody-synthesizing mechanisms are initiated. This phase is highly radiosensitive. The dose–response curve for this criterion of damage is a typical sigmoid type.

Induction period: This period is relatively less radiosensitive. Irradiation does not affect the total quantities of antibodies produced, but their rates of production are slower.

Production period: The production period of antibody formation is highly radioresistant. The antibodies produced by irradiated rabbits are similar to those produced by nonirradiated animals.

Secondary response: The secondary response is relatively more resistant to irradiation than the primary response. A dose of 1.53 Gy causes a 50% reduction in the primary response, whereas a dose of 4.37 Gy is needed for a similar decrease in the secondary response.

Radiosensitivity of immune cells: Peripheral lymphocytes are one of the most radiosensitive cells in the human body, and this may account for the high radiosensitivity of immune response. Among the various types of lymphocytes, B-lymphocytes, and unprimed T-lymphocytes are very radiosensitive, whereas macrophages, primed T-lymphocytes, and plasma cells are radioresistant. Natural killer cells (NK cells) destroy tumor cells. A dose of 30 Gy totally destroys the activity of NK cells.

Changes in cytokines: Irradiation of human umbilical endothelial cells in culture increased the production of IL-6 and IL-8 and upregulated the expression of intercellular adhesion molecule 1 (ICAM-1).[107] Irradiation of murine splenocytes in culture with 5 Gy reduced the expression of a T helper cell type 1 (Th1-type) cytokine, interferon (IFN)-gamma, whereas the levels of IL-2 expression did not change.[108] Irradiation also significantly reduced the expression of IL-2 and INF-gamma in concanavalin-activated splenocytes. On the other hand, the expression of TH-2 (T helper cell type 2) cytokines, such as IL-4, IL-5, and IL-10, was increased in both naïve and activated splenocytes after irradiation. In addition, irradiation increased the levels of pro-inflammatory cytokines in naïve splenocytes. It has been reported that murine splenocytes inhibited IFN-gamma expression, phosphorylation of signal transducer and activator of transcription (STAT1) and cell-mediated immunity.[109] Whole-body irradiation of rats with 10 Gy increased the expression of pro-inflammatory cytokines, such as IL-6, IL-8, and TNF-alpha in the ileal muscularis layer of the small intestine, whereas the expression of anti-inflammatory cytokine IL-10 was markedly reduced.[110] A dose of 4 Gy markedly enhanced the level of IL-6 proteins in human umbilical vascular endothelial cells in culture.[111] This effect of radiation is mediated through mitogen-activated protein kinase/p38-mediated NF-kappaB activation. It has been demonstrated that thorax irradiation with a dose of 12 Gy in mice induced expression of several pro-inflammatory cytokines, such as TNF-alpha, IL-1alpha, and IL-6 in the bronchial epithelium soon after irradiation. A long-lasting release of these cytokines by the bronchiolar and alveolar epithelium and by inflammatory cells was observed at the onset of acute pneumonitis. Thus, pro-inflammatory cytokines promote inflammation in the lung through recruitment and activation of inflammatory cells.[112]

It has been shown that irradiated hematopoietic tissue exhibits increased macrophage activation and lysosomal and nitric oxide synthase activities.[113] Activation of macrophages after irradiation was associated with enhanced respiratory burst activities and neutrophil infiltration. Using p53-null mice, it was demonstrated that macrophage activation and neutrophil infiltration were not due to the direct effects of irradiation, but they were a consequence of the recognition and clearance of radiation-induced apoptotic cells. Increased phagocytic activity and inflammatory responses continued after clearance of apoptotic cell debris.

Irradiation of skin cells in vivo and in vitro increased the expression of ICAM-1 and its protein level, which can contribute to cutaneous radiation syndrome.[114] Protein kinase C, mitogen-activated protein (MAP) ERK kinase, p38MAP kinase, and phosphatidylinositol-3 kinase appear to be involved in induction of radiation-induced ICAM-1, because their inhibitors significantly reduced this effect of radiation on the skin. The anti-inflammatory agent dexamethasone suppressed the radiation-induced elevation of ICAM-1 expression. It has been demonstrated that the expression of ICAM-1 and E-selectin increased after irradiation in a dose-dependent manner, whereas the expression of P-selectin and vascular cell adhesion molecule-1(VACM-1) did not change.[115] These changes in cell adhesion molecules after irradiation occur in the absence of TNF-alpha and interleukin-1 production. It has been reported that irradiation of mice with a dose of 10 Gy gamma radiation induced iNOS-mediated epithelial dysfunction in the absence of an inflammatory response.[116] Irradiation of human umbilical vein endothelial cells in culture markedly increased IL-32 expression through multiple pathways, including activation of NF-kappaB, induction of cytosolic phospholipase A2 (cPLA2) and lysophosphatidylcholine (LPC), as well as induction of Cox-2 activity, which leads to conversion of arachidonic acid to prostacyclin. Additionally, IL-32 significantly enhanced radiation-induced expression of VCAM-1 and leukocyte adhesion on endothelial cells. These results suggest that IL-32 acts as a positive regulator in radiation-induced vascular inflammation.[117]

RADIATION EFFECTS ON FETUS

Most studies with high doses of ionizing radiation on fetuses have been performed on animal models. Data on the effects of irradiation on human fetuses come from the survivors of the atomic bombing of Hiroshima and Nagasaki, and from pregnant women who received diagnostic doses of ionizing radiation. These studies have been described in detail in a book.[1]

Developing organisms constitute a highly dynamic system in which rapid cell proliferation, cell migration, and cell differentiation occur. Therefore, it is expected that the radiation response of the embryo as a whole, or of specific tissue, would markedly differ, depending upon the stage of development. Several animal studies have suggested that radiation-induced damage is similar in different species when they are irradiated with the same dose and at the same stage of development.[118,119]

In humans, the fetuses are most radiosensitive to radiation-induced damage between 8 and 15 weeks. The most sensitive period in the etiology of a cell is during its transformation from the embryonic state to the adult state, whether it be neuroblast, myoblast, or erythroblast. In humans, such transformation occurs between days 18 and 38 of gestation. Therefore, during this period, irradiation of the fetus causes the highest incidence of congenital anomalies. It is also clear that, during this period, a fractionated dose would produce a greater number of organ abnormalities than a single dose, because a great variety of formative cells would be exposed to radiation; therefore, more organ primordia would be damaged at their critical stage of radiosensitivity. It should be pointed out that this is the period during which neither the woman nor her physician might suspect a pregnancy. The first three weeks of human embryonic development is the most radiosensitive period on the criterion of lethality. Cells during this period are undifferentiated and are rapidly dividing. Depending upon the dose, the irradiated cells could either recover or die. The recovered cells may undergo normal differentiation and maturation; therefore, radiation damage may not be reflected in terms of organ abnormalities. However, these cells may carry functional damage or mutational changes that may not manifest during the F1 generation. The LD_{50} dose of radiation for the embryo observed at the term increases as a function of gestational age at the time of radiation exposure.[120,121]

A dose of 5.1 Gy delivered in a single dose before 18 weeks of gestation causes 100% mortality in the human fetus. Radiation-induced congenital abnormalities are highest during days 18–45 of gestation in humans. After the organs become differentiated and are largely involved in further growth and maturation, radiation-induced congenital anomalies or lethality are greatly reduced, and the effects of radiation are largely expressed as pathologic damage to the tissues, organs, and associated functional impairment. These radiation-induced defects may become detectable at some time during one's lifespan. Irradiation of the fetus at any time can cause some loss of neural connections, resulting in functional deficiencies.

TABLE 3.6
Effect of X-Irradiation on the Implantation of the Zygote

	Average % of Normal Implantation			
Age (hours)	Control	10 R	25 R	50 R
2	85	83	81	76
6	85	79	71	61
7	85	76	71	59
24	85	87	77	83

Source: Data are presented in a simplified form from R. Rugh, M. Wohlfromm, and A. Varma, "Low-Dose X-Ray Effects on the Precleavage Mammalian Zygote," *Radiat Res* 37, no. 2 (1969): 401–414.

RADIATION EFFECTS ON EMBRYOS DURING PREIMPLANTATION OF ZYGOTES

The studies on the radiation effects on the preimplantation of the zygote have been performed primarily on animal models. The period of preimplantation includes the cleavage, morula, and blastocyte stages of the embryo. These stages of the embryo are very sensitive to radiation and can either die or survive after irradiation, depending upon the doses. Those irradiated embryos that survive do not show any abnormality with respect to morphology, size, short- and long-term survival, and reproductive ability.[122–124] The effect of x-irradiation on the implantation of the zygote in mice is presented in Table 3.6 .[125] It appears that the most reduction in implantation occurred when the zygote was 7 hours old at the time of irradiation with a radiation exposure of 50 R (about 0.5 Gy), although a significant reduction in implantation also occurred when the zygote was 6–7 hours old at the time of irradiation with an exposure of 25 R (about 0.25 Gy). It has been reported that when pregnant mice are irradiated during the preimplantation period (soon after sperm entry), loss of the sex chromosome occurs, which may result in XO females (Turner's syndrome in humans).[126] The frequency of this effect in pregnant mice irradiated with an x-ray exposure of 100 R (about 1 Gy) is about 4%.

RADIATION EFFECTS ON EMBRYOS AFTER IMPLANTATION OF ZYGOTES

The studies on radiation effects after implantation of the zygote have been performed primarily on animal models (Table 3.7). The results showed that the number of abnormal implantations of the zygote irradiated at 7 hours after implantation increased in a dose-dependent manner. When irradiated at 2 hours after implantation with a radiation exposure of 50 R (about 0.5 Gy), the number of abnormal

TABLE 3.7

Effect Irradiation after Implantation of the Zygote in Animals

Age	Radiation Exposure	No. of Implantations	No. of Abnormal Implantations (% of Implanted Zygotes)
7 hours	0	596	15
7 hours	10 R	583	23
7 hours	25 R	583	31
7 hours	50 R	583	47
2 hours	50 R	581	24

Source: Data from R. Rugh, M. Wohlfromm, and A. Varma, "Low-Dose X-Ray Effects on the Precleavage Mammalian Zygote," *Radiat Res* 37, no. 2 (1969): 401–414.

implantations was 24% compared to 47% among those that were irradiated with the same radiation exposure at 7 hours after implantation. This suggested that the period of 7 hours after implantation of a zygote was very sensitive to radiation on the criterion of abnormal implantation.

RADIATION EFFECTS ON EMBRYOS DURING ORGANOGENESIS

Irradiation of embryos during organogenesis produced an increased number of organ defects. Protraction of radiation doses (continuous radiation at a very low dose rate) reduces the incidence of gross organ abnormalities. This may be due to the fact that the threshold dose received within the sensitive period may be less than that needed to produce detectable injury to the organ. On the other hand, fractionation of a single dose is more effective than a single dose. This may be due to the fact that more precursors of organs are damaged by fractionated doses. The effect of irradiation on organogenesis is summarized in Table 3.8. A dose of less than 5 rads (0.05 Gy) delivered during organogenesis can significantly increase the risk of brain defects in humans.

On the criterion of brain abnormalities, neuroblasts (not dividing, one neuroblast matures to one neuron) are more sensitive to radiation than neuroectoderms (dividing), whereas neurons are highly radioresistant. This is due to the fact that the reduced number of neuroblasts after irradiation would proportionally decrease the number of neurons in the adult brain. On the other hand, dividing neuroectoderms may repopulate after initial damage, and thus may not change the number of neurons in the adult brain. X-irradiation of human fetuses at any time before the completion of neurogenesis induces severe CNS abnormalities. Among CNS defects, microcephaly (often associated with mental retardation) is most common. These defects may not be apparent upon histological examination, regardless of the degree of neuronal loss. About 64% of the children exposed in utero to atomic bomb radiation within 1200 meters, and who appeared normal at delivery, showed microcephaly with mental retardation by 4.5 years of age. Microcephaly is particularly associated with irradiation delivered during the early stages of pregnancy. It has

TABLE 3.8
Effect of Irradiation during Organogenesis

Radiation Exposure	Species	Effects on Organs
25 R	Mouse	Skeletal defects
<10 R	Mouse	Mitotic delay in brain
5 rads	Mouse	Kill 50% oocytes
5–10 rads	Humans	17% microcephaly
<5 rads	Humans	11% microcephaly

Source: Data were summarized from Committee on the Biological Effects of Ionizing Radiation, *The Effects on Populations of Exposure to Low Levels of Ionizing Radiation* (Washington, DC: National Academic Press, 1980) and L. B. Russell, "Effects of Low Doses of X-Rays on Embryonic Development in the Mouse, *Proc Soc Exp Biol Med* 95, no. 1 (1957): 174–178.

Note: Period of organogenesis: 7.5–2.5 days in mice; 14–50 days in humans.

been reported that when fetuses were exposed to radiation between 4 and 13 weeks of gestation, the incidence of microcephaly was 28%, whereas it was only 7% when irradiation was delivered after 13 weeks of gestation.[127]

Table 3.9 describes the incidence of microcephaly among children of Hiroshima who were exposed to atomic bomb radiation during the most radiosensitive fetal phase (6–11weeks).[128] It has been reported that 11% of fetuses of less than 18 weeks of age who received irradiation with doses between 0.10 to 0.19 Gy developed microcephaly. In a BEIR Report of 1990, it was proposed that a threshold dose for mental retardation may lie between 0.2 to 0.4 Gy. The incidence of seizure was highest among subjects who were exposed to radiation between 8 and 15 weeks of gestation with doses higher than 0.10 Gy. The severity of seizure increased linearly as a function of dose to fetuses.[129] No seizure was observed among individuals who were exposed to irradiation from 0 to 7 weeks of gestation with doses higher than 0.10 Gy. Another study reported that radiation doses greater than 500 mGy may significantly increase the risk of growth retardation and CNS damage.[130]

RADIATION EFFECTS ON THE EYE OF EMBRYOS

Among eye defects, anophthalmia and microphthalmia are common responses in rodents exposed to radiation exposure of 100–300 R (about 1–3 Gy) at a particular stage during fetal lifetime. Although developing retina is very sensitive to radiation, it exhibits a remarkable capacity to reconstitute and repair the damaged area. Even after reconstitution of the retina, there is evidence of persistence damage in the form of microphthalmia. A dose of 0.15 Gy to the rat fetuses produces brain and eye abnormalities.[131]

TABLE 3.9
Effect of Irradiation during the Most
Radiosensitive Period (6–11 weeks)
on the Incidence of Microcephaly

Air Dose[a]	Incidence of Microcephaly (%)
0.01–0.9 Gy	11
0.10–0.19 Gy	17
0.20–0.29 Gy	30
0.30–0.49 Gy	40
0.50–0.99 Gy	70
<1.0 Gy	100
0.0 Gy	4

Source: Data were summarized from Committee on the Biological Effects of Ionizing Radiation, *The Effects on Populations of Exposure to Low Levels of Ionizing Radiation* (Washington, DC: National Academic Press, 1980) and W. J. Blot, "Growth and Development following Prenatal and Childhood Exposure to Atomic Radiation," *J Radiat Res* (Tokyo) 16 Suppl (1975): 82–88.

[a] 0.01–0.09 Gy air dose = average fetus dose of 0.013 Gy of gamma rays plus 0.001 Gy of neutron radiation, totaling 0.014 Gy.

RADIATION EFFECTS ON THE GONADS OF EMBRYOS

Animals: It has been reported that when pregnant rats were exposed to whole-body radiation with 100 R (about 1 Gy) of x-rays at days 13.5, 15.5, 17.5, and 19.5 of gestation, and gonads were analyzed 100 days after birth, the radiosensitivity of germinal elements of the female was maximal at 15.5 days after gestation. In contrast, the male germ cells became increasingly radiosensitive between 17.5 to 19.5 days of gestation. In females, severe depopulation of oocytes appeared to be associated with a slight decline in the secretory activity of the ovary. On the other hand, in males there was complete elimination of germ cells without affecting the endocrine function of the testis.[132] In another study, female mice were whole-body irradiated with radiation exposures of 20 R (0.2 Gy) and 80 R (about 0.8 Gy). The female offspring showed reduction in oocytes; however, the reduction was less at the age of 50 days than at 28 days after birth.[133] It has been shown that radiation-induced fetal germ cell apoptosis in testis of mice is dependent on activation of p53. This effect of p53 requires an upregulation of Fas.[134]

Humans: The effect of gamma irradiation on the function of human fetal testis during the first trimester of gestation was investigated, using an organ culture system.

The results showed that germ cells are very sensitive to radiation even at doses of 0.1 and 0.2 Gy. At these doses, about one-third of germ cells died by apoptosis, while others were blocked in their cycle. Although sertoli cells were less sensitive to radiation than germ cells, their proliferation rate and the level of anti-Müllerian hormone were reduced. Irradiation had no effect on the production of testosterone. The apoptosis of germ cells is mediated through activation of p53, which was associated with increased expression of the pro-apoptotic genes Bax and PUMA , whereas the expression of anti-apoptotic gene Bcl2 remained unaffected. The expression of p21, which is responsible for cell cycle arrest, was also elevated after irradiation.[135]

It has been reported that both male and female germ cells display a similar number of phosphorylated histone protein gamma H2AX foci in response to irradiation.[136] In organ culture of human fetal ovaries, doses 1.5 Gy or above induced apoptosis. Accumulation of p53 was found only in male human germ cells, but not in female cells. This suggested that p53-mediated apoptosis in germ cells is involved in males, but not in females. This was further confirmed by the fact that pifithrin-alpha, an inhibitor of p53, did not affect the oogonia apoptosis. Using p53 knockout mice, it was demonstrated that germ cell survival in testis was better in comparison to that obtained using wild type mice, whereas female germ cell survival was unaffected by p53 or p53 knockout.

IDENTIFICATION OF A NOVEL MECHANISM OF ACTION OF RADIATION

Investigators from the Russian Federation have identified large molecular weight (200–250 kDa) antigenic molecules comprised of 50% protein, 38% lipid, 10% carbohydrate, and 1.3% mineral residue from the lymphatic vessels shortly after high doses (6–10 Gy) of whole-body gamma irradiation.[137] These doses of radiation produced 100% lethality in animals within seven days. The antigenic molecules, when injected subcutaneously into different unirradiated species of animals (cattle, pigs, sheep, horses, dogs, mice, and rats), produce syndromes similar to those produced by high doses of radiation. These radiation toxin molecules have been referred to as specific radiation determinants (SRDs). At least seven variants of SRDs are associated with the specific syndrome of ARS that they produce rather than with the species of animal. Although multiple variants of SRDs are produced following high doses of irradiation, a predominant variant of SRD is responsible for producing specific ARS syndrome. For example, SRD 1 produced cerebral ARS, SRD2 produced nonspecific toxic ARS, SRD3 produced gastrointestinal ARS, and SRD 4 produced BM syndrome.

Radiation-induced SRDs appear in the lymphatic fluid within hours after irradiation and enter the blood circulation affecting various organs. The precise time of appearance of SRDs varies from one species to another. The SRDs generated from the lymphatic fluid of animals irradiated with a dose that produces GI syndrome when injected subcutaneously, produced GI syndrome in nonirradiated animals similar to that produced by radiation. On the other hand, if the SRD has been generated from animals irradiated with a dose that produces CNS syndrome, administration of this SRD into nonirradiated animals produced CNS syndromes similar to those caused by radiation. In my opinion, SRDs can be considered a product of acute inflammation of lymphatic tissue that is released into the lymphatic fluids. The antigenic nature of SRDs was confirmed by the fact that the antibody of SRD has been successfully developed and tested. Investigators from other countries have not attempted to confirm the existence of SRD in lethally irradiated animals and its antigenic function. In my opinion, this observation is very novel and should be investigated extensively because of its importance in prevention of ARS.

SUMMARY

The effects of high doses of ionizing radiation have been performed extensively in animal models; however, data on the effects of high doses of radiation on humans primarily come from populations exposed to atomic explosions in Hiroshima and Nagasaki during World War II. Ionizing radiation causes injury by generating excessive amounts of free radicals and producing acute inflammatory reactions. Low LET radiation, such as x-rays or gamma rays, initiates radiation damage primarily by generating excessive amounts of free radicals, whereas high LET radiation, such as proton and alpha particles, produces damage primarily through ionization. In response to tissue damage, acute inflammatory reactions are set in motion; therefore, inflammation contributes to the progression of damage caused by both low and high LET radiation. For the same total dose, high LET radiation is more effective than low LET radiation. Generally, the higher the dose rate, the greater the damage. Dividing cells are more radiosensitive than nondividing cells.

The signs and symptoms produced by high doses of radiation in a single dose to the whole body are referred to as *radiation syndromes*. Radiation syndromes have been divided into three major categories: (1) bone marrow syndrome (BM), (2) gastrointestinal (GI) syndrome, and (3) central nervous system (CNS) syndrome. Each radiation syndrome is characterized by specific doses, survival, survival time, and signs and symptoms. Acute radiation sickness (ARS) includes both BM syndrome and GI syndrome.

Increased expressions of certain oncogenes, such as c-myc, H-ras, and raf, appear to be associated with enhanced radioresistance. The combination of high expression of v-myc and H-ras produces a synergistic effect on radioresistance. The role of p53, a tumor suppressor protein and a regulator of the cell cycle, in regulating BM syndrome is totally different from that in GI syndrome. It has been reported that the deficiency of PUMA in mice increases radioresistance of hematopoietic cells. Ionizing radiation activates p53 in GI epithelium, which induces PUMA to kill cells by the intrinsic pathway of apoptosis. Deletion of either p53 or PUMA blocks radiation-induced apoptosis of most epithelial cells of the intestine. Therefore, p53-mediated apoptosis has been implicated in the GI syndrome.

It is now established that DNA DSBs are critical radiation-induced cellular lesions which, if left unrepaired, can lead to gene mutations, chromosomal aberrations, and cell death, or can promote the onset of chronic diseases associated with genomic instability, such as cancer. One of the most important cellular responses to DSBs is the accumulation and localization of a plethora of DNA damage signaling and repair proteins in the vicinity of the lesion, initiated by ATM-mediated phosphorylation of histone H2AX (gamma-H2AX). This results in the formation of microscopically visible nuclear domains referred to as radiation-induced foci (RIF).

X-irradiation is known to induce mutations that may result from a change in nucleotides, called gene mutation, and/or they may result from damage to chromosomes. The frequency of gene mutation increases linearly with dose, and is independent of dose rate (at least in fruit fly), fractionation, LET, and temperature. The dose–response curve for a single-break chromosomal damage is linear, whereas for a double-break chromosomal damage, it is of the sigmoid type. Ionizing radiation often causes terminal deletion, which requires one break, and interstitial deletion, which requires two breaks in a chromosome. Ionizing radiation can cause duplication of a chromosome, which requires one break. The translocation type of chromosomal damage requires two breaks, and occurs when an exchange of segments between two chromosomes occurs without any loss of genetic materials. The inversion type of chromosomal damage requires two breaks, and occurs when the two breaks are on the same chromosome, and the resulting piece is inverted and reinserted into the same chromosome. The dicentric ring type of chromosomal damage occurs when the rearrangement of chromosomes can lead to the configuration of a dicentric ring. The nondisjunction type of chromosomal damage may produce trisomy (can cause Down's syndrome) and monosomy (can cause Turner's syndrome). Unlike *Drosophila melanogaster*, the frequency of mutation in mice is dependent upon dose rate. In spermatogonia, the frequency of mutations from a high dose rate to a low dose rate is reduced by a factor of 3. There is no evidence of a threshold dose rate for male mice. In female mice, the effect of dose rate on the frequency of mutations is different from that found in males. The incidence of radiation (400 R)-induced dominant lethal mutations in spermatozoa is about 50% higher for rats than in mice.

Ionizing radiation suppresses the immune response of mammals, including humans. It does so either by inhibiting innate immunity, which includes destruction of mechanical barrier, cellular defense mechanisms, clearance mechanisms, bactericidal efficiency of the blood, and the properdin system, or by inhibiting the formation of antibodies. Excessive amounts of products of acute inflammation, such as pro-inflammatory and anti-inflammatory cytokines, are released. The persistence of pro-inflammatory cytokines after irradiation contributes to the progression of radiation injury.

Most studies with high doses of ionizing radiation on fetuses have been performed on animal models. Data on the effects of irradiation on human fetuses come from the survivors of the atomic bombing of Hiroshima and Nagasaki, and from pregnant women who received diagnostic doses of ionizing radiation. Developing organisms constitute a highly dynamic system in which rapid cell proliferation, cell migration, and cell differentiation occur. Therefore, it is expected that the radiation response of the embryo as a whole, or of specific tissue, would markedly differ, depending upon the stage of development. Several animal studies have suggested that radiation-induced damages are similar in different species when they are irradiated with the same dose and at the same stage of development. In humans, the fetuses are most radiosensitive to radiation-induced damage between 8 and 15 weeks. The most sensitive period in the etiology of a cell is during its transformation from the embryonic state to the adult state, whether it be neuroblast, myoblast, or erythroblast. In humans, such transformation occurs between days 18 and 38 of gestation. Therefore, during this period, irradiation of the fetus causes the highest incidence of congenital anomalies. The first three weeks of human embryonic development is the most radiosensitive period on the criterion of lethality. Cells during this period are undifferentiated and are rapidly dividing. Depending upon the dose, the irradiated cells could either recover or die. The recovered cells may undergo normal differentiation and maturation; therefore, radiation damage may not be reflected in terms of organ abnormalities. However, these cells may carry functional damage or mutational changes that may not manifest during the F1 generation. The LD_{50} dose of radiation for the embryo observed at the term increases as a function of gestational age at the time of radiation exposure. A dose of 5.1 Gy delivered in a single dose before 18 weeks of gestation causes 100% mortality in the human fetus. Radiation-induced congenital abnormalities are highest during days 18–45 of gestation in humans. After the organs become differentiated and are largely involved in further growth and maturation, radiation-induced congenital anomalies or lethality are greatly reduced, and the effects of radiation are largely expressed as pathologic damage to the tissues, organs, and associated functional impairment. Irradiation of the fetus at any time can cause some loss of neural connections, resulting in functional deficiencies. The studies on the radiation effects on the preimplantation of the zygote have been performed primarily on animal models. The period of preimplantation includes the cleavage, morula, and blastocyte stages of the embryo. These stages of the embryo are very sensitive to radiation and can either die or survive after irradiation, depending upon the doses. Those irradiated embryos that survive do not show any abnormality with respect to morphology, size, short- and long-term survival, and reproductive ability. The studies on the radiation effects after implantation of the zygote have been performed primarily on animal models. The results showed that the number of abnormal implantations of the zygote irradiated at 7 hours after implantation increased in a dose-dependent manner. Irradiation of embryos during organogenesis produced an increased number of organ defects. A dose of less than 5 rads delivered during organogenesis can significantly increase the risk of brain defects in humans.

Among eye defects, anophthalmia and microphthalmia are common responses in rodents exposed to radiation exposure to 100–300 R (about 1–3 Gy) at a particular stage during fetal lifetime. Although the developing retina is very sensitive to radiation, it exhibits a remarkable capacity to reconstitute and repair the damaged area. A dose of 0.15 Gy to rat fetuses produces brain and eye abnormalities.

The radiosensitivity of germinal elements of the female was maximal at 15.5 days after gestation. In contrast, the male germ cells became increasingly radiosensitive between 17.5 and 19.5 days of gestation. In females, severe depopulation of oocytes appeared to be associated with a slight decline in the secretory activity of the ovary. On the other hand, in males, complete elimination of germ cells without affecting the endocrine function of the testis irradiation had no effect on the production of testosterone. It has been reported that both male and female germ cells display a similar number of histone protein gamma-H2AX foci in response to irradiation. In organ culture of human fetal ovaries, doses 1.5 Gy or above induced apoptosis.

Accumulation of p53 was found only in male human germ cells, but not in female cells. This suggested that p53-mediated apoptosis in germ cells is involved in males, but not in females.

Investigators from the Russian Federation have identified large molecular weight (200–250 kDa) antigenic molecules comprising 50% protein, 38% lipid, 10% carbohydrate, and 1.3% mineral residue from the lymphatic vessels shortly after high doses (6–10 Gy) of whole-body gamma irradiation. The antigenic molecules, when injected subcutaneously into different species of unirradiated animals, produced syndromes similar to those produced by high doses of radiation. These radiation toxin molecules have been referred to as SRDs. Although multiple variants of SRDs are produced following high doses of irradiation, a predominant variant of SRD is responsible for producing specific ARS syndromes. For example, SRD 1 produced cerebral ARS, SRD2 produced nonspecific toxic ARS, SRD3 produced gastrointestinal ARS, and SRD 4 produced BM syndrome.

REFERENCES

1. Prasad, K., *Handbook of Radiobiology,* 2nd Ed., CRC Press, Boca Raton, FL, 1995.
2. Elkind, M. M., and. Whitmore, G. F., *The Radiobiology of Cultured Mammalian Cells,* Gordon and Breach, New York, 1967.
3. Hall, E. J., *Radiobiology for the Radiologists,* J. B. Lippincott, Philadelphia, 1994.
4. Hamada, N., Matsumoto, H., Hara, T., and Kobayashi, Y., Intercellular and intracellular signaling pathways mediating ionizing radiation-induced bystander effects, *J Radiat Res (Tokyo)* 48 (2), 87–95, 2007.
5. Herok, R., Konopacka, M., Polanska, J., Swierniak, A., Rogolinski, J., Jaksik, R., Hancock, R., and Rzeszowska-Wolny, J., Bystander effects induced by medium from irradiated cells: Similar transcriptome responses in irradiated and bystander K562 cells, *Int J Radiat Oncol Biol Phys* 77 (1), 244–252, 2010.
6. Gerashchenko, B. I., and Howell, R. W., Cell proximity is a prerequisite for the proliferative response of bystander cells co-cultured with cells irradiated with gamma-rays, *Cytometry A* 56 (2), 71–80, 2003.
7. Baskar, R., Balajee, A. S., Geard, C. R., and Hande, M. P., Isoform-specific activation of protein kinase C in irradiated human fibroblasts and their bystander cells, *Int J Biochem Cell Biol* 40 (1), 125–134, 2008.
8. Yang, H., Asaad, N., and Held, K. D., Medium-mediated intercellular communication is involved in bystander responses of x-ray-irradiated normal human fibroblasts, *Oncogene* 24 (12), 2096–2103, 2005.
9. He, M., Zhao, M., Shen, B., Prise, K. M., and Shao, C., Radiation-induced intercellular signaling mediated by cytochrome-C via a p53-dependent pathway in hepatoma cells, *Oncogene* 30, 1947–1955, 2010.
10. Pasi, F., Facoetti, A., and Nano, R., IL-8 and IL-6 bystander signaling in human glioblastoma cells exposed to gamma radiation, *Anticancer Res* 30, 2769–2772, 2010.
11. Ivanov, V. N., Zhou, H., Ghandhi, S. A., Karasic, T. B., Yaghoubian, B., Amundson, S. A., and Hei, T. K., Radiation-induced bystander signaling pathways in human fibroblasts: A role for interleukin-33 in the signal transmission, *Cell Signal* 22 (7), 1076–1087, 2010.
12. Kulkarni, R., Marples, B., Balasubramaniam, M., Thomas, R. A., and Tucker, J. D., Mitochondrial gene expression changes in normal and mitochondrial mutant cells after exposure to ionizing radiation, *Radiat Res* 173 (5), 635–644, 2010.

13. Pajonk, F., and McBride, W. H., Ionizing radiation affects 26s proteasome function and associated molecular responses, even at low doses, *Radiother Oncol* 59 (2), 203–212, 2001.
14. Weisserger, E., and Okada, S., Radiosensitivity of single- and double-stranded deoxyribonucleic acid, *Int J Radiat Biol* 3, 331–335., 1961.
15. Altman, K. I., and S. Okada, *Radiation Biochemistry*, Academic Press, New York, 1970.
16. Little, J. B., Irradiation of primary human amnion cell cultures: Effects on DNA synthesis and progression through the cell cycle, *Radiat Res* 44 (3), 674–699, 1970.
17. Teresima, T., and Tolmach, L. J., Variation in several responses of Hela cells to x-irradiation during the division cycle, *Biophys. J.* 3 (11–18), 1963.
18. Beltz, E., Van Lancker, J., and Potter, V. R., Nucleic acid metabolism in regenerating liver. IV The effect of x-irradiation of the whole-body on nucleic acid synthesis in vivo, *Cancer Res* 17, 688–696, 1957.
19. Diaz Borges, J. M., and Drujan, B. D., The effects of gamma-irradiation upon monoamine oxidase activity, *Radiat Res* 45 (3), 589–597, 1971.
20. Berry, H. K., Saenger, E. L., Perry, H., Friedman, B. I., Kereiakes, J. G., and Scheel, C., Deoxycytidine in urine of humans after whole-body irradiation, *Science* 142, 396–398, 1963.
21. Gerber, G. B., Gertler, P., Altman, K. I., and Hempelmann, L. H., Dose dependency of radiation-induced creatine excretion in rat urine, *Radiat Res* 15, 307–313, 1961.
22. Gerber, G., Kurohara, S., Altman, K. I., and Hempelmann, L. H., Urinary excretion of several metabolites in persons accidentally exposed to ionizing radiation, *Radiat Res* 15, 314–318, 1961.
23. Bond, V. P. Fliedner, T. M., and Archambeau, J. O., *Mammalian Radiation Lethality*, Academic Press, New York, 1965.
24. Rubin, P., and Casarett, G. W., *Clinical Radiation Pathology*, W. B. Saunders, Philadelphia, 1968.
25. Boche, R. D., et al., Studies on the effects of massive doses of x-radiation on mortality in animals, in *Biological Effects of External Radiation, Vol 2*, ed. H. A. Blair, McGraw Hill, New York, 1954, 3.
26. Jones, D. C., Osborn, G. K., and Kimeldorf, D. J., Age at x-irradiation and acute radiation mortality in the adult male rat, *Radiat Res* 38 (3), 614–621, 1969.
27. Greenberger, J. S., and Epperly, M. Bone marrow-derived stem cells and radiation response, *Semin Radiat Oncol* 19 (2), 133–139, 2009.
28. Acharya, M. M., Lan, M. L., Kan, V. H., Patel, N. H., Giedzinski, E., Tseng, B. P., and Limoli, C. L., Consequences of ionizing radiation-induced damage in human neural stem cells, *Free Radic Biol Med* 49 (12), 1846–1855, 2010.
29. Bond, V. P., Fliedner, T. M., and Aychambeau, J. O., *Mammalian Radiation Lethality*, Academic Press, New York, 1965.
30. Freeman, S. L., Hossain, M., and MacNaughton, W. K., Radiation-induced acute intestinal inflammation differs following total-body versus abdominopelvic irradiation in the ferret, *Int J Radiat Biol* 77 (3), 389–395, 2001.
31. MacNaughton, W. K., Aurora, A. R., Bhamra, J., Sharkey, K. A., and Miller, M. J., Expression, activity and cellular localization of inducible nitric oxide synthase in rat ileum and colon post-irradiation, *Int J Radiat Biol* 74 (2), 2255–264, 1998.
32. Mettler, F. A. Jr., Gus'kova, A. K., and Gusev, I., Health effects in those with acute radiation sickness from the Chernobyl accident, *Health Phys* 93 (5), 462–469, 2007.
33. Gottlober, P., Steinert, M., Weiss, M., Bebeshko, V., Belyi, D., Nadejina, N., Stefani, F. H., Wagemaker, G., Fliedner, T. M., and Peter, R. U., The outcome of local radiation injuries: 14 years of follow-up after the Chernobyl accident, *Radiat Res* 155 (3), 409–416, 2001.

34. Diczfalusy, E. N., et al., Influence of ovarian irradiation on urinary estrogens in breast cancer patients, in *Effects of Ionizing Radiation on the Reproductive System*, ed. W. D. Carlson, McMillan, New York, 1964, 393.

35. Johnston, C. J., Williams, J. P., Okunieff, P., and Finkelstein, J. N., Radiation-induced pulmonary fibrosis: Examination of chemokine and chemokine receptor families, *Radiat Res* 157 (3), 256–265, 2002.

36. Samid, D., Miller, A. C., Rimoldi, D., Gafner, J., and Clark, E. P., Increased radiation resistance in transformed and nontransformed cells with elevated ras proto-oncogene expression, *Radiat Res* 126 (2), 244–250, 1991.

37. Sklar, M. D., The ras oncogenes increase the intrinsic resistance of NIH 3T3 cells to ionizing radiation, *Science* 239 (4840), 645–647, 1988.

38. Pirollo, K. F., Garner, R., Yuan, S. Y., Li, L., Blattner, W. A., and Chang, E. H., Raf involvement in the simultaneous genetic transfer of the radioresistant and transforming phenotypes, *Int J Radiat Biol* 55 (5), 783–796, 1989.

39. McKenna, W. G., Weiss, M. C., Endlich, B., Ling, C. C., Bakanauskas, V. J., Kelsten, M. L., and Muschel, R. J., Synergistic effect of the v-myc oncogene with H-ras on radio-resistance, *Cancer Res* 50 (1), 97–102, 1990.

40. Miller, A. C., Kariko, K., Myers, C. E., Clark, E. P., and Samid, D., Increased radioresis-tance of E ras-transformed human osteosarcoma cells and its modulation by lovastatin, an inhibitor of P21ras isoprenylation, *Int J Cancer* 53 (2), 302–307, 1993.

41. Kidd, A. R. III, Snider, J. L., Martin, T. D., Graboski, S. F., Der, C. J., and Cox, A. D., Ras-related small GTPases RalA and RalB regulate cellular survival after ionizing radi-ation, *Int J Radiat Oncol Biol Phys* 78 (1), 2205–212, 2010.

42. Chang, I. Y., Youn, C. K., Kim, H. B., Kim, M. H., Cho, H. J., Yoon, Y., Lee, Y. S., Chung, M. H., and You, H. J., Oncogenic H-Ras up-regulates expression of Ku80 to protect cells from gamma-ray irradiation in NIH3T3 cells, *Cancer Res* 65 (15), 6811–6819, 2005.

43. Kim, I. A., Bae, S. S., Fernandes, A., Wu, J., Muschel, R. J., McKenna, W. G., Birnbaum, M. J., and Bernhard, E. J., Selective inhibition of ras, phosphoinositide 3 kinase, and Akt isoforms increases the radiosensitivity of human carcinoma cell lines, *Cancer Res* 65 (17), 7902–7910, 2005.

44. Chautard, E., Loubeau, G., Tchirkov, A., Chassagne, J., Vermot-Desroches, C., Morel, L., and Verrelle, P., Akt signaling pathway: A target for radiosensitizing human malig-nant glioma, *Neuro Oncol* 12 (5), 434–443, 2010.

45. Park, H. S., Yun, Y., Kim, C. S., Yang, K. H., Jeong, M., Ahn, S. K., Jin, Y. W., and Nam, S. Y., A critical role for Akt activation in protecting cells from ionizing radiation-induced apoptosis and the regulation of acinus gene expression, *Eur J Cell Biol* 88 (10), 563–575, 2009.

46. Bucci, B., D'Agnano, I., Amendola, D., Citti, A., Raza, G. H., Miceli, R., De Paula, U., Marchese, R., Albini, S., Felsani, A., Brunetti, E., and Vecchione, A., Myc down-regulation sensitizes melanoma cells to radiotherapy by inhibiting MLH1 and MSH2 mismatch repair proteins, *Clin Cancer Res* 11 (7), 2756–2767, 2005.

47. Tang, W. Y., Chau, S. P., Tsang, W. P., Kong, S. K., and Kwok, T. T., The role of Raf-1 in radiation resistance of human hepatocellular carcinoma Hep G2 cells, *Oncol Rep* 12 (6), 1349–1354, 2004.

48. Pfeifer, A., Mark, G., Leung, S., Dougherty, M., Spillare, E., and Kasid, U., Effects of c-raf-1 and c-myc expression on radiation response in an in vitro model of human small-cell-lung carcinoma, *Biochem Biophys Res Commun* 252 (2), 481–486, 1998.

49. Pruschy, M., Shi, Y. Q., Crompton, N. E., Steinbach, J., Aguzzi, A., Glanzmann, C., and Bodis, S., The proto-oncogene c-fos mediates apoptosis in murine T-lymphocytes induced by ionizing radiation and dexamethasone," *Biochem Biophys Res Commun* 241 (2), 519–524, 1997.

50. Sherman, M. L., Datta, R., Hallahan, D. E., Weichselbaum, R. R., and Kufe, D. W., Ionizing radiation regulates expression of the c-jun protooncogene, *Proc Natl Acad Sci USA* 87 (15), 5663–5666, 1990.

51. Collart, F. R., Horio, M., and Huberman, E., Heterogeneity in c-jun gene expression in normal and malignant cells exposed to either ionizing radiation or hydrogen peroxide, *Radiat Res* 142 (2), 188–196, 1995.

52. Syljuasen, R. G., Hong, J. H., and McBride, W. H., Apoptosis and delayed expression of c-jun and c-fos after gamma irradiation of Jurkat T cells, *Radiat Res* 146 (3), 276–282, 1996.

53. Choi, E. K., Rhee, Y. H., Park, H. J., Ahn, S. D., Shin, K. H., and Park, K. K., Effect of protein kinase C inhibitor (PKCI) on radiation sensitivity and c-fos transcription, *Int J Radiat Oncol Biol Phys* 49 (2), 397–405, 2001.

54. Sakakura, C., Sweeney, E. A., Shirahama, T., Igarashi, Y., Hakomori, S., Nakatani, H., Tsujimoto, H., Imanishi, T., Ohgaki, M., Ohyama, T., Yamazaki, J., Hagiwara, A., Yamaguchi, T., Sawai, K., and Takahashi, T., Overexpression of Bax sensitizes human breast cancer MCF-7 cells to radiation-induced apoptosis, *Int J Cancer* 67 (1), 101–105, 1996.

55. Santucci, M. A., Anklesaria, P., Laneuville, P., Das, I. J., Sakakeeny, M. A., Fitzgerald, T. J., and Greenberger, J. S., Expression of p210 bcr/abl increases hematopoietic progenitor cell radiosensitivity, *Int J Radiat Oncol Biol Phys* 26 (5), 831–836, 1993.

56. Yacoub, A., Mitchell, C., Lister, A., Lebedeva, I. V., Sarkar, D., Su, Z. Z., Sigmon, C., McKinstry, R., Ramakrishnan, V., Qiao, L., Broaddus, W. C., Gopalkrishnan, R. V., Grant, S., Fisher, P. B., and Dent, P., Melanoma differentiation-associated 7 (interleukin 24) inhibits growth and enhances radiosensitivity of glioma cells in vitro and in vivo, *Clin Cancer Res* 9 (9), 3272–3281, 2003.

57. von Holzen, U., Pataer, A., Raju, U., Bocangel, D., Vorburger, S. A., Liu, Y., Lu, X., Roth, J. A., Aggarwal, B. B., Barber, G. N., Keyomarsi, K., Hunt, K. K., and Swisher, S. G., The double-stranded RNA-activated protein kinase mediates radiation resistance in mouse embryo fibroblasts through nuclear factor KappaB and Akt activation, *Clin Cancer Res* 13 (20), 6032–6039, 2007.

58. Anai, S., Goodison, S., Shiverick, K., Hirao, Y., Brown, B. D., and Rosser, C. J., Knockdown of Bcl-2 by antisense oligodeoxynucleotides induces radiosensitization and inhibition of angiogenesis in human PC-3 prostate tumor xenografts, *Mol Cancer Ther* 6 (1), 101–111, 2007.

59. Skvara, H., Thallinger, C., Wacheck, V., Monia, B. P., Pehamberger, H., Jansen, B., and Selzer, E., Mcl-1 blocks radiation-induced apoptosis and inhibits clonogenic cell death, *Anticancer Res* 25 (4), 2697–2703, 2005.

60. Kim, R., Inoue, H., and Toge, T., Bax is an important determinant for radiation sensitivity in esophageal carcinoma cells, *Int J Mol Med* 14 (4), 697–706, 2004.

61. Adhami, V. M., Siddiqui, I. A., Sarfaraz, S., Khwaja, S. I., Hafeez, B. B., Ahmad, N., and Mukhtar, H., Effective prostate cancer chemopreventive intervention with green tea polyphenols in the TRAMP model depends on the stage of the disease, *Clin Cancer Res* 15 (6), 1947–1953, 2009.

62. Wang, J., Wakeman, T. P., Lathia, J. D., Hjelmeland, A. B., Wang, X. F., White, R. R., Rich, J. N., and Sullenger, B. A., Notch promotes radioresistance of glioma stem cells, *Stem Cells* 28 (1), 17–28, 2010.

63. Purow, B. W., Sundaresan, T. K., Burdick, M. J., Kefas, B. A., Comeau, L. D., Hawkinson, M. P., Su, Q., Kotliarov, Y., Lee, J., Zhang, W., and Fine, H. A., Notch-1 regulates transcription of the epidermal growth factor receptor through p53, *Carcinogenesis* 29 (5), 918–925, 2008.

64. Platta, C. S., Greenblatt, D. Y., Kunnimalaiyaan, M., and Chen, H., Valproic acid induces notch1 signaling in small cell lung cancer cells, *J Surg Res* 148 (1), 31–37, 2008.

65. Shao, L., Sun, Y., Zhang, Z., Feng, W., Gao, Y., Cai, Z., Wang, Z. Z., Look, A. T., and Wu, W. S., Deletion of proapoptotic puma selectively protects hematopoietic stem and progenitor cells against high-dose radiation, *Blood* 115 (23), 4707–4714, 2010.

66. Yu, H., Shen, H., Yuan, Y., XuFeng, R., Hu, X., Garrison, S. P., Zhang, L., Yu, J., Zambetti, G. P., and Cheng, T., Deletion of PUMA protects hematopoietic stem cells and confers long-term survival in response to high-dose gamma-irradiation, *Blood* 115 (17), 3472–3480, 2010.

67. Komarova, E. A., Kondratov, R. V., Wang, K., Christov, K., Golovkina, T. V., Goldblum, J. R., and Gudkov, A. V., Dual effect of p53 on radiation sensitivity in vivo: p53 promotes hematopoietic injury, but protects from gastro-intestinal syndrome in mice, *Oncogene* 23 (19), 3265–3271, 2004.

68. Pawlik, A., Alibert, O., Baulande, S., Vaigot, P., and Tronik-Le Roux, D., Transcriptome characterization uncovers the molecular response of hematopoietic cells to ionizing radiation, *Radiat Res* 175 (1), 66–82, 2011.

69. Merritt, A. J., Potten, C. S., Kemp, C. J., Hickman, J. A., Balmain, A., Lane, D. P., and Hall, P. A., The role of p53 in spontaneous and radiation-induced apoptosis in the gastrointestinal tract of normal and p53-deficient mice, *Cancer Res* 54 (3), 614–617, 1994.

70. Potten, C. S., Extreme sensitivity of some intestinal crypt cells to X and gamma irradiation, *Nature* 269 (5628) 518–521, 1977.

71. Qiu, W., Carson-Walter, E. B., Liu, H., Epperly, M., Greenberger, J. S., Zambetti, G. P., Zhang, L., and Yu, J., PUMA regulates intestinal progenitor cell radiosensitivity and gastrointestinal syndrome, *Cell Stem Cell* 2 (6), 576–583, 2008.

72. Paris, F., Fuks, Z., Kang, A., Capodieci, P., Juan, G., Ehleiter, D., Haimovitz-Friedman, A., Cordon-Cardo, C., and Kolesnick, R., Endothelial apoptosis as the primary lesion initiating intestinal radiation damage in mice, *Science* 293 (5528), 293–297, 2001.

73. Santana, P., Pena, L. A., Haimovitz-Friedman, A., Martin, S., Green, D., McLoughlin, M., Cordon-Cardo, C., Schuchman, E. H., Fuks, Z., and Kolesnick, R., Acid sphingomyelinase-deficient human lymphoblasts and mice are defective in radiation-induced apoptosis, *Cell* 86 (2), 189–199, 1996.

74. Garcia-Barros, M., Paris, F., Cordon-Cardo, C., Lyden, D., Rafii, S., Haimovitz-Friedman, A., Fuks, Z., and Kolesnick, R., Tumor response to radiotherapy regulated by endothelial cell apoptosis, *Science* 300 (5622), 1155–1159, 2003.

75. Rotolo, J. A., Maj, J. G., Feldman, R., Ren, D., Haimovitz-Friedman, A., Cordon-Cardo, C., Cheng, E. H., Kolesnick, R., and Fuks, Z., Bax and Bak do not exhibit functional redundancy in mediating radiation-induced endothelial apoptosis in the intestinal mucosa, *Int J Radiat Oncol Biol Phys* 70 (3), 804–815, 2008.

76. Kirsch, D. G., Santiago, P. M., di Tomaso, E., Sullivan, J. M., Hou, W. S., Dayton, T., Jeffords, L. B., Sodha, P., Mercer, K. L., Cohen, R., Takeuchi, O., Korsmeyer, S. J., Bronson, R. T., Kim, C. F., Haigis, K. M., Jain, R. K., and Jacks, T., p53 controls radiation-induced gastrointestinal syndrome in mice independent of apoptosis, *Science* 327 (5965), 593–596, 2010.

77. Bristow, R. G., Benchimol, S., and Hill, R. P., The p53 gene as a modifier of intrinsic radiosensitivity: Implications for radiotherapy, *Radiother Oncol* 40 (3), 197–223, 1996.

78. Matsui, Y., Tsuchida, Y., and Keng, P. C., Effects of p53 mutations on cellular sensitivity to ionizing radiation, *Am J Clin Oncol* 24 (5), 486–490, 2001.

79. Lee, J. M. and Bernstein, A., p53 mutations increase resistance to ionizing radiation, *Proc Natl Acad Sci USA* 90 (12), 5742–5746, 1993.

80. Segreto, H. R., Oshima, C. T., Franco, M. F., Silva, M. R., Egami, M. I., Teixeira, V. P., and Segreto, R. A., Phosphorylation and cytoplasmic localization of MAPK p38 during apoptosis signaling in bone marrow granulocytes of mice irradiated in vivo and the role of amifostine in reducing these effects, *Acta Histochem* 113 (3), 300–307, 2010.

81. Fournier, C., Wiese, C., and Taucher-Scholz, G., Accumulation of the cell cycle regulators TP53 and CDKN1A (p21) in human fibroblasts after exposure to low- and high-LET radiation, *Radiat Res* 161 (6), 675–684, 2004.

82. Jakob, B., Scholz, M., and Taucher-Scholz, G., Immediate localized CDKN1A (p21) radiation response after damage produced by heavy-ion tracks, *Radiat Res* 154 (4), 398–405, 2000.

83. Xiao, M., Inal, C. E., Parekh, V. I., Li, X. H., and Whitnall, M. H., Role of NF-kappaB in hematopoietic niche function of osteoblasts after radiation injury, *Exp Hematol* 37 (1), 52–64, 2009.

84. Sokolov, M., Panyutin, I. G., and Neumann, R., Genome-wide gene expression changes in normal human fibroblasts in response to low-LET gamma-radiation and high-LET-like 125IUdR exposures, *Radiat Prot Dosimetry* 122 (1–4), 195–201, 2006.

85. Boerma, M., van der Wees, C. G., Vrieling, H., Svensson, J. P., Wondergem, J., van der Laarse, A., Mullenders, L. H., and van Zeeland, A. A., Microarray analysis of gene expression profiles of cardiac myocytes and fibroblasts after mechanical stress, ionizing or ultraviolet radiation, *BMC Genomics* 6 (1), 6, 2005.

86. Morgan, W. F., Day, J. P., Kaplan, M. I., McGhee, E. M., and Limoli, C. L., Genomic instability induced by ionizing radiation, *Radiat Res* 146 (3), 247–58, 1996.

87. Burlakova, E. B., Mikhailov, V. F., and Azurik, V. K., The redox homeostasis system in radiation-induced genomic instability [in Russian], *Radiats Biol Radioecol* 41 (5), 489–499, 2001.

88. Lassmann, M., Hanscheid, H., Gassen, D., Biko, J., Meineke, V., Reiners, C., and Scherthan, H., In vivo formation of gamma-H2AX and 53BP1 DNA repair foci in blood cells after radioiodine therapy of differentiated thyroid cancer, *J Nucl Med* 51 (8), 1318–1325, 2010.

89. Bekker–Jensen, S., and Mailand, N., Assembly and function of DNA double–strand break repair foci in mammalian cells, *DNA Repair (Amst)* 9 (12), 1219–28, 2010.

90. Costes, S. V., Chiolo, I., Pluth, J. M., Barcellos–Hoff, M. H., and Jakob, B., Spatiotemporal characterization of ionizing radiation induced DNA damage foci and their relation to chromatin organization, *Mutat Res* 704 (1–3), 78–87, 2010.

91. Koike, M., Mashino, M., Sugasawa, J., and Koike, A., Histone H2AX phosphorylation independent of ATM after x-irradiation in mouse liver and kidney in situ, *J Radiat Res (Tokyo)* 49 (4), 445–449, 2008.

92. Magic, Z., Matic-Ivanovic, S., Savic, J., and Poznanovic, G., Ionizing radiation-induced expression of the genes associated with the acute response to injury in the rat, *Radiat Res* 143 (2), 187–193, 1995.

93. Chen, D. J., and Nirodi, C. S., The epidermal growth factor receptor: A role in repair of radiation-induced DNA damage, *Clin Cancer Res* 13 (22 Pt 1), 6555–60, 2007.

94. Rodemann, H. P., Dittmann, K., and Toulany, M., Radiation-induced EGFR-signaling and control of DNA-damage repair, *Int J Radiat Biol* 83 (11–12), 781–791, 2007.

95. Dittmann, K., Mayer, C., Kehlbach, R., Rothmund, M. C., and Peter Rodemann, H., Radiation-induced lipid peroxidation activates src kinase and triggers nuclear EGFR transport, *Radiother Oncol* 92 (3), 379–382, 2009.

96. Dittmann, K., Mayer, C., Kehlbach, R., and Rodemann, H. P., Radiation-induced caveolin-1 associated EGFR internalization is linked with nuclear EGFR transport and activation of DNA-PK, *Mol Cancer* 7, 69, 2008.

97. Thacker, J., The RAD51 gene family, genetic instability and cancer, *Cancer Lett* 219 (2), 125–135, 2005.

98. Davies, A. A., Masson, J. Y., McIlwraith, M. J., Stasiak, A. Z., Stasiak, A., Venkitaraman, A. R., and West, S. C., Role of BRCA2 in control of the RAD51 recombination and DNA repair protein, *Mol Cell* 7 (2), 273–282, 2001.

99. van Veelen, L. R., Essers, J., van de Rakt, M. W., Odijk, H., Pastink, A., Zdzienicka, M. Z., Paulusma, C. C., and Kanaar, R., Ionizing radiation-induced foci formation of mammalian Rad51 and Rad54 depends on the Rad51 paralogs, but not on Rad52, *Mutat Res* 574 (1–2), 34–49, 2005.

100. Liu, J., Yuan, Y., Huan, J., and Shen, Z., Inhibition of breast and brain cancer cell growth by BCCIPalpha: An evolutionarily conserved nuclear protein that interacts with BRCA2, *Oncogene* 20 (3), 336–345, 2001.

101. Ono, T., Kitaura, H., Ugai, H., Murata, T., Yokoyama, K. K., Iguchi-Ariga, S. M., and Ariga, H., TOK-1, a novel p21Cip1-binding protein that cooperatively enhances p21-dependent inhibitory activity toward CDK2 kinase, *J Biol Chem* 275 (40), 31145–32254, 2000.

102. Lu, H., Guo, X., Meng, X., Liu, J., Allen, C., Wray, J., Nickoloff, J. A., and Shen, Z., The BRCA2-interacting protein BCCIP functions in RAD51 and BRCA2 focus formation and homologous recombinational repair, *Mol Cell Biol* 25 (5), 1949–1957, 2005.

103. Wray, J., Liu, J., Nickoloff, J. A., and Shen, Z., Distinct RAD51 associations with RAD52 and BCCIP in response to DNA damage and replication stress, *Cancer Res* 68 (8), 2699–2707, 2008.

104. Wolf, S., Radiation genetics, in *Mechanisms in Radiobiology*, ed. Errera, M. and Forssberg, A. Academy Press, New York, 1961, 419–430.

105. Russell, W. L., The effect of radiation dose rate and fractionation on mutation in mice, in *Repair from Genetic Radiation Damage and Differential Radiosensitivity in Germ Cells*, ed. Ashburner, M., Macmillan, New York, 1963, pp. 205–210.

106. Anderson, R. E., and Warner, N. L., Ionizing radiation and the immune response, *Adv Immunol* 24, 215–335, 1976.

107. Van Der Meeren, A., Squiban, C., Gourmelon, P., Lafont, H., and Gaugler, M. H., Differential regulation by IL-4 and IL-10 of radiation-induced IL-6 and IL-8 production and ICAM-1 expression by human endothelial cells, *Cytokine* 11 (11), 831–838, 1999.

108. Han, S. K., Song, J. Y., Yun, Y. S., and Yi, S. Y., Effect of gamma radiation on cytokine expression and cytokine-receptor mediated STAT activation, *Int J Radiat Biol* 82 (9), 686–697, 2006.

109. Han, S. K., Song, J. Y., Yun, Y. S., and Yi, S. Y., Gamma irradiation-reduced IFN-gamma expression, STAT1 signals, and cell-mediated immunity, *J Biochem Mol Biol* 35 (6), 583–589, 2002.

110. Linard, C., Marquette, C., Mathieu, J., Pennequin, A., Clarencon, D., and Mathe, D., Acute induction of inflammatory cytokine expression after gamma-irradiation in the rat: Effect of an NF-kappaB inhibitor, *Int J Radiat Oncol Biol Phys* 58 (2), 427–434, 2004.

111. Chou, C. H., Chen, S. U., and Cheng, J. C., Radiation-induced interleukin-6 expression through MAPK/p38/NF-kappaB signaling pathway and the resultant antiapoptotic effect on endothelial cells through Mcl-1 expression with sIL6-Ralpha, *Int J Radiat Oncol Biol Phys* 75 (5), 1553–1561, 2009.

112. Rube, C. E., Uthe, D., Wilfert, F., Ludwig, D., Yang, K., Konig, J., Palm, J., Schuck, A., Willich, N., Remberger, K., and Rube, C., The bronchiolar epithelium as a prominent source of pro-inflammatory cytokines after lung irradiation, *Int J Radiat Oncol Biol Phys* 61 (5), 1482–1492, 2005.

113. Lorimore, S. A., Coates, P. J., Scobie, G. E., Milne, G., and Wright, E. G., Inflammatory-type responses after exposure to ionizing radiation in vivo: A mechanism for radiation-induced bystander effects?, *Oncogene* 20 (48), 7085–7095, 2001.

114. Muller, K., Kohn, F. M., Port, M., Abend, M., Molls, M., Ring, J., and Meineke, V., Intercellular adhesion molecule-1: A consistent inflammatory marker of the cutaneous radiation reaction both in vitro and in vivo, *Br J Dermatol* 155 (4), 670–679, 2006.

115. Hallahan, D., Kuchibhotla, J., and Wyble, C., Cell adhesion molecules mediate radiation-induced leukocyte adhesion to the vascular endothelium, *Cancer Res* 56 (22), 5150–5155, 1996.

116. Freeman, S. L., and MacNaughton, W. K., Ionizing radiation induces iNOS-mediated epithelial dysfunction in the absence of an inflammatory response, *Am J Physiol Gastrointest Liver Physiol* 278 (2), G243–250, 2000.

117. Kobayashi, H., Yazlovitskaya, E. M., and Lin, P. C., Interleukin-32 positively regulates radiation-induced vascular inflammation, *Int J Radiat Oncol Biol Phys* 74 (5), 1573–1579, 2009.

118. Committee on the Biological Effects of Ionizing Radiation, *The Effects on Populations of Exposure to Low Levels of Ionizing Radiation*, National Academic Press, Washington, DC, 1980.

119. Rugh, R., X-irradiation effects on the human fetus, *J Pediatr* 52 (5), 531–538, 1958.

120. Brent, R. L., The effect of radiation on mammalian fetus, *Clin Obstet Gynecol* 3, 928–935, 1960.

121. Wilson, J. G., Differentiation and the reaction of rat embryos to radiation, *J. Cell Comp. Physiol, Suppl* 43, 11–16, 1954.

122. Brent, R. L., and Bolden, B. T., The indirect effect of irradiation on embryonic development. 3. The contribution of ovarian irradiation, uterine irradiation, oviduct irradiation, and zygote irradiation to fetal mortality and growth retardation in the rat, *Radiat Res* 30 (4), 759–773, 1967.

123. Russell, L. B., X-ray-induced development abnormalities in the mouse and their use in the analysis of embryological patterns. II Abnormalities of the vertebral column and thorax, *J Exp Zool*, 329–337, 1956.

124. Russell, L. B., and Russell, W. L., An analysis of the changes radiation response of the developing mouse embryo, *J Cell Comp Physiol* 43 (Suppl. 1), 103–149, 1954.

125. Rugh, R., Wohlfromm, M., and Varma, A., Low-dose x-ray effects on the precleavage mammalian zygote, *Radiat Res* 37 (2), 401–414, 1969.

126. Russell, L. B., and Montgomery, D. S., Radiation-sensitivity differences within cell division cycles during cleavages, *Int J Radiat Biol* 10, 151–156, 1966.

127. Miller, R. W., and Mulvihill, J. J., Small head size after atomic irradiation, *Teratology* 14 (3), 355–357, 1976.

128. Blot, W. J., Growth and development following prenatal and childhood exposure to atomic radiation, *J Radiat Res (Tokyo)* 16 (Suppl.), 82–88, 1975.

129. Dunn, K., Yoshimaru, H., Otake, M., Annegers, J. F., and Schull, W. J., Prenatal exposure to ionizing radiation and subsequent development of seizures, *Am J Epidemiol* 131 (1), 114–123, 1990.

130. Timins, J. K., Radiation during pregnancy, *N J Med* 98 (6), 29–33, 2001.

131. Rugh, R., and Vandyke, R. H., The fate and effect of x-irradiated neuroblasts (presumptive neural ectoderm) in a normal environment, *J Exp Zool* 157, 197–216, 1964.

132. Beaumat, H. M., Effect of irradiation during fetal life on the subsequent structures and secretory activity of the gonads, *J Endocrinol* 24, 325–339, 1962.

133. Henricson, B., and Nilsson, A., Roentgen ray effects on the ovaries of fetal mice, *Acta Radiol* 9, 443–450, 1970.

134. Embree-Ku, M., Venturini, D., and Boekelheide, K., Fas is involved in the p53-dependent apoptotic response to ionizing radiation in mouse testis, *Biol Reprod* 66 (5), 1456–4561, 2002.

135. Lambrot, R., Coffigny, H., Pairault, C., Lecureuil, C., Frydman, R., Habert, R., and Rouiller-Fabre, V., High radiosensitivity of germ cells in human male fetus, *J Clin Endocrinol Metab* 92 (7), 2632–2639, 2007.

136. Guerquin, M. J., Duquenne, C., Coffigny, H., Rouiller-Fabre, V., Lambrot, R., Bakalska, M., Frydman, R., Habert, R., and Livera, G., Sex-specific differences in fetal germ cell apoptosis induced by ionizing radiation, *Hum Reprod* 24 (3), 670–678, 2009.
137. Maliev, V. P., Popov, D., Jones, J. A., Casey, R. C., Mechanisms of action of anti-radiation vaccine in reducing the biological impact of high-dose gamma-irradiation, *J Adv Space Res* 40 (4), 586–590, 2007.
138. Bond, V. P., Radiation mortality in different species, in *Comparative Cellular and Species Radiosensitivity*, ed. Bond, V. P., and Sagahara, T., Igaku Shoin, Tokyo, 1969, 5.
139. Russell, L. B., Effects of low doses of x-rays on embryonic development in the mouse, *Proc Soc Exp Biol Med* 95 (1), 174–178, 1957.

4 Long-Term Damages among Survivors of High Doses of Ionizing Radiation

INTRODUCTION

Radiation-induced neoplasms cannot be distinguished from those that occur spontaneously or that are induced by chemical carcinogens. The structural and functional changes in carcinogenesis may be similar irrespective of causative agents. Cancer cells result from accumulation of multiple genetic changes in normal dividing cells, which occur over a long period of time. The latent period (time interval between exposure to carcinogens and detection of cancer) of a neoplasm depends upon the radiation dose, type of organ exposed, and experimental models (tissue culture, animals, or humans). Three models are used to study radiation carcinogenesis: the tissue culture model, the animal model, and the human model. The latent period for radiation-induced cancer is about a few weeks in the tissue culture model, about a few months in the animal model, and several years in the human model. In addition to cancer, irradiation can also increase the risk of nonneoplastic diseases. At present, there are no effective preventive strategies to reduce the risk of either radiation-induced cancer or nonneoplastic diseases. Before discussing the studies on radiation-induced cancer using each experimental model, it is important to understand the incidence of spontaneous cancer per year in order to evaluate the impact of radiation-induced increases in the incidence of cancer on society in terms of cost and well-being.

SPONTANEOUS CANCER INCIDENCE, MORTALITY, AND COST

The American Cancer Society has estimated that in 2010 about 1,529,560 new cases of cancer were diagnosed in the United States; this number shows an increase of 50,210 new cases of cancer in comparison to that observed in 2009. The pattern of changes in cancer incidence is described in Table 4.1. It appears that the incidence of new cancer has increased from about 1.2 million in 2000 to about 1.5 million in 2010. The American Cancer Society has estimated that lifetime probability of developing cancer for all sites is 1 in 2 in men and 1 in 3 in women; for prostate cancer it is 1 in 6, and for breast cancer, 1 in 8; for lung and bronchus cancer 1 in 13 in men

TABLE 4.1

Pattern of Changes in Cancer Incidence per Year during the Last 10 Years in the United States

Year	Number of New Cancers	Increase/Decrease from Previous Year
2000	1,220,100	
2008	1,437,180	Increase of 217,080
2009	1,479,350	Increase of 42,170
2010	1,529,560	Increase of 50,210

Source: Data were summarized from the publications of the American Cancer Society.

TABLE 4.2

Pattern of Changes in Cancer Mortality between 1950 and 2006 in the US Population

Year	Death Rate per 100,000 Persons
1950	194
1991	251.1
2005	184
2006	180.7

and 1 in 16 in women; and for colon and rectal cancer 1 in 18 in men and 1 in 20 in women.

The US mortality rate from cancer has not changed significantly since 1950 in spite of extensive research and development of new treatment modalities. In 1950, the death rate was about 194 per 100,000 persons; however in 1991, this value increased to about 251.1 per 100,000 persons. In 2005, this value dropped to about 184, and in 2006, it remained at about 180.7 per 100,000 persons (Table 4.2). These data suggest that in 1991 the cancer death rate was markedly enhanced compared to that observed in 1950. Thus, it appears that cancer death in 2006 was reduced compared to that observed in 1991, but it remains similar to that observed in 1950. It is more likely that cancer deaths, which were increasing during 1950–1991, were prevented due to advancement of tumor treatment modalities. Table 4.3 shows that there is a slight increase in cancer deaths between 2009 and 2010; this may not be statistically significant.

In 2009, the National Institutes of Health (NIH) estimated that the overall annual direct and indirect cost of cancer in 2008 was $228.1 billion. This cost included $93.2 billion for medical, $18.8 billion for lost productivity due to illness, and $116.1 billion for lost productivity due to premature death.

TABLE 4.3
Pattern of Changes in Cancer Death per Year during Last
10 Years in the United States

Year	Number of Cancer Deaths	Increase/Decrease from Previous Year
2000	552,200	
2008	565,650	Increase of 13,450
2009	562,340	Decrease of 3,310
2010	569,490	Increase of 7,150

Source: Data were summarized from the publications of the American Cancer Society.

STUDY OF CARCINOGENESIS IN EXPERIMENTAL
MODELS (TISSUE CULTURE AND ANIMALS)

Experiments on tissue culture and animals utilizing chemicals and ionizing radiation have established two broad stages in carcinogenesis: (1) initiation and (2) promotion.[1] Based on the above model of carcinogenesis, several distinct agents that initiate or promote carcinogenesis have been identified. Agents that initiate carcinogenesis are called *tumor initiators* and those that promote cancer are called *tumor promoters*. Tumor initiators at any doses can cause cancer, whereas tumor promoters by themselves do not cause cancer, but they in combination with tumor initiators increase the risk of cancer, and may reduce the latent period. There are two types of tumor initiators (carcinogens), direct-acting carcinogens such as ionizing radiation, and indirect-acting carcinogens such as benzo (a) pyrene that require conversion to an active form in the liver. Examples of tumor promoters are phorbol ester,12-O-tetradecanoyl-13- acetate (TPA), excessive fat consumption, and high estrogen levels in women.

Tumor progression that can lead to metastasis may be the result of accumulation of additional mutations, some of which may make cancer cells more aggressive. The nature of mutated genes that make cancer cells more aggressive has not been adequately defined.

The following hypotheses have been proposed for carcinogenesis:

1. Activation of proto-oncogenes, such as c-myc and c-ras[2,3]
2. Loss of anti-oncogenes, such as the retinoblastoma gene (rb)[4]
3. Infection with certain viruses[5]
4. Substitution of normal promoters of proto-oncogenes with strong promoters of viruses[6]
5. Chromosomal aberrations[7]

Radiation-induced neoplastic transformation in vitro and in vivo is also associated with alterations in the expression of certain oncogenes.[8–13] Activation of k-ras and the presence of k-ras[8] and N-ras mutation[9] have been demonstrated in some of the x-ray-induced lymphoma in mice. K-ras mutation and increased expression of c-myc were also present in radiation-induced skin cancer in rats.[10] Neutron-induced k-ras mutation is different from that produced by gamma irradiation.[11,12] Although the ras gene abnormality was detected in some cases of radiation-induced tumors, it is less frequent in tumors produced by certain chemical carcinogens. The frequency of mutation or overexpression of oncogenes is seldom over 50% in human tumors or in chemical-induced animal tumors. Therefore, the involvement of mutated ras or overexpression of cellular oncogenes in radiation-induced cancer in humans may be incidental.

The previously proposed concepts of human carcinogenesis have been critically reviewed.[14–16] Although the proposed concepts are intriguing, none of them alone is sufficient to explain the initial events in human carcinogenesis. For example, chromosomal aberrations that can occur spontaneously or be induced by ionizing radiation can be observed in dividing normal cells. These cells may or may not transform to cancer cells, depending upon the subsequent specific genetic changes. Similarly, overexpression and/or mutation of cellular oncogenes are not sufficient to convert normal cells to cancer cells, nor is the induced expression of anti-oncogenes sufficient to reverse cancer cells to a normal phenotype. Recently, polymorphism of certain genes appears to be associated with increased risk of cancer.[17–19]

PROPOSED STAGES OF CARCINOGENESIS

Humans are exposed to several tumor initiators and promoters daily from environment-, diet-, and lifestyle-related factors. Radiation represents one of the tumor initiators. Human carcinogenesis is a very complex process with a long latent period (3–30 years) between exposures to carcinogens, such as ionizing radiation, and clinically detectable cancer. This implies that preventive strategy can be implemented at any time before cancer detection. The identification of biochemical events that can alter activities of genes responsible for cancer formation during the latent period can help to select agents that can attenuate the risk of cancer. The radiation-induced tumors, spontaneous tumors, or familial tumors cannot be distinguished histologically. Despite profound advances in molecular carcinogenesis, the primary genes that initiate the development of human cancer remain elusive in most cases.

Normal cells go through at least two identifiable stages of carcinogenesis, immortalization and cancerous. Immortalized cells continue to divide without undergoing differentiation leading to the formation of benign tumors, whereas cancerous cells continue to divide and metastasize to distant organs. Certain oncogenic viruses can induce immortalization when inserted into normal cells in culture. For example, rat and human brain cells can be immortalized by inserting large T-antigen gene from SV40 and polyoma virus, respectively.[16,20] Viral oncogenes E6 and E7 of human papilloma virus (HPV) increase the risk of several cancers, including benign and malignant cervical cancer.[21–23] HPV can also act as a cocarcinogen in combination with tobacco smoking in increasing the risk of oral squamous cell carcinoma.[24] In the case of HPV-induced cervical cancer, herpes viruses may act as a cofactor in the development of cancer.[25]

When immortalized hepatocytes are transfected with the oncogenic c-Has in culture, they gain the capacity to form tumors in appropriate hosts.[26] This suggests that activation of cellular oncogenes can constitute a cancer risk factor, secondary to a critical step leading to immortalization.

A generic model of carcinogenesis suggests that cancer cells are the result of multiple mutations (gene defects) due to exposure to environment-, diet-, and lifestyle-related carcinogenic agents. Based on the histological progression of cancer formation, we have proposed a three-stage model of human carcinogenesis.[15] A diagrammatic representation of this model, which is also applicable to radiation-induced cancer, is shown. This model shows that intervention can be made at any time before the cancer cells are formed in order to reduce the risk of cancer. The latent period for each of these stages in humans could vary from a few to several years.

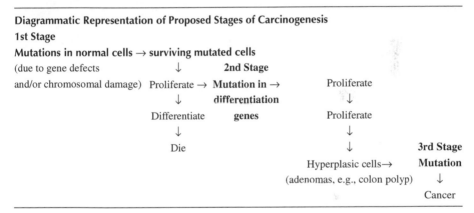

Diagrammatic Representation of Proposed Stages of Carcinogenesis

1st Stage

Mutations in normal cells → surviving mutated cells

(due to gene defects ↓ **2nd Stage**

and/or chromosomal damage) Proliferate → **Mutation in** → Proliferate

↓ **differentiation** ↓

Differentiate **genes** Proliferate

↓ ↓

Die ↓ **3rd Stage**

Hyperplasic cells→ **Mutation**

(adenomas, e.g., colon polyp) ↓

Cancer

The first stage involves induction of random mutations (due to gene defects or chromosomal damage) in normal dividing cells due to exposure to ionizing radiation and other cancer-causing substances associated with the environment, diet, and lifestyle, a deficiency in the natural repair system, or a deficiency in protective substances, such as antioxidants. These mutations can also occur spontaneously due to random errors during replication, and increased endogenous oxidative stress and chronic inflammation. The mutated cells may die or survive, depending upon the severity of genetic damage. The surviving mutated cells continue to divide and differentiate and die similar to the patterns observed in normal cells that do not have mutations. The mutated cells continue to accumulate additional mutations at a higher rate, but continue to divide and differentiate like unmutated normal dividing cells for a long period of time.

The second stage of carcinogenesis involves induction of random mutations in specific genes that are responsible for inducing differentiation in normal cells. As a result, the mutated cells continue to divide without achieving differentiation and subsequent cell death. Such cells become immortal and form precancerous or benign growths such as polyps in the colon or a cyst in the female breast and ovary. They continue to proliferate while accumulating additional mutations for a long period of time.

The third stage of carcinogenesis involves induction of random mutations in the immortal cells. Most such mutations play no role in converting immortal cells to cancer cells; however, when mutations occur in specific cellular genes, oncogenes, or anti-oncogenes, immortal cells become cancerous. This is well demonstrated in colon polyps and female breast and ovarian cysts which remain noncancerous for a long time, but if not removed, could become cancerous. Because mutation occurs randomly, the colon polyp may carry defects in more than one oncogene. The multiple, heterogeneous foci of cancer cells found in colon polyps are not necessarily clonal with respect to a given oncogene. This heterogeneity may be the reason why, in spite of extensive research in molecular carcinogenesis, it has not been possible to establish any direct relationship between the presence of one defective oncogene or other cellular genes and tumor type or tumor behavior, although some associations between oncogene or anti-oncogene and tumor behavior have been documented.

The induction of random mutations in cancer cells may lead to aggressive behavior of cancer cells. Again, most mutations may not have any significant impact on tumor behavior; however, when mutations occur in certain specified genes, the cancer cells may become very aggressive and invasive, and, as a consequence, cause distant metastasis. Although several studies have tried to establish relationships between defects in a particular gene and aggressive behavior of tumors, the results have not been consistent for the same tumor type.

In order to develop a novel biological strategy to reduce the risk of cancer, it is essential to know the potential stages that normal cells go through before they become cancer cells, and what risk factors contribute to the development of cancer at each stage. Irrespective of the types of mutagenic or carcinogenic agents, increased oxidative stress and chronic inflammation[27–31] play a central role in inducing gene mutations and/or chromosomal changes that initiate carcinogenic changes. Therefore, agents that can attenuate oxidative stress and chronic inflammation may reduce the risk of cancer.

Tissue culture models to study carcinogenesis: Monolayer cultures of normal embryonic cells have been used to establish dose–effect relationships for the induction of cancer by radiation. Generally, the presence of transformed cells in a culture after irradiation has been considered an index of potentially carcinogenic cells. The transformed cells grow and form colonies in tissue culture dishes. They can be easily identified by their growth pattern (they grow in clumps). If the transformed cells when injected into syngeneic animals produce tumors, they are considered cancerous. On the other hand, if they do not produce cancers, they are considered immortalized cells. There is no doubt that one can study the mechanisms of radiation-induced carcinogenesis processes in cell culture models with great ease and precision and with a short latent period; however, the dose–effect relationship or the latent period cannot be extrapolated to whole organisms. Nevertheless, one can develop new principles and concepts that may be applicable to humans in a qualitative sense. The following three types of dose responses have been observed in tissue culture models:[32]

1. Above 1 Gy: The transformation frequency may exhibit a quadratic dependence on doses.
2. Between 0.30 and 1 Gy: The transformation frequency may not vary with dose.
3. Below 0.30 Gy: The transformation frequency may be directly proportional to dose.

These results suggest that a linear extrapolation from the results at high doses cannot provide an accurate estimation of the transformation incidence at low doses.[32] Another type of dose response suggests that the transformation frequency increases up to 4 Gy, after which it decreases. The latter phenomenon is due to the fact that cell survival decreases significantly after 4 Gy, and thus fewer cells remain available for transformation.

It has been reported that fresh explants of hamster embryo cells are ten times more sensitive to x-ray-induced transformation than established cell lines (C3H10T1/2 cells). Radiation doses as low as 0.01 and 0.001 Gy of neutrons can produce transformation in the fresh explants of hamster embryo cells. It has been reported that radiation-induced transformation can be produced in embryos of Syrian hamsters irradiated with 3 Gy in utero and assayed in vitro.[33] However, the frequency of transformation is at least tenfold lower than that obtained when the cells are irradiated in vitro. This suggests that some repair of radiation damage may occur when the embryos are irradiated in utero. This frequency of transformation (0.07–0.1%) is close to the incidence of radiation-induced tumors in animals.[33] A dose of 8 Gy induced morphological neoplastic transformation, which was heritable in mouse embryo fibroblasts (C3H 101/2 cells).[34]

It has been reported that human epidermal keratinocytes immortalized by transfection with adenovirus type 12 and SV40 can be transformed to a malignant state by x-irradiation with doses of 4–8 Gy.[35] Several other types of human and rodent embryonic normal cells have been immortalized by inserting mutated oncogenes or genes from carcinogenic viruses. These immortalized cell lines can be very useful in the study of the mechanisms of radiation-induced cancers. Radiation-induced human mammary cancer is not well understood. It has been found that a single dose of 3 Gy gamma rays reduced the levels of c-myc, c-jun, and c-fos proteins by 23% to 80% in spontaneously immortalized breast epithelial cells in culture. When these cells were treated with benzo(a)pyrene and then transfected with c-Ha-ras, the same dose of radiation decreased the levels of oncoproteins by 50% to 100% compared to their unirradiated controls.[36] These observations are interesting with respect to the mechanisms of mammary carcinogenesis, because overexpression of the c-myc oncogene is commonly observed in breast cancer.

The effect of fractionated doses on the frequency of transformation depends upon the total dose, number of fractions, and time interval between fractionated doses. At doses of about 1.5 Gy, the fractionation of a single dose reduced the transformation frequency in comparison to that produced by a single dose. However, at doses between 0.3 and 1.0 Gy, the fractionation of a dose enhances the frequency of transformation. Furthermore, at 1.0 Gy, the increase in the number of fractions leads to a progressive enhancement in transformation frequency.[32]

High linear energy transfer (LET) radiation (helium-3 ions) is more effective in causing transformation in surviving rodent fibroblasts than low LET radiation (gamma rays).[37] However, the transformation frequency decreases proportional to a decrease in the survival of cells irradiated with both high LET and low LET radiation.

The effects of tumor promoters in combination with radiation on the transformation frequency have been investigated on tissue culture models. The active principal in croton oil is a class of compounds called the *phorbol esters* that act as tumor promoters. Among them, 12-o-tetradecanoyl phorbol-13-acetate (TPA) is the most active tumor promoter for chemically initiated tumors. It has been observed that TPA also markedly enhances the frequency of x-ray-induced transformation.[38] It is interesting to note that TPA is particularly effective at low radiation doses. For example, a dose of 0.25 Gy by itself does not produce detectable levels of transformation; however, when this dose of radiation is followed by TPA treatment, the transformation frequency increases to (5×10^{-4} per viable cell). This level of transformation is generally achieved by radiation doses of 2–3 Gy without any tumor promoters.[38] It has been reported that the irradiated cells retain the ability to respond to TPA treatment even after many generation times.[39] This observation is similar to that observed in mouse skin, where application of croton oil many months after carcinogen exposure still leads to a high incidence of tumors.[40] It has been observed that an increase in the iron contents of cells and tissues may increase the risk of radiation-induced cancer.[41] This may be due to the fact that free iron may increase the levels of free radicals. A human epidemiologic study has revealed that increased iron stores in the body may enhance cancer risk.[42]

It is established that ionizing radiation interacts with chemical and biological carcinogens, and tumor promoters, resulting in increased cancer incidence. For example, x-radiation enhances chemical carcinogen-induced transformation in Syrian hamster cells in culture by about ninefold[22] and UV-induced transformation by about twelvefold.[43] X-irradiation also enhances the level of ozone-[44] and viral-induced[38] transformation in cell culture. Radiation doses that alone do not transform normal fibroblasts do so when combined with a tumor promoter.[45] Ionizing radiation in combination with tobacco smoking increases the risk of lung cancer by about 50%.[46] Estrogen also increases the frequency of radiation-induced transformation in cell culture.[47] The combination of high LET radiation and asbestos fibers increases the transformation frequency in mouse embryo cells by more than additive.[48]

Hyperthermia at high temperatures enhances the frequency of radiation-induced transformation.[49–51] For example, when mouse embryo cells (C3H 101/2 cells) were exposed to heat at 45°C for 15 minutes or 43°C for 60 minutes, no transformation was observed; however, when the heat treatment was preceded by an x-ray dose of 2 Gy, the frequency of x-ray-induced transformation increased by 4–5 times.

Recently, it has been reported that when normal breast epithelial cells in culture are exposed to a radiation dose of 0.1 Gy in combination with cigarette smoke condensate, they acquired characteristics of a transformed phenotype. In addition, changes in gene expression include the upregulation of genes encoding proteins involved in metabolic pathways and inflammation.[52]

Animal models to study carcinogenesis: Cancer arises from an organ containing dividing cells or cells having potential to divide upon a proper stimulus. The frequency of radiation-induced cancer has been extensively studied in animal models. However, the dose–response curves vary from one tumor type to another and from one species to another.[53] This may be due to the fact that environmental and cellular factors that may influence the process of carcinogenesis may be different. The factors that may influence radiation-induced cancer may include (1) age, (2) gender, (3) family history, (4) capacity to repair DNA damage, (5) chemical and biological carcinogens, (6) tumor promoters, (7) immunological status, and (8) dietary factors.

Leukemia: Several studies show that the incidence of thymic leukemia in RFM mice is dependent upon the dose rate: the higher the dose rate, the greater is the frequency of this cancer.[54,55] It has been reported that in mice, the smallest dose that causes leukemia is below 200 R (about 2 Gy).[56] The susceptibility of mice to the induction of radiation-induced leukemia depends upon the gender and strain of mice. In most strains of mice, spontaneous leukemia is of thymic origin. Surgical removal of the thymus decreases the incidence of spontaneous leukemia or its induction by irradiation. On the other hand, removal of the adrenal gland increases the susceptibility to the induction of leukemia by x-irradiation. It has been observed that continuous gamma irradiation throughout life at the rate of 0.062 rad/day failed to induce leukemia in A/Jax mice, but did induce leukemia in 55% of irradiated Rap mice.[57] Although C57BL mice rarely develop spontaneous lymphosarcoma or lymphatic leukemia, they are very sensitive to radiation-induced leukemia. Radiation-induced leukemia in these mice can be transmitted by a cell-free extract. The agent responsible for inducing tumor was found to be a virus, called *radiation leukemia virus*. It was concluded that whole-body irradiation with an exposure of 300 R (about 3 Gy) caused activation and/or release of this virus or made the cellular environment compatible for virus replication, and thereby transformed normal cells to malignant cells.[58] The relative risk of mortality from gamma ray–induced thymic lymphoma, myeloid leukemia, and reticulum cell sarcoma was predominantly linear as a function of dose between 0.11 Gy to 4 Gy.[59] High doses of radiation (3 Gy or above) induced acute myeloid leukemia (AML) in C3H mice. The PU.1 gene is a member of the E-twenty sis (ETS) family of transcriptional factors and is exclusively expressed in cells of hematopoietic lineage. This gene is considered a critical factor for initiation of leukemogenesis. It has been found that radiation-induced acute myeloid leukemia is associated with upregulation of c-myc and PU.1 in PU.1-deficient AML compared with AML.[60] It appears that deletion of the PU.1 gene on chromosome 2 is a crucial event in the initiation of AML in mice. At one month after irradiation with x-rays or ^{56}Fe ions, the proportions of cells with PU.1 deletion were similar.[61,62] The effectiveness of 1 GeV ^{56}Fe-ion radiation and gamma rays was similar for the induction of AML;[63] however, the incidence of hepatocellular carcinoma was much higher in the ^{56}Fe ion–irradiated group than that in the gamma-irradiated group. The estimated relative biological effectiveness (RBE) of ^{56}Fe ion radiation for the induction of hepatocellular carcinoma was about 50. Generally, tumor development requires prevention of apoptosis. However, gamma radiation–induced thymic lymphomagenesis needs PUMA-driven apoptosis of leukocytes.[64] The suppression of PUMA caused inhibition of tumor incidence. On the other hand, loss of NOXA, a pro-apoptotic member of the Bcl2 family, accelerated the development of tumors.

NRH: Quinine oxidoreductase 2, also called NAD(P)H:quinine oxidoreductase (NQO2) is known to protect against myelogenous hyperplasia. It has been reported that 72% of NQO2-null mice developed B-cell lymphomas in multiple tissues compared with 11% in wild-type mice after irradiation with 3 Gy gamma rays; however, only 22% of NQO2-null mice showed myeloproliferation compared with none in wild-type mice.[65] It was further revealed that bone marrow from NQO2-null mice contained lower levels of p53 compared with wild-type mice due to rapid degradation of p53. In addition, irradiation of NQO2-null mice caused lower induction of p53 and Bax and decreased apoptosis of B-cell lymphocytes. It was suggested that the lack of significant induction of p53 and Bax and decrease in B-cell apoptosis by irradiation contributed to the development of lymphoma. These data suggest that NQO2 in wild-type mice plays an important role in preventing against radiation-induced B-cell lymphomas.

Lung cancer: Induction of lung cancer in mice requires high doses of radiation, and is dependent upon the dose rate. Chronic gamma irradiation at a dose rate of 8 R/day, for a total of 2400 R (about 24 Gy) over a period of 10 months, produced a 50% increase in lung cancer. An exposure of 3,000 R (about 30 Gy) or 4,000 R (about 40 Gy) delivered in a single dose to the chest cause 43% bronchial carcinoma in irradiated rats, but only 2% in hamsters.[66] Administration of radioactive plutonium dioxide increased the incidence of pulmonary tumors in control and thymectomized rats to similar levels (52–54%); however, the incidence of nonpulmonary tumors in thymectomized animals was fourfold higher (17%) than that observed in the control group (4%).[67] The same study showed that in thymectomized rats, the size and invasiveness of the tumor and the frequency of regional metastasis were much higher in comparison to the control group. These results suggest that the suppression of immune function may not affect the frequency of PuO_2-induced lung cancer, but it may affect the rate of progression of this tumor. The relative risk of mortality from gamma ray–induced lung cancer was predominantly linear as a function of dose between 0.11 Gy and 4 Gy.[59] It has been shown that the RBE of 5.5 MeV alpha particle–induced preneoplastic transformation in rat tracheal epithelial cells in culture is 2.4.[68]

Bone tumor: Administration of [226]Radium-, [90]Sr-, and [239]Pu induced osteosarcoma as well as bone carcinoma in animals as well as in humans. The incidence of these tumors increases linearly as a function of dose in beagles. No tumor was detected in these animals at doses below 0.5 rad (0.005 Gy).[69–72] Administration of ^{32}P induced a high incidence of osteosarcoma. Estimated skeletal doses were in the range of 100 Gy to 200 Gy. A single dose of 10–16 Gy produced a high incidence of osteosarcoma in rabbits with a latent period of greater than one year.[73] The same study revealed that the RBE of fission neutrons compared with gamma rays was about 3 for the induction of osteosarcoma, fibrosarcoma, and basal cell carcinoma.

Breast tumors: The incidence of mammary tumors in rats and mice is dependent upon the strain, tumor type, and irradiation conditions.[74] It has been reported that the incidence of breast tumors in rats appears to be linearly related to dose for both x-rays and gamma rays between radiation exposure of 25 R and 400 R (about 0.25 to 4 Gy).[75] The RBE of fission neutrons for the production of mammary tumors was at least as high as 20; however, when the incidence of tumors was considered 11 months after irradiation, the RBE value was only about 8.[76] The average latent period was 218 days. The dose rate may have slight or no effect on the incidence of mammary tumors, depending upon the type of tumors. Fractionation of radiation exposure of 400–500 R (about 4–5 Gy) of x-rays into as many as 32 fractions in a 16-week period produced no significant change in the total incidence of mammary tumors in rats, compared to single doses; however, an increase in the number of adenocarcinomas was observed.[77]

Ovarian cancer: Most studies suggest that doses of 50–100 rads (0.5 to 1 Gy) of x-rays produced the maximum incidence of ovarian tumors in RFM and BALB/c mice.[55,78,79] The incidence of tumors in these mice appears to be dependent upon the dose rate within a certain range of dose rates. It has been reported that the incidence of tumors decreased at low dose rates.[80,81]

Uterine cancer: A high incidence of uterine cancer was reported in rabbits after radiation exposures of 1400–2900 R (about 14–29 Gy) at dose rates of 1.1 to 8.8 R delivered in 8 hours. The latent period at a high dose rate of 8.8 R/day in 8 hours was 36 months; however, it was 57.5 months at a low dose rate of 1.1 R/day in 8 hours. It has been suggested that radiation-induced uterine carcinoma results from a hormonal imbalance.[82]

Skin Cancer: A high incidence of skin cancers (about 97%) in rats receiving whole-body proton irradiation (1–25 Gy) with a maximum penetration of 1.4 mm was observed. Doses at 2 Gy produced a high incidence of skin cancer.[83] Rats surviving a single whole-body radiation exposure of 660 R (about 6.6 Gy) showed a high incidence of skin cancer. The incidence of skin tumors was about 25% after s.c. injection of ^{90}Y at concentrations of 10 and 30 µCi per mouse. The RBE of fission neutrons for the production of basal carcinoma is about 3.[73]

Tumors of the alimentary tract: We have reported that the irradiation of the exteriorized small intestine by x-ray at radiation exposures of 1400 R and above (about 14 Gy and above) produced a high incidence (67% of the survivors) of low-grade adenocarcinoma of the small intestine.[84] Irradiation with a dose of 2.7 Gy of fast neutrons also produced a high incidence of intestinal tumors.[85] It has been reported that irradiation of mice with gamma and/or high-energy charge particles radiation, the main component of galactic cosmic rays, significantly enhanced development of and progression of intestinal tumors.[86] An increased incidence of hepatoma in mice after chronic gamma irradiation has been reported.[82] An injection of 40–160 µCi of ^{198}Au produced 15% hepatoma in C57BL mice and 32% in RF mice, but cirrhosis occurred in 90% of the mice.[87] Clonal expansion of tumor-initiated cells appears to be a necessary process for the formation of cancer. It is possible that loss of p53 protein function in preneoplastic cells may favor growth of these cells over normal cells. Indeed, it has been reported that preneoplastic hepatocytes have a lower level of p53 protein and are not able to increase its level following x-irradiation.[88]

Thyroid cancer: The injection of radioactive iodine increased the incidence of thyroid cancer in animals.[89,90] A concentration of 32 µCi of ^{125}I increased the incidence of medullary thyroid carcinoma from 13% to 32% and follicular cysts from 4% to 43% in rats, whereas ^{131}I at a lower concentration of 3.2 µCi produced about 15% medullary carcinoma and 53% of follicular cysts.[91]

Pituitary and adrenal tumors: In mice, destruction of thyroid by ^{131}I (200–300 µCi) may be followed by the development of pituitary tumors.[92,93] This type of tumor appears to have a very long latent period. An intravenous administration of radioactive polonium at a concentration of 1 µCi/kg of body weight induced adenomas of the adrenal medulla in surviving rats one year later.

Induction of tumors in infant mice: Mice during infancy are highly sensitive to the induction of solid tumors. Irradiation in the neonatal period in female mice with gamma radiation doses of 0.48–5.70 Gy increased the incidence of several solid tumors. It was found that the incidence of liver, pituitary, ovarian, and lung cancer is proportional to radiation doses, whereas the incidence of bone tumors was proportional to the square of the dose. A significant increase in the incidence of gastrointestinal tumors, kidney tumors, adrenal tumors, and hemangiomas of the spleen was also detected in the previous study.[94] In order to investigate the effect of age on radiation-induced cancer, female mice (B6C3F1 strain) were irradiated with gamma radiation doses of 0.10, 0.48, or 0.95 Gy at 7 days of age. The results showed that infant mice were susceptible to solid tumors, especially endocrine tumors. The high doses of radiation increased the incidence of these tumors more than that produced by lower doses of radiation.[95]

Radiation-induced tumors in adult mice irradiated in utero: In order to investigate the effect of irradiation on tumor induction, pregnant Swiss albino mice were irradiated with gamma-ray doses of 0.1–1.5 Gy on days 14 or 17 of gestation. The incidence of tumors in adult mice showed a linear-quadratic dose response. The organs showing tumors include ovary, uterus, liver, and spleen; however, the highest incidence was observed for ovarian tumors. Tumors in the ovary and uterus developed at an earlier age than in the liver and spleen.[96]

Effect of irradiation before conception on tumor incidence in animal offspring: In order to evaluate the effect of irradiation before conception on the incidence of tumors in the offspring, female mice were irradiated with an x-ray dose of 4 Gy (divided into two fractions) 2 weeks before mating with untreated males. After weaning, half of the offspring were exposed to an immunomodulating and tumor-promoting cyclosporine A through diet for a period of six months, and the incidence of tumors was evaluated at 28 months after irradiation in all groups. Fertility and the lifetime of the irradiated mice were reduced and the incidence of tumors was also increased in these mice compared to nonirradiated mice. The incidence of hematopoietic and lymphatic tumors increased in the offspring of the irradiated mother. These tumors become detectable 6 months after treatment with cyclosporine A. A higher incidence of lung and liver tumors was observed in male offspring of mothers irradiated before conception.[97] These data suggest that radiation-induced gene alterations in female germ cells responsible for carcinogenesis can be transmitted to offspring. It has been demonstrated that prenatal exposure to radiation or chemical carcinogens causes adverse effects including increased incidence of tumors in mice. In addition, the tumor-susceptibility phenotype is transmittable beyond the first 1-generation.[98] The incidence of tumors was 100 times higher than that produced by mouse-specific locus mutations. In humans, a higher risk of leukemia and birth defects has been reported in the children of a father who had been exposed to radionuclides in nuclear reprocessing plants and diagnostic radiation.

Induction of radiation-induced tumors in nonhuman primates: Increasing use of whole-body irradiation in combination with chemotherapy in the treatment of hematological malignancies and refractory autoimmune disease has produced significant improvement in the survival rate; however, the risk of a second new cancer exists among survivors. The induction of tumors in Rhesus monkeys irradiated whole-body either with neutron doses ranging from 2.34–4..4 Gy (average dose 3 Gy) or with x-ray doses ranging from 2.8–8.6 Gy (average dose 7.1 Gy) was compared with age-matched nonirradiated control animals. The results showed that 90% of neutron-irradiated animals had one or more malignant tumors, whereas 50% of x-irradiated animals had malignant tumors at the time of their death.[99] The incidence of brain tumors, especially glioblastoma multiforme or grade IV astrocytoma, increased among Rhesus monkeys whole-body irradiated with doses between 4.0–8.9 Gy of 55-MeV protons, a component of solar particles. The latent period for tumor induction ranged from 14 months to 20 years.[100] It has been shown that RBE of 55-MeV proton radiation for the induction of brain tumors may be about 1, suggesting that proton radiation–induced genetic alterations may not be different from those produced by x-irradiation or gamma irradiation.[101] This observation is unusual because protons on other criteria of radiation damage produce an RBE value greater than 1.

EFFECT OF INTERACTION BETWEEN RADIATION AND CHEMICALS ON CARCINOGENESIS

It has been reported that ionizing radiation in combination with several chemicals, such as benzo(a)pyrene, smoking, urethane, diethylstilbestrol, estrogen, ethionine, and procarbazine, produced more tumors than that produced by irradiation alone.[102-111] It has been found that application of repeated radiation exposure of 20 R (about 0.2 Gy) of x-rays during and following exposure to 7,12-dimethylbenz (a) anthracene (DMBA) increased DMBA-induced cheek pouch cancer in the Syrian hamster. Irradiation prior to DMBA treatment was ineffective.[112] It is unknown whether a similar phenomenon is observed when the interaction of radiation with other types of carcinogens occurs.

RADIATION-INDUCED CANCER INCIDENCE IN HUMANS

The primary sources of radiation-induced cancer in humans include the following:

1. Populations of Hiroshima and Nagasaki who were exposed to single whole-body irradiation during atomic bombardment in World War II (primarily leukemia and most solid tumors)
2. Marshall Islanders who were exposed to atomic fallout radiation during testing of an atom bomb (primarily thyroid cancer)
3. Uranium miners (primarily lung cancer)
4. Nuclear accidents (primarily leukemia)
5. Patients who were treated with fractionated doses of ionizing radiation locally for nonneoplastic diseases, such as ankylosing spondylitis (primarily leukemia), tuberculosis (primarily breast cancer), and tinea capititis (primarily thyroid cancer)
6. Cancer survivors following treatment with ionizing radiation and/or chemotherapy (primarily leukemia and breast cancer)
7. Patients who receive diagnostic x-rays (primarily leukemia, breast cancer, and thyroid cancer)

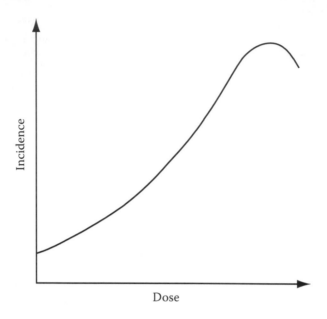

FIGURE 4.1   General radiation dose-response curve for cancer incidence.[113]

Although the effect of a single dose of ≤ 0.1 Gy can be estimated with a certain degree of accuracy, it is difficult to estimate the carcinogenic effect of fractionated low doses of x-rays (<0.1 Gy). Any estimation of the effect of fractionated low doses of radiation involves certain assumptions and extrapolations; therefore, any value of the incidence of cancer must be considered crude estimates.[113] Tissue culture and animal studies on carcinogenesis suggest that the incidence of tumors in humans can be modified by several factors, including tissue and organ types, cellular origin, gender, age, total dose, dose rate, fractionation, LET, co-carcinogens, and tumor promoters. In addition, diet, lifestyle, environmental, and genetic factors also can influence the incidence of tumors in humans.

Models used for the estimation of cancer risk: Generally, the incidence of cancer increases as a function of radiation dose. Cell killing at higher doses decreases the incidence of cancer (Figure 4.1). Three dose–response models, linear dose response, quadratic dose response, and linear-quadratic dose response, have been used to estimate the cancer risk in humans[114] (Figure 4.2). The values of cancer risk estimates depend upon the particular dose–response model, and they may vary at least by a factor of 2. The linear dose response is primarily observed at low doses, whereas the proportional to square of dose is found at higher doses. Below 4 Sv, the incidence is a linear function of dose for all cancers except leukemia. The incidence of leukemia reveals a linear-quadratic dose response.

Methods of cancer risk estimation: Three methods, *absolute risk estimate, relative risk estimate*, and *lifetime risk estimate*, can be used to estimate cancer risk in humans. The excess cancer above the baseline that can be induced by irradiation alone is referred to as the *absolute cancer risk*. Since it is impossible to separate the effect of radiation from other co-carcinogens and tumor-promoting factors on cancer

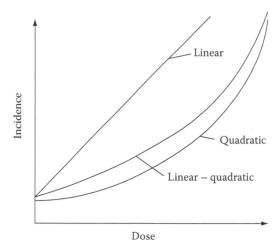

FIGURE 4.2   Radiation dose-response curve for cancer incidence.[113]

incidence, it is not meaningful to express radiation-induced cancer risk in absolute terms. The relative risk estimate is the excess of cancer risk above baseline that is due to the interaction between radiation and other factors. This is commonly used in estimating radiation-induced cancer risk. Lifetime risk is determined after the end of the life of the individual in a population and is expressed as the number of cases per Gy.

INCIDENCE OF RADIATION-INDUCED CANCER IN HUMANS

The survivors of the atomic bombing in Hiroshima and Nagasaki represent a population of all ages and sexes who received a wide range of single doses; therefore, they have been used by several committees to determine cancer risk, including the Biological Effects of Ionizing Radiation (BEIR), the International Commission on Radiological Protection (ICRP), and the United Nations Scientific Committees on the Effects of Atomic Radiation (UNSCEAR). The incidence of total cancer has been updated by a series of BEIR reports. Other sources of radiation-induced cancer come from survivors of cancer therapy and patients who received x-rays for the treatment of nonneoplastic diseases.

Hematopoietic cancer: The incidence of leukemia among atomic bomb survivors in Hiroshima and Nagasaki was analyzed for the period between October 1, 1950 to September 30, 1966.[115] The total air-dose at 1000 meters was 4.48 Gy of neutrons + gamma radiation in Hiroshima, and 9.25 Gy of primarily gamma radiation in Nagasaki. The rate of leukemia increased sharply with these doses in both Hiroshima and Nagasaki; however, the rates were higher in Hiroshima than in Nagasaki. The difference in the rates of leukemia between two cities may be due to differences in the quality of radiation. The population of Hiroshima was exposed to both neutrons and gamma rays, whereas the population of Nagasaki was exposed to only gamma rays. The RBE of neutrons for the induction of leukemia was considered to be about 5. The relative risk was greater in males than in females, especially at higher doses. Younger individuals were more sensitive to radiation-induced leukemia than older individuals; however, the incidence of leukemia at the fractionated low-dose range increased with age. In Hiroshima, the incidence of lymphocytic leukemia and chronic granulocytic leukemia increased in individuals receiving doses between 0.05 to 0.99 Gy, whereas the incidence of acute and chronic granulocytic leukemia increased in individuals receiving higher doses. The BEIR report showed that radiation-induced leukemia generally appears within 10 years after irradiation. In contrast, radiation-induced solid tumors can appear even 30 years or more after irradiation.

Another study examined the relative risk of cancer among survivors from 1950 to 1985.[115] The relative risk factor for leukemia increased fivefold, whereas this varied from 1.2 to 3.0 for other cancers. The risk of leukemia was higher among individuals who were under the age of 20 years at the time of irradiation than those who were older than 20 years. Excess relative risk of leukemia per Sv was 4.24 to 5.21. There appears to be prevalence of Hodgkin's disease, lymphosarcoma, reticulum cell sarcoma, and multiple myeloma among survivors of Hiroshima who received an estimated dose greater than 1 Gy. Males who were less than 25 years of age at the time of the atomic explosion showed relatively high risk in comparison with those who were older than 25 years of age.[116] No such effect of irradiation on the incidence of lymphoma was observed in the population of Nagasaki. This difference between the populations of Hiroshima and Nagasaki may be due to the fact that the Hiroshima atom bomb released both neutrons and gamma rays, whereas the Nagasaki atom bomb released only gamma rays. The genetic difference between the two populations may also account for the difference in lymphoma incidence. It has been estimated that if 100,000 people of all ages were exposed to 0.1 Gy (whole-body, single dose), 800 extra deaths due to radiation exposure may occur; however, 20,000 cancer deaths will occur without radiation exposure.[114]

A recent analysis of survivors of the Hiroshima and Nagasaki atomic bombings revealed some new information regarding the dose–effect relationship with respect to induction of cancers.[117] It is interesting to note that the risk of chronic lymphocytic leukemia, pancreatic carcinoma, prostate cancer, and uterine carcinoma did not increase in these populations. The dose–response curves for most solid tumors were linear, although analysis of mortality data suggests that there may be an upward curvature for all solid tumors. For leukemia, the dose–response curve is of quadratic type within the dose range of 0–2 Sv.

The German Hodgkin Lymphoma Study Group evaluated the incidence of second solid tumors among 5,367 patients with early, intermediate, and advanced Hodgkin's disease who received chemotherapy in combination with radiation therapy. The results showed that the most frequent tumors were 23.6% lung cancer, 20.5% colorectal cancer, and 10.2% breast cancer. After a median follow-up of 6 years, the cumulative risk of developing solid cancer was 2% with an overall relative risk of 2.4% for lung cancer, 3.8% for colorectal cancer, and 3.2% for breast cancer.[118] In another study, the effect of radiation therapy and chemotherapy on the incidence of contralateral breast cancer among 7,221 treated patients was evaluated. The results showed that radiotherapy-associated risk of contralateral breast cancer increased with decreasing age at first treatment (age below 35 years).[119] In addition, women treated before the age of 45 years with postlumpectomy radiotherapy had 1.5-fold increased risk of contralateral breast cancer compared with those who received radiotherapy postmastectomy. It has been reported that the incidence of second cancer among patients with hereditary retinoblastoma was much higher than those with nonhereditary retinoblastoma after radiation therapy.[120] Cumulative incidence of a second cancer (primarily sarcoma) at 50 years after diagnosis was about 51.0% for hereditary retinoblastoma, and only 5% for nonhereditary retinoblastoma. Treatment of breast cancer with radiation therapy–induced lung cancer at 10–14 years and 15 years or more after initial breast cancer diagnosis with a relative risk of 1.62, myeloid leukemia at 1–5 years with a relative risk of 2.99, second breast cancer at 5–10 years with a relative risk of 1.34, and esophageal cancer at 15 years or more with a relative risk of 2.19.[121]

An analysis of 13,352 patients who received fractionated doses of x-rays for the treatment of ankylosing spondylitis showed that the incidence of leukemia is proportional to dose with no threshold dose. Nine cases of leukemia occurred among 3,224 men who participated in military maneuvers during the 1957 nuclear test explosion. This value was about three-fold higher than that of the expected value of 3.5 cases. The average latent period was about 14.2 years.[122] Others have reported that that incidence of leukemia increased among those who were 10–14 years of age at the time of nuclear testing, and who received an average dose of 4.662 mSv from fallout radiation in Utah.[123]

Solid cancer: Analysis of solid cancer incidence among survivors of the Hiroshima and Nagasaki atomic bombing showed that individuals irradiated at the age of 30 years, and cancer incidence determined at the age of 70 years, the incidence of solid cancer increased by 35% per Gy for men and 58% per Gy for women.[124] Significant increases in radiation-induced cancer was observed for most sites, including oral cavity, esophagus, stomach, colon, liver, lung, nonmelanoma skin, breast, ovary, bladder, nervous system, and thyroid. The analysis of solid tumor incidence among survivors of the Hiroshima and Nagasaki atomic bombings who were in utero ($n = 2452$) or younger than 6 years of age at the time of the explosion showed that at the age of 50 years, estimated absolute rates per 10,000 person-year per Sv were 6.8 for those exposed in utero, and 5.6 for those exposed as young children.[125]

Lung cancer: The more proximal regions of the bronchial tree are most radiosensitive to the induction of bronchiogenic carcinoma. An analysis of the Hiroshima and Nagasaki data revealed that a dose greater than 1.28 Gy increased the incidence of lung cancer by a factor of 2 among survivors.[126] The induction of lung cancer depended upon the age at the time of irradiation and the duration of observation period.[127] The excess death rate was 2.1 cases per 10^6/year/rad for those individuals aged 35–49 years at the time of the atomic explosion, and it was 4.9 cases for those 50 years of age or older, when the values obtained from both Hiroshima and Nagasaki were combined. The mean absorbed dose to lungs of survivors of the Hiroshima bombing was 0.88 Gy of gamma rays and 0.0095 Gy of neutrons.

The incidence of lung cancer among miners working in mines involving radioactive substances is high.[128–130] The average dose received at the site of the tumor was about 30 Gy and the average tumor induction time was about 17 years. The risk of lung cancer among miners is about 22–45 cases per 10^6/year/rad. The RBE value of alpha particles is between 8 and 15, whereas the RBE value of neutrons is only 5.

Smoking enhanced the incidence of lung cancer by about 50% and shortened the latent period.[129] The latent period appears to be dependent upon the age of the individual. The latent period appears to decrease with an increase in age.

Irradiated before the age 15 years = 25 years latent period

Irradiated between 15 and 34 years = 15–20 years latent period

Irradiated at ≥ 35 years = 10 years latent period

Bone cancer: No increase in the incidence of bone cancer was found among the survivors of the Hiroshima and Nagasaki bombings. This may be due to the fact that the skeletal radiation doses of survivors were lower than needed to induce osteosarcoma.[131] Osteosarcomas are the most common forms of bone cancer. Therapeutic radiation doses of 3000 R (about 30 Gy) can increase the incidence of bone cancer.[132] The threshold skeletal dose from ^{226}Ra for the induction of osteosarcoma is between 0.67–0.90 Gy.[133] Another study reported that the threshold for a cumulative radiation dose to induce bone cancer was 0.5–1.10 Gy for mice, dogs, and humans.[72] The risk of radiation-induced bone cancer is independent of age and gender.

Breast cancer: The female breast is very sensitive to radiation-induced cancer. A number of papers have been published on this issue, including those listed here.[134–143] The following radiobiological principles have emerged from these studies:

1. The incidence of breast tumor is primarily related to dose at low dose levels.
2. Fractionated doses of x-rays are approximately as effective as a single dose.
3. Breast tissue becomes more radiosensitive during pregnancy when the cells have increased proliferation rates. Increased levels of estrogen may enhance the incidence of radiation-induced breast cancer.
4. The latent period does not depend upon the dose, but upon the age at the time of irradiation. Younger women have a shorter latent period than older women.

Ovarian, uterine, and cervical cancer: X-rays can induce both benign and malignant tumors in the ovaries. Among 731 gynecological patients treated with intracavitory ^{226}Ra or x-rays (7–17 Gy), the incidence of ovarian tumor was 3.1 times higher than the expected value.[144] The latent period was 10.1 years. The above study also revealed that the incidence of uterine carcinoma was 5.9 times higher than the expected value, with a latent period of about 9.7 years; however, the incidence of cervical cancer in this group of patients was only 1.7 times higher than the expected value, with a mean latent period of 8.5 years. The data on survivors of the Hiroshima and Nagasaki bombings have failed to show any definite relationship between radiation dose and cancer of the uterus and cervix. This may be due to the fact that these survivors did not receive radiation doses high enough to induce these tumors.

Skin Cancer: Ionizing radiation is known to induce skin cancer, primarily basal cell carcinoma. It has been reported that two patients who were treated for malignant disease by surgery and gamma radiation (36–40 Gy fractionated doses) developed melanoma in tattoo sites used for marking the radiation field.[145] It has been shown that 19% of skin cancer patients had a previous history of radiation exposure, whereas only 4.5% of nonirradiated subjects developed skin cancer.[146] Therapeutic doses of 500–2,000 R (about 5–20 Gy) induced a 4.5% incidence of skin cancer in the Japanese population, and only 0.6% developed skin cancer in nonirradiated individuals.[147] Fractionation of radiation doses decreases the incidence of skin cancer. The risk of cancer appears to increase with age.

Cancer of the alimentary canal: Radiation-induced cancer of the esophagus has been observed but with less frequency. It has been estimated that the risk of this type of cancer among survivors of the Hiroshima atomic bombing was only 0.39 excess cases per 10^6/year/rad. The risk may be higher for persons who were 35 years or older at the time of radiation exposure.[114] The same BEIR report showed that an estimated risk of stomach cancer among patients with ankylosing spondylitis may be 0.59 excess cases per 10^6/year/rad. The dose estimate to the stomach was about 2.5 Gy. Among the survivors of the Hiroshima bombing, excess deaths from stomach cancer may be 1.57 cases/10^6/year/rad. The mean dose to the stomach is estimated to be about 0.37 Gy. However, among the survivors of the Nagasaki atomic bombing, the excess deaths from stomach cancer may be 1.05 cases/10^6/year/rad. The same BEIR report revealed that the risk of colon cancer among patients with ankylosing spondylitis is estimated to be 1.7 cases/10^6/year/rad, with a latent period of about 15–25 years. The average estimated dose to the colon was 0.57 Gy.

Cancer of liver and pancreas: Ionizing radiation can induce cancers of liver and pancreas. It has been estimated that the risk for induction of liver cancer by gamma irradiation would be 0.7 cases/10^6/year/rad, whereas it was 13 cases/10^6/year/rad.[114] The risk of ^{239}PU-induced liver cancer is four times higher than that of ^{239}Pu-induced bone sarcoma.[148]

Among 14,554 patients with ankylosing spondylitis who received a pancreatic dose of 0.9 Gy, excess deaths from pancreatic cancer were 0.7 cases/10^6/year/rad, with a latent period of 7–9 years.[149] The incidence of pancreatic cancer increased among nuclear workers at the Hanford nuclear plant.[150,151] An estimation of excess deaths from pancreatic cancer in this population was 10 cases/10^6/year/rad.[152] Increased risk of pancreatic cancer has been observed after radiation therapy in patients with cervical cancer and lymphoma.[153,154]

Cancer of the thyroid and salivary glands: A tenfold increase in incidence of thyroid cancer was observed in children living in Belarus, Russia, and Ukraine who were exposed to radioactive fallout following the Chernobyl nuclear accident after 10 years of exposure.[155] X-irradiation is about 10–80 times more effective than beta irradiation.[156] Radiation-induced thyroid cancers are primarily of the papillary and follicular types.[156,157] The BEIR report suggested that females are about 2.3 time more sensitive to radiation for the induction of thyroid cancer than males.[114] The incidence among females of Jewish heritage may be 17 times more sensitive to radiation for the induction of thyroid cancer than those of non-Jewish heritage. A single dose of 0.06 Gy can produce thyroid tumors; lower fractionated doses of x-rays can increase the risk of thyroid cancer. The latent period varies from 10 to 35 years after irradiation.

Both benign and malignant cancers of the salivary gland occur after therapeutic doses of irradiation.[158–160] The BEIR report estimated that the induction rate for both benign and malignant salivary gland tumors was about no more than 10 excess cases/10^6 children/year/rad.[114]

Central nervous system (CNS) tumors: Ionizing radiation can induce brain tumors. Radiation therapy is commonly used in the treatment of various intracranial tumors. It has been reported that increased incidence of meningioma occurred following irradiation of cranial tumors. The time interval between the irradiation and the onset of meningioma was significantly less in younger patients.[161] Among children surviving the treatment of leukemia and brain tumor with standard therapy, the risk of developing secondary primary CNS tumors is high. Radiation exposure was associated with the increased risk of glioma and meningioma. Glioma occurred after a median of 9 years and meningioma after 17 years from initial cancer treatment.[162] The higher risk of secondary primary glioma in child survivors who were irradiated at an early age may suggest greater susceptibility of the developing brain to radiation. In addition to tumors of the central nervous system, childhood survivors of cancer treatment appear to be associated with more than a ninefold increase in the risk of secondary sarcomas compared to the general population.[163] An epidemiologic study reported that the survivors of childhood cancer treatment had increased risk for the second primary cancer, including nonmelanoma skin cancer.[164]

An epidemiologic study on children who received an average dose of 0.08 Gy in utero, and who died of CNS tumors between 0 and 14 years of age, suggests that the absolute risk for excess deaths was about 6.1 cases/10^6/year/rad.[165] Another study estimated that the absolute risk for excess deaths due to CNS tumors was 6.3 cases/10^6/year/rad.[166] In children 8 years of age with ringworm of the scalp who were treated with 1.4 Gy, the absolute risk for developing cancer was about 1.3 excess deaths/10^6/year/rad on the basis of a follow-up period of 15–34 years. The minimum latent period was 5 years, but the tumors were observed 25 years after irradiation.[167]

LATE EFFECTS OF HIGH-DOSES OF RADIATION ON THE RISK OF NON-NEOPLASTIC DISEASES

In the survivors of childhood cancer, ovarian and testicular failure has been reported after treatment with tumor therapeutic agents.[168,169] Patients who received radiation doses of 10–40 Gy fractionated doses for the treatment of other malignant diseases developed hypothyroidism a few months to several years after radiation therapy.[170] Other non-neoplastic diseases include cataract and delayed necrosis in brain, muscle, auditory ossicles, and bone.[113] There are excess risks of nonneoplastic diseases, particularly cardiovascular, respiratory, and digestive diseases, among the survivors of the Hiroshima and Nagasaki atomic bombing.[117] Because of many confounding factors, the dose–effect relationships for these diseases have not been established. It has been reported that the immune system of survivors of the Hiroshima and Nagasaki bombings was damaged proportional to the dose of radiation at the time of the explosion.[171] Although the immune system in these radiation-exposed populations tended to repair and regenerate as the hematopoietic system recovered, significant residual injury persists as shown by the presence of abnormalities in lymphoid cell composition and function. These abnormalities include: (1) abnormal T-cell function as shown by reduction in mitogen-dependent proliferation and IL-2 production; (2) decrease in helper T-cell populations; and (3) increase in inflammatory cytokine levels. It has been reported that among survivors of the Hiroshima and Nagasaki bombings, the peripheral blood showed a decrease in the T-cell population and an increase in the B-cell population.[172] The impairment in T-cell function and composition may lead to acceleration of immunological aging, which could be associated with some chronic diseases. Indeed, radiation-induced T-cell immunosenescence may increase inflammatory responses that may be partly involved in the development of age-associated and inflammation-related chronic diseases that are commonly observed among survivors of atomic bombing.[173]

SUMMARY

Radiation-induced neoplasms cannot be distinguished from those that occur spontaneously or that are induced by chemical carcinogens. Cancer cells result from accumulation of multiple genetic changes in normal dividing cells, which occur over a long period of time. The latent period (time interval between exposure to carcinogens and detection of cancer) of a neoplasm depends upon the radiation dose, type of organ exposed, and experimental models (tissue culture, animals, or humans). The latent period for radiation-induced cancer is about a few weeks in the tissue culture model, about a few months in the animal model, and several years in the human model. The US mortality rate from cancer has not changed significantly since 1950 in spite of extensive research and development of new treatment modalities. In 1950, the death rate was about 194 per 100,000 persons; however in 1991, it increased to about 251.1 per 100,000 persons. In 2006, it dropped to about 180.7 per 100,000 persons. These data suggest that the cancer death, which was increasing between 1950 and 1991, was prevented due to advancement of current cancer treatment modalities. The establishment of two-stage carcinogenesis, initiation, and promotion, helped to identify agents that can initiate carcinogenesis (tumor initiators) and those that can promote cancer (tumor promoters). Tumor initiators at any dose can cause cancer, whereas tumor promoters by themselves do not cause cancer, but they, in combination with tumor initiators, increase the risk of cancer, and may reduce the latent period. Tumor progression that can lead to metastasis may be the result of accumulation of additional mutations, some of which may make cancer cells more aggressive. The nature of mutated genes that make cancer cells more aggressive has not been adequately defined.

The proposed hypotheses for carcinogenesis include activation of proto-oncogenes, loss of anti-oncogenes, infection with certain viruses, substitution of normal promoters of proto-oncogenes with strong promoters of viruses, and chromosomal aberrations. Radiation-induced neoplastic transformation in vitro and in vivo is also associated with alterations in the expression of certain oncogenes. Although the ras gene abnormality was detected in some cases of radiation-induced tumors, it is less frequent compared to cancer induced by certain chemical carcinogens. The frequency of mutation or overexpression of oncogenes is seldom over 50% in human tumors or in chemical-induced animal tumors. Therefore, the involvement of mutated ras or overexpression of cellular oncogenes in radiation-induced cancer in humans may be incidental. None of the proposed hypotheses alone is sufficient to explain the initial events in human carcinogenesis.

Humans are exposed to several tumor initiators and promoters daily from environment-, diet-, and lifestyle-related factors. Radiation represents one of the tumor initiators. Normal cells undergo at least two identifiable stages of carcinogenesis, immortalization and cancerous. Immortalized cells continue to divide without undergoing differentiation leading to the formation of benign tumors, whereas cancerous cells continue to divide and metastasize to distant organs. Certain oncogenic viruses can induce immortalization when inserted into normal cells in culture. When immortalized hepatocytes are transfected with the oncogenic c-Has in culture, they gain the capacity to form tumors in appropriate hosts. This suggests that activation of cellular oncogenes can constitute a cancer risk factor, secondary to a critical step leading to immortalization. The studies on tissue culture models have established the following radiation dose–effect relationships: above 1 Gy, the transformation frequency may exhibit a quadratic dependence on doses; between 0.30 and 1 Gy, the transformation frequency may not vary with dose; and below 0.30 Gy, the transformation frequency may be directly proportional to dose. Animal studies have revealed that high-dose ionizing radiation can induce tumors in most adult organs, which contain dividing cells or cells that have the potential to decide, such as hematopoietic tumors (leukemia and lymphoma) and solid tumors, such as lung cancer, bone cancer, breast cancer, ovarian cancer, uterine cancer, skin cancer, tumors of the alimentary canal, thyroid cancer, and pituitary and adrenal gland tumors. Fetuses and children are more sensitive to radiation for the induction of cancer. Irradiation can induce the risk of cancer in the offspring of females who received radiation before conception.

The incidence of radiation-induced cancer in humans comes from the following sources: (1) populations of Hiroshima and Nagasaki who were exposed to a single whole-body irradiation during atomic bombardment in World War II (primarily leukemia and most solid tumors); (2) Marshall Islanders who were exposed to atomic fallout radiation during the testing of an atom bomb (primarily thyroid cancer); (3) uranium miners (primarily lung cancer); (4) nuclear accidents (primarily leukemia); (5) patients who were treated with fractionated doses of ionizing radiation locally for nonneoplastic diseases, such as ankylosing spondylitis (primarily leukemia), tuberculosis (primarily breast cancer), and tinea capititis (primarily thyroid cancer); (6) cancer survivors following treatment with ionizing radiation and/or chemotherapy (primarily leukemia and breast cancer); and (7) patients who receive diagnostic x-rays (primarily cancer, including leukemia, breast cancer, and thyroid cancer). The last item will be discussed in detail in Chapter 7.

It has been reported that ionizing radiation in combination with several chemicals, such as benzo(a)pyrene, smoking, urethane, diethylstilbestrol, estrogen, ethionine, and procarbazine, produced more tumors than were produced by irradiation alone.

Ionizing radiation can also induce nonneoplastic diseases, such as ovarian and testicular failure, hypothyroidism, cataract, and delayed necrosis in brain, muscle, auditory ossicles, and bone. There are excess risks of nonneoplastic diseases, particularly cardiovascular, respiratory, and digestive diseases, among the survivors of the Hiroshima and Nagasaki atomic bombings. Because of many confounding factors, the dose–effect relationships for these diseases have not been established. It has been reported that the immune systems of survivors of the Hiroshima and Nagasaki bombings were damaged proportional to the dose of radiation at the time of the explosion. Although the immune systems in these radiation-exposed populations tended to repair and regenerate as the hematopoietic system recovered, significant residual injury persists as shown by the presence of abnormalities in lymphoid cell composition and function. The impairment in T-cell function and composition may lead to acceleration of immunological aging, which increases the risk of some chronic diseases. At present, there are no effective strategies to reduce the risk of radiation-induced cancer or nonneoplastic diseases.

REFERENCES

1. Boutwell, R., Biology and biochemistry of two-steps carcinogenesis, in *Modulation and Mediation of Cancer by Vitamins*, ed. Meyskens, F. J., and Prasad, K. N., Karger, Basel, 1983, 2.
2. Bishop, J. M., Cellular oncogenes and retroviruses, *Annu Rev Biochem* 52, 301–354, 1983.
3. Varmus, H. E., The molecular genetics of cellular oncogenes, *Annu Rev Genet* 18, 553–612, 1984.
4. Knudson, A. G. Jr., Hereditary cancer, oncogenes, and antioncogenes, *Cancer Res* 45 (4), 1437–1443, 1985.
5. Evans, A. S., Viral infection of humans, in *Epidemiology and Control*, Plenum Press, New York, 1989.
6. Duesberg, P. H., Cancer genes: Rare recombinants instead of activated oncogenes (a review), *Proc Natl Acad Sci USA* 84 (8), 2117–2124, 1987.
7. Heim, S., *Cancer Cytogenetics*, Alan R. Liss, New York, 1987.
8. Guerrero, I., Villasante, A., Corces, V., and Pellicer, A., Activation of a c-K-ras oncogene by somatic mutation in mouse lymphomas induced by gamma radiation, *Science* 225 (4667), 1159–1162, 1984.
9. Diamond, L. E., Guerrero, I., and Pellicer, A., Concomitant K- and N-ras gene point mutations in clonal murine lymphoma, *Mol Cell Biol* 8 (5), 2233–2236, 1988.
10. Sawey, M. J., Hood, A. T., Burns, F. J., and Garte, S. J., Activation of c-myc and c-K-ras oncogenes in primary rat tumors induced by ionizing radiation, *Mol Cell Biol* 7 (2), 932–935, 1987.
11. Sloan, S. R., Newcomb, E. W., and Pellicer, A., Neutron radiation can activate K-ras via a point mutation in codon 146 and induces a different spectrum of ras mutations than does gamma radiation, *Mol Cell Biol* 10 (1), 405–408, 1990.
12. Sloan, S. R., and Pellicer, A., Activation of the ras oncogene in gamma radiation and neutron radiation induced thymic lymphomas, *Prog Clin Biol Res* 374, 1–18, 1992.
13. Endlich, B., Salavati, R., Zhang, J., Weiss, H., and Ling, C. C., Molecular analysis of rat embryo cell transformants induced by alpha-particles, *Int J Radiat Biol* 64 (6), 715–726, 1993.
14. Duesberg, P. H., and Schwartz, J. R., Latent viruses and mutated oncogenes: No evidence for pathogenicity, *Prog Nucleic Acid Res Mol Biol* 43, 135–204, 1992.
15. Prasad, K. N., Cole, W., and Hovland, P., Cancer prevention studies: Past, present, and future directions, *Nutrition* 14 (2), 197–210; discussion 237–238, 1998.
16. Prasad, K. N., Carvalho, E., Edwards-Prasad, J., La Rosa, F. G., Kumar, R., and Kumar, S., Establishment and characterization of immortalized cell lines from rat parotid glands, *In Vitro Cell Dev Biol Anim* 30A (5), 321–328, 1994.

17. Zhai, R., Liu, G., Asomaning, K., Su, L., Kulke, M. H., Heist, R. S., Nishioka, N. S., Lynch, T. J., Wain, J. C., Lin, X., and Christiani, D. C., Genetic polymorphisms of VEGF, interactions with cigarette smoking exposure and esophageal adenocarcinoma risk, *Carcinogenesis* 29 (12), 2330–2334, 2008.

18. Xu, T., Zhu, Y., Wei, Q. K., Yuan, Y., Zhou, F., Ge, Y. Y., Yang, J. R., Su, H., and Zhuang, S. M., A functional polymorphism in the miR-146a gene is associated with the risk for hepatocellular carcinoma, *Carcinogenesis* 29 (11), 2126–2131, 2008.

19. Zhang, Z., Wang, S., Wang, M., Tong, N., and Fu, G., Genetic variants in RUNX3 and risk of bladder cancer: A haplotype-based analysis, *Carcinogenesis* 29 (10), 1973–1978, 2008.

20. La Rosa, F. G., Adams, F. S., Krause, G. E., Meyers, A. D., Edwards-Prasad, J., Kumar, R., Freed, C. R., and Prasad, K. N., Inhibition of proliferation and expression of T-antigen in SV40 large T-antigen gene-induced immortalized cells following transplantations, *Cancer Lett* 113 (1–12), 55–60, 1997.

21. Cullmann, C., Hoppe-Seyler, K., Dymalla, S., Lohrey, C., Scheffner, M., Durst, M., and Hoppe-Seyler, F., Oncogenic human papillomaviruses block expression of the B-cell translocation gene-2 (BTG2) tumor suppressor gene, *Int J Cancer* 125 (9), 2014–2010, 2009.

22. DiPaolo, J. A., Popescu, N. C., Alvarez, L., and Woodworth, C. D., Cellular and molecular alterations in human epithelial cells transformed by recombinant human papillomavirus DNA, *Crit Rev Oncog* 4 (4), 337–360, 1993.

23. Franceschi, S., Munoz, N., Bosch, X. F., Snijders, P. J., and Walboomers, J. M., Human papillomavirus and cancers of the upper aerodigestive tract: A review of epidemiological and experimental evidence, *Cancer Epidemiol Biomarkers Prev* 5 (7), 567–575, 1996.

24. Chocolatewala, N. M., and Chaturvedi, P., Role of human papilloma virus in the oral carcinogenesis: An Indian perspective, *J Cancer Res Ther* 5 (2), 71–77, 2009.

25. Szostek, S., Zawilinska, B., Kopec, J., and Kosz-Vnenchak, M., Herpesviruses as possible cofactors in HPV-16-related oncogenesis, *Acta Biochim Pol* 56 (2), 337–342, 2009.

26. Jacob, J. R., and Tennant, B. C., Transformation of immortalized woodchuck hepatic cell lines with the c-Ha-ras proto-oncogene, *Carcinogenesis* 17 (4), 631–636, 1996.

27. Li, Y., Ambrosone, C. B., McCullough, M. J., Ahn, J., Stevens, V. L., Thun, M. J., and Hong, C. C., Oxidative stress-related genotypes, fruit and vegetable consumption and breast cancer risk, *Carcinogenesis* 30 (5), 777–784, 2009.

28. Matsui, H., and Rai, K., Oxidative stress in gastric carcinogenesis [in Japanese], *Gan To Kagaku Ryoho* 35 (9), 1451–1456, 2008.

29. Nelson, W. G., De Marzo, A. M., DeWeese, T. L., and Isaacs, W. B., The role of inflammation in the pathogenesis of prostate cancer, *J Urol* 172 (5 Pt 2), S6–11; discussion S11–12, 2004.

30. Sugar, L. M., Inflammation and prostate cancer, *Can J Urol* 13 (Suppl. 1), 46–47, 2006.

31. Walser, T., Cui, X., Yanagawa, J., Lee, J. M., Heinrich, E., Lee, G., Sharma, S., and Dubinett, S. M., Smoking and lung cancer: The role of inflammation, *Proc Am Thorac Soc* 5 (8), 811–815, 2008.

32. Hall, E. J., and Miller, R. C., The how and why of in vitro oncogenic transformation, *Radiat Res* 87 (2), 208–223, 1981.

33. Borek, C., Pain, C., and Mason, H., Neoplastic transformation of hamster embryo cells irradiated in utero and assayed in vitro, *Nature* 266 (5601), 452–454, 1977.

34. Crompton, N. E., Sigg, M., and Jaussi, R., Genome lability in radiation-induced transformants of C3H 10T1/2 mouse fibroblasts, *Radiat Res* 138 (Suppl. 1), S105–108, 1994.

35. Thraves, P., Salehi, Z., Dritschilo, A., and Rhim, J. S., Neoplastic transformation of immortalized human epidermal keratinocytes by ionizing radiation, *Proc Natl Acad Sci USA* 87 (3), 1174–1177, 1990.

36. Calaf, G. M., and Hei, T. K., Ionizing radiation induces alterations in cellular proliferation and c-myc, c-jun and c-fos protein expression in breast epithelial cells, *Int J Oncol* 25 (6), 1859–1866, 2004.
37. Hall, E. J., Oncogenic transformation by radiation and chemicals, in *Proceedings of the VII International Congress of Radiation Research*, ed. Fielden, E. M. F., et al., Taylor and Francis, London, 1987.
38. Little, J. B., Influence of noncarcinogenic secondary factors on radiation carcinogenesis, *Radiat Res* 87 (2), 240–250, 1981.
39. Kennedy, A. R., Murphy, G., and Little, J. B., Effect of time and duration of exposure to 12-O-tetradecanoylphorbol-13-acetate on x-ray transformation of C3H 10T 1/2 cells, *Cancer Res* 40 (6), 1915–1920, 1980.
40. Berenblum, I., The carcinogenic action of vroton resin, *Cancer Res* 1, 44–50, 1941.
41. Stevens, R. G., and Kalkwarf, D. R., Iron, radiation, and cancer, *Environ Health Perspect* 87, 291–300, 1990.
42. Knekt, P., Reunanen, A., Takkunen, H., Aromaa, A., Heliovaara, M., and Hakulinen, T., Body iron stores and risk of cancer, *Int J Cancer* 56 (3), 379–382, 1994.
43. Borek, C., Zaider, M., Ong, A., Mason, H., and Witz, G., Ozone acts alone and synergistically with ionizing radiation to induce in vitro neoplastic transformation, *Carcinogenesis* 7 (9), 1611–1613, 1986.
44. Pollock, E. J., and Todaro, G. J., Radiation enhancement of SV40 transformation in 3T3 and human cells, *Nature* 219 (5153), 520–521, 1968.
45. Puck, T. T., Morse, H., Johnson, R., and Waldren, C. A., Caffeine enhanced measurement of mutagenesis by low levels of gamma-irradiation in human lymphocytes, *Somat Cell Mol Genet* 19 (5), 423–429, 1993.
46. Committee on the Biological Effects of Ionizing Radiation, *Biological Effects of Ionizing Radiation (BEIR V)*, National Academic Press, Washington, DC, 1990.
47. Holtzman, S., Stone, J. P., and Shellabarger, C. J., Synergism of estrogens and x-rays in mammary carcinogenesis in female ACI rats, *J Natl Cancer Inst* 67 (2), 455–459, 1981.
48. Hall, E. J., and Hei, T. K., Modulating factors in the expression of radiation-induced oncogenic transformation, *Environ Health Perspect* 88, 149–155, 1990.
49. Clark, E. P., Hahn, G. M., and Little, J. B., Hyperthermic modulation of x-ray-induced oncogenic transformation in C3H 10T1/2 cells, *Radiat Res* 88 (3), 619–622, 1981.
50. Azzam, E. I., George, I., and Raaphorst, G. P., Alteration in thermal sensitivity of Chinese hamster cells by D2O treatment, *Radiat Res* 90 (3), 644–648, 1982.
51. Raaphorst, G. P., Szekely, J., Lobreau, A., and Azzam, E. I., A comparison of cell killing by heat and/or x-rays in Chinese hamster V79 cells, friend erythroleukemia mouse cells, and human thymocyte MOLT-4 cells, *Radiat Res* 94 (2), 340–349, 1983.
52. Botlagunta, M., Winnard, P. T. Jr., and Raman, V., Neoplastic transformation of breast epithelial cells by genotoxic stress, *BMC Cancer* 10, 343, 2010.
53. Fry, R. J. M., Principles of carcinogenesis, in *Principles and Practice of Oncology, 3rd Edition*, ed. DeVita, V., Lippincott, Philadelphia, 1989, 136–148.
54. Ullrich, R. L., and Storer, J. B., Influence of gamma irradiation on the development of neoplastic disease in mice. III. Dose-rate effects, *Radiat Res* 80 (2), 325–342, 1979.
55. Upton, A. C., Randolph, M. L., Conklin, J. W., Kastenbaum, M. A., Slater, M., Melville, G. S. Jr., Conte, F. P., and Sproul, J. A. Jr., Late effects of fast neutrons and gamma-rays in mice as influenced by the dose rate of irradiation: Induction of neoplasia, *Radiat Res* 41 (3), 467–491, 1970.
56. Upton, A. C., The dose-response relation in radiation-induced cancer, *Cancer Res* 21, 717–729, 1961.

57. Warren, S., and Gates, O., The induction of leukemia and life shortening in mice by continuous low-level external gamma radiation, *Radiat Res* 47 (2), 480–490, 1971.

58. Kaplan, H. S., On the natural history of the murine leukemias: Presidential address, *Cancer Res* 27 (8), 1325–1340, 1967.

59. Prentice, R. L., Peterson, A. V., and Marek, P., Dose-mortality relationships in RFM mice following 137Cs gamma-ray irradiation, *Radiat Res* 90 (1), 57–76, 1982.

60. Hirouchi, T., Takabatake, T., Yoshida, K., Nitta, Y., Nakamura, M., Tanaka, S., Ichinohe, K., Oghiso, Y., and Tanaka, K., Upregulation of c-myc gene accompanied by PU.1 deficiency in radiation-induced acute myeloid leukemia in mice, *Exp Hematol* 36 (7), 871–885, 2008.

61. Peng, Y., Borak, T. B., Bouffler, S. D., Ullrich, R. L., Weil, M. M., and Bedford, J. S., Radiation leukemogenesis in mice: Loss of PU.1 on chromosome 2 in CBA and C57BL/6 mice after irradiation with 1 GeV/nucleon 56Fe ions, x-rays or gamma rays. Part II. Theoretical considerations based on microdosimetry and the initial induction of chromosome aberrations, *Radiat Res* 171 (4), 484–493, 2009.

62. Peng, Y., Brown, N., Finnon, R., Warner, C. L., Liu, X., Genik, P. C., Callan, M. A., Ray, F. A., Borak, T. B., Badie, C., Bouffler, S. D., Ullrich, R. L., Bedford, J. S., and Weil, M. M., Radiation leukemogenesis in mice: Loss of PU.1 on chromosome 2 in CBA and C57BL/6 mice after irradiation with 1 GeV/nucleon 56Fe ions, x-rays or gamma rays. Part I. Experimental observations, *Radiat Res* 171 (4), 474–483, 2009.

63. Weil, M. M., Bedford, J. S., Bielefeldt-Ohmann, H., Ray, F. A., Genik, P. C., Ehrhart, E. J., Fallgren, C. M., Hailu, F., Battaglia, C. L., Charles, B., Callan, M. A., and Ullrich, R. L., Incidence of acute myeloid leukemia and hepatocellular carcinoma in mice irradiated with 1 GeV/nucleon (56)Fe ions, *Radiat Res* 172 (2), 213–219, 2009.

64. Michalak, E. M., Vandenberg, C. J., Delbridge, A. R., Wu, L., Scott, C. L., Adams, J. M., and Strasser, A., Apoptosis-promoted tumorigenesis: Gamma-irradiation-induced thymic lymphomagenesis requires PUMA-driven leukocyte death, *Genes Dev* 24 (15), 1608–1613, 2010.

65. Iskander, K., Barrios, R. J., and Jaiswal, A. K., NRH:quinone oxidoreductase 2-deficient mice are highly susceptible to radiation-induced B-cell lymphomas, *Clin Cancer Res* 15 (5), 1534–1542, 2009.

66. Gross, P., Pfitzer, E. A., Watson, J., De Treville, R. T., Kaschak, M., Tolker, E. B., and Babyak, M. A., Experimental carcinogenesis. Bronchial intramural adenocarcinomas in rats from x-ray irradiation of the chest, *Cancer* 23 (5), 1046–1060, 1969.

67. Nolibe, D., Masse, R., and Lafuma, J., The effect of neonatal thymectomy on lung cancers induced in rats by plutonium dioxide, *Radiat Res* 87 (1), 90–99, 1981.

68. Thomassen, D. G., Seiler, F. A., Shyr, L. J., and Griffith, W. C., Alpha-particles induce preneoplastic transformation of rat tracheal epithelial cells in culture, *Int J Radiat Biol* 57 (2), 395–405, 1990.

69. Whittemore, A. S., and McMillan, A., Osteosarcomas among beagles exposed to [239]Pu, *Radiat Res* 90 (1), 41–56, 1982.

70. Peterson, A. V., Prentice, R. L., and Marek, P., Relationship between dose of injected [239]Pu and bone sarcoma mortality in young adult beagles, *Radiat Res* 90 (1), 77–89, 1982.

71. White, R. G., Raabe, O. G., Culbertson, M. R., Parks, N. J., Samuels, S. J., and Rosenblatt, L. S., Bone sarcoma characteristics and distribution in beagles injected with radium-226, *Radiat Res* 137 (3), 361–370, 1994.

72. Raabe, O. G., Book, S. A., and Parks, N. J., Bone cancer from radium: Canine dose response explains data for mice and humans, *Science* 208 (4439), 61–64, 1980.

73. Hulse, E. V., Osteosarcoma, fibrosarcoma, and basal cell carcinoma in rabbits after irradiation with gamma-rays or fission neutrons, an interim report on incidence, site of tumors and RBE, *Int J Radiat Biol* 16, 27–30, 1969.

74. Shellabarger, C. J., Modifying factors in rat mammary gland carcinogenesis, in *Biology of Radiation Carcinogenesis*, ed. Yuhas, J. M., Tennant, R. W., and Regans, J. B., Raven Press, New York, 1976, 31–43.

75. Bond, V. P., Cronkite, E. P., Lippincott, S. W., and Shellabarger, C. J., Studies on radiation-induced mammary gland neoplasia in the rat. III. Relation of the neoplastic response to dose of total-body radiation, *Radiat Res* 12, 276–285, 1960.

76. Vogel, H. H., Experimental mammary neoplasms: A comparison of effectiveness between neutrons, x-ray and gamma-radiation, in *Neutron in Radiobiology*, U.S. Atomic Energy Commission, Washington, DC, 1970, 207–229.

77. Shellabarger, C. J., Bond, V. P., Aponte, G. E., and Cronkite, E. P., Results of fractionation and protraction of total-body radiation on rat mammary neoplasia, *Cancer Res* 26 (3), 509–513, 1966.

78. Yuhas, J. M., Recovery from radiation-carcinogenic injury to the mouse ovary, *Radiat Res* 60 (2), 321–332, 1974.

79. Ullrich, R. L., Jernigan, M. C., Cosgrove, G. E., Satterfield, L. C., Bowles, N. D., and Storer, J. B., The influence of dose and dose rate on the incidence of neoplastic disease in RFM mice after neutron irradiation, *Radiat Res* 68 (1), 115–131, 1976.

80. Kaplan, H. S., Influence of ovarian function on incidence of radiation-induced ovarian tumors in mice, *J Natl Cancer Inst* 11 (1), 125–132, 1950.

81. Bonser, A. M., Tumor of ovary, in *The Ovary*, ed. Zuckerman, L., and Weir, B. J., Academic Press, New York, 1977, 129–184.

82. Lorenz, E., Some biologic effects of long continued irradiation, *Am J Roentgenol Radium Ther* 63 (2), 176–185, 1950.

83. Casey, H. W., Prince, J. E., Hinkle, D. K., Caraway, B. L., and Williams, W. T., Neoplastic skin response of rats after 13 Mev proton irradiation, *Aerosp Med* 39 (4), 360–365, 1968.

84. Osborne, J. W., Nicholson, D. P., and Prasad, K. N., Induction of intestinal carcinoma in the rat by x-irradiation of the small intestine, *Radiat Res* 18, 76–85, 1963.

85. Cole, L. J., Ellis, M. E., and Nowell, P. C., Induction of intestinal carcinoma in the mouse by whole-body fast-neutron irradiation, *Cancer Res* 16 (9), 873–876, 1956.

86. Trani, D., Datta, K., Doiron, K., Kallakury, B., and Fornace, A. J. Jr., Enhanced intestinal tumor multiplicity and grade in vivo after HZE exposure: Mouse models for space radiation risk estimates, *Radiat Environ Biophys* 49 (3), 389–396, 2010.

87. Upton, A. C., Historical perspective on radiation carcinogenesis, in *Radiation Carcinogenesis*, ed. Burns, F. J., Upton, A. C., and Silini, G., Elsevier, New York, 1986, 1–10.

88. van Gijssel, H. E., Stenius, U., Mulder, G. J., and Meerman, J. H., Lack of p53 protein expression in preneoplastic rat hepatocytes in vitro after exposure to N-acetoxy-acetylaminofluorene, x-rays or a proteasome inhibitor, *Eur J Cancer* 36 (1), 106–112, 2000.

89. Lindsay, S., and Chaikoff, I. L., The effects of irradiation on the thyroid gland with particular reference to the induction of thyroid neoplasms, *Cancer Res* 24, 1099–1107, 1964.

90. Jongejan, W. J., et al., The effect of I-131, and I-125 on mouse and rat thyroid, *Int J Radiat Biol* 22, 489–499, 1972.

91. deRuiter, J. H., Boorman, C. F., Hennemann, G., Dector, R., and Van Putten, L. M., Comparison of carcinogenicity of I-131 and I-125 in thyroid gland of the rat, International Atomic Energy Agency, V. Vienna, 1976, 31–39.

92. Haran-Ghera, N., Furth, J., Buffett, R. F., and Yokoro, K., Studies on the pathogenesis of neoplasms by ionizing radiation. II. Neoplasms of endocrine organs, *Cancer Res* 19, 1181–1187, 1959.

93. Furth, J., Haran-Ghera, N., Curtis, H. J., and Buffett, R. F., Studies on the pathogenesis of neoplasms by ionizing radiation. I. Pituitary tumors, *Cancer Res* 19 (5), 550–556, 1959.

94. Sasaki, S. and Fukuda, N., Dose–response relationship for induction of solid tumors in female B6C3F1 mice irradiated neonatally with a single dose of gamma rays, *J Radiat Res (Tokyo)* 40 (3), 229–241, 1999.

95. Sasaki, S., and Fukuda, N., Dose-response relationship for life-shortening and carcinogenesis in mice irradiated at day 7 postnatal age with dose range below 1 Gy of gamma rays, *J Radiat Res (Tokyo)* 47 (2), 135–145, 2006.

96. Uma Devi, P., and Hossain, M., Induction of solid tumours in the Swiss albino mouse by low-dose foetal irradiation, *Int J Radiat Biol* 76 (1), 95–99, 2000.

97. Dasenbrock, C., Tillmann, T., Ernst, H., Behnke, W., Kellner, R., Hagemann, G., Kaever, V., Kohler, M., Rittinghausen, S., Mohr, U., and Tomatis, L., Maternal effects and cancer risk in the progeny of mice exposed to x-rays before conception, *Exp Toxicol Pathol* 56 (6), 351–360, 2005.

98. Nomura, T., Transgenerational effects of radiation and chemicals in mice and humans, *J Radiat Res (Tokyo)* 47 (Suppl. B), B83–97, 2006.

99. Hollander, C. F., Zurcher, C., and Broerse, J. J., Tumorigenesis in high-dose total body irradiated rhesus monkeys: A life span study, *Toxicol Pathol* 31 (2), 209–213, 2003.

100. Wood, D. H., Yochmowitz, M. G., Hardy, K. A., and Salmon, Y. L., Occurrence of brain tumors in rhesus monkeys exposed to 55-MeV protons, *Adv Space Res* 6 (11), 213–216, 1986.

101. Dalrymple, G. V., Leichner, P. K., Harrison, K. A., Cox, A. B., Hardy, K. A., Salmon, Y. L., and Mitchell, J. C., Induction of high grade astrocytoma (HGA) by protons: Molecular mechanisms and RBE considerations, *Adv Space Res* 14 (10), 267–270, 1994.

102. Motran, J. E., Production of epithelial tumor by a combination of beta-radiation and painting with benzopyrene, *Am J. Cancer* 32, 76–80, 1938.

103. Bock, F. G., and Moore, G. E., Carcinogenic activity of cigarette-smoke condensate. I. Effect of trauma and remote x irradiation, *J Natl Cancer Inst* 22 (2), 401–411, 1959.

104. Goldfeder, A., Urethan and x-ray effects on mice of a tumor-resistant strain, X-Gf, *Cancer Res* 32 (12), 2771–2777, 1972.

105. Shellabarger, C. J., Stone, J. P., and Holtzman, S., Synergism between neutron radiation and diethylstilbestrol in the production of mammary adenocarcinomas in the rat, *Cancer Res* 36 (3), 1019–1022, 1976.

106. Segaloff, A., and Pettigrew, H. M., Effect of radiation dosage on the synergism between radiation and estrogen in the production of mammary cancer in the rat, *Cancer Res* 38 (10), 3445–3452, 1978.

107. Little, J. B., McGandy, R. B., and Kennedy, A. R., Interactions between polonium-210 alpha-radiation, benzo(a)pyrene, and 0.9% NaCl solution instillations in the induction of experimental lung cancer, *Cancer Res* 38 (7), 1929–1935, 1978.

108. Stone, J. P., Holtzman, S., and Shellabarger, C. J., Synergistic interactions of various doses of diethylstilbestrol and x-irradiation on mammary neoplasia in female ACI rats, *Cancer Res* 40 (11), 3966–3972, 1980.

109. Telles, N. C., and Ward, B. C., The effects of radiation and ethionine on rat mammary tumor incidence, *Radiat Res* 37 (3), 577–589, 1969.

110. Arseneau, J. C., Synergistic tumorgenic effect of procarbazine and ionizing radiation in (BALB/CXDBA/2)F mice, *J Natl Cancer Inst* 59, 423–425, 1977.

111. Vesselinovitch, S. D., Simmons, E. L., Mihailovich, N., Lombard, L. S., and Rao, K. V., Additive leukemogenicity of urethan and x-irradiation in infant and young adult mice, *Cancer Res* 32 (2), 222–225, 1972.

112. Lurie, A. G., Interactions between 7,12-dimethylbenz(a)anthracene (DMBA) and repeated low-level X radiation in hamster cheek pouch carcinogenesis: Dependence on the relative timing of DMBA and radiation treatments, *Radiat Res* 90 (1), 155–164, 1982.
113. Prasad, K., *Handbook of Radiobiology*, 2nd Ed., CRC Press, Boca Raton, FL, 1995.
114. Committee on the Biological Effects of Ionizing Radiation, *The Effects on Populations of Exposure to Low Levels of Ionizing Radiation*, National Academic Press, Washington, DC, 1980.
115. Shimizu, Y., Schull, W. J., and Kato, H., Cancer risk among atomic bomb survivors. The RERF Life Span Study. Radiation Effects Research Foundation, *JAMA* 264 (5), 601–604, 1990.
116. Anderson, R. E., Malignant lymphoma, *Hum Pathol* 2 (4), 515–519, 1971.
117. Little, M. P., Cancer and non-cancer effects in Japanese atomic bomb survivors, *J Radiol Prot* 29 (2A), A43–59, 2009.
118. Behringer, K., Josting, A., Schiller, P., Eich, H. T., Bredenfeld, H., Diehl, V., and Engert, A., Solid tumors in patients treated for Hodgkin's disease: A report from the German Hodgkin Lymphoma Study Group, *Ann Oncol* 15 (7), 1079–1085, 2004.
119. Hooning, M. J., Aleman, B. M., Hauptmann, M., Baaijens, M. H., Klijn, J. G., Noyon, R., Stovall, M., and van Leeuwen, F. E., Roles of radiotherapy and chemotherapy in the development of contralateral breast cancer, *J Clin Oncol* 26 (34), 5561–5568, 2008.
120. Wong, F. L., Boice, J. D. Jr., Abramson, D. H., Tarone, R. E., Kleinerman, R. A., Stovall, M., Goldman, M. B., Seddon, J. M., Tarbell, N., Fraumeni, J. F. Jr., and Li, F. P., Cancer incidence after retinoblastoma: Radiation dose and sarcoma risk, *JAMA* 278 (15), 1262–1267, 1997.
121. Roychoudhuri, R., Evans, H., Robinson, D., and Moller, H., Radiation-induced malignancies following radiotherapy for breast cancer, *Br J Cancer* 91 (5), 868–872, 2004.
122. Caldwell, G. G., Kelley, D. B., and Heath, C. W., Jr., Leukemia among participants in military maneuvers at a nuclear bomb test. A preliminary report, *JAMA* 244 (14), 1575–1578, 1980.
123. Lyon, J. L., Klauber, M. R., Gardner, J. W., and Udall, K. S., Childhood leukemias associated with fallout from nuclear testing, *N Engl J Med* 300 (8), 397–402, 1979.
124. Preston, D. L., Ron, E., Tokuoka, S., Funamoto, S., Nishi, N., Soda, M., Mabuchi, K., and Kodama, K., Solid cancer incidence in atomic bomb survivors: 1958–1998, *Radiat Res* 168 (1), 1–64, 2007.
125. Preston, D. L., Cullings, H., Suyama, A., Funamoto, S., Nishi, N., Soda, M., Mabuchi, K., Kodama, K., Kasagi, F., and Shore, R. E., Solid cancer incidence in atomic bomb survivors exposed in utero or as young children, *J Natl Cancer Inst* 100 (6), 428–436, 2008.
126. Cihak, R. W., Radiation and lung cancer, *Hum Pathol* 2 (4), 525–528, 1971.
127. Beebe, G. W., Kato, H., and Land, C. E., Lifespan Study Report 8: Mortality experience of atomic bomb survivors, 1950–1974, in *Radiation Effect Res. Found, Tech. Rep. TR-177* Radiation Effect Research Foundation, Hiroshima, 1978.
128. Archer, V. E., Saccomanno, G., and Jones, J. H., Frequency of different histologic types of bronchogenic carcinoma as related to radiation exposure, *Cancer* 34 (6), 2056–2060, 1974.
129. Archer, V. E., Wagoner, J. K., and Lundin, F. E. Jr., Uranium mining and cigarette smoking effects on man, *J Occup Med* 15 (3), 204–211, 1973.
130. Horacek, J., Placek, V., and Sevc, J., Histologic types of bronchogenic cancer in relation to different conditions of radiation exposure, *Cancer* 40 (2), 832–835, 1977.
131. Yamamoto, T., and Wakabayashi, T., Bone tumors among the atomic bomb survivors of Hiroshima and Nagasaki, *Acta Pathol Jpn* 19 (2), 201–212, 1969.

132. Soloway, H. B., Radiation-induced neoplasms following curative therapy for retinoblastoma, *Cancer* 19 (12), 1984–1988, 1966.
133. Mays, C. W., Skeletal effects following 224Ra injections into humans, *Health Phys* 35 (1), 83–90, 1978.
134. Mackenzie, I., Breast cancer following multiple fluoroscopies, *Br J Cancer* 19, 1–8, 1965.
135. Myrden, J. A., and Hiltz, J. E., Breast cancer following multiple fluoroscopies during artificial pneumothorax treatment of pulmonary tuberculosis, *Can Med Assoc J* 100 (22), 1032–1034, 1969.
136. Wanebo, C. K., Johnson, K. G., Sato, K., and Thorslund, T. W., Breast cancer after exposure to the atomic bombings of Hiroshima and Nagasaki, *N Engl J Med* 279 (13), 667–671, 1968.
137. Curtin, C. T., McHeffy, B., and Kolarsick, A. J., Thyroid and breast cancer following childhood radiation, *Cancer* 40 (6), 2911–2913, 1977.
138. Boice, J. D. Jr., Rosenstein, M., and Trout, E. D., Estimation of breast doses and breast cancer risk associated with repeated fluoroscopic chest examinations of women with tuberculosis, *Radiat Res* 73 (2), 373–390, 1978.
139. Shore, R. E., Hempelmann, L. H., Kowaluk, E., Mansur, P. S., Pasternack, B. S., Albert, R. E., and Haughie, G. E., Breast neoplasms in women treated with x-rays for acute postpartum mastitis, *J Natl Cancer Inst* 59 (3), 813–822, 1977.
140. McGregor, H., Land, C. E., Choi, K., Tokuoka, S., Liu, P. I., Wakabayashi, T., and Beebe, G. W., Breast cancer incidence among atomic bomb survivors, Hiroshima and Nagasaki, 1950–69, *J Natl Cancer Inst* 59 (3), 799–811, 1977.
141. Tokunaga, M., Norman, J. E. Jr., Asano, M., Tokuoka, S., Ezaki, H., Nishimori, I., and Tsuji, Y., Malignant breast tumors among atomic bomb survivors, Hiroshima and Nagasaki, 1950–74, *J Natl Cancer Inst* 62 (6), 1347–1359, 1979.
142. Baral, E., Larsson, L. E., and Mattsson, B., Breast cancer following irradiation of the breast, *Cancer* 40 (6), 2905–2910, 1977.
143. Boice, J. D. Jr., Land, C. E., Shore, R. E., Norman, J. E., and Tokunaga, M., Risk of breast cancer following low-dose radiation exposure, *Radiology* 131 (3), 589–597, 1979.
144. Palmer, J. P., and Spratt, D. W., Pelvic carcinoma following irradiation for benign gynecological diseases, *Am J Obstet Gynecol* 72 (3), 497–505, 1956.
145. Bartal, A. H., Cohen, Y., and Robinson, E., Malignant melanoma arising at tattoo sites used for radiotherapy field marking, *Br J Radiol* 53 (633), 913–914, 1980.
146. Martin, H., Strong, E., and Spiro, R. H., Radiation-induced skin cancer of the head and neck, *Cancer* 25 (1), 61–71, 1970.
147. Takahashi, S., Kitabatake, T., Wakabayashi, M., Koga, Y., Miyakawa, T., Yamashita, H., Masuyama, M., Hibino, S., Miyakawa, M., Okajima, S., Kaneda, H., Tachiiri, H., Anno, Y., and Irie, H., A statistical study on human cancer induced by medical irradiation, *Nippon Igaku Hoshasen Gakkai Zasshi* 23, 1510–1530, 1964.
148. Mays, C. W., and Spiess, H., Bone sarcoma risks to man from Ra-224, Ra-226 and Pu-239, in *Biological Effects of Ra224: Benefits and Risk of Therapeutic Application*, Muller, W. A., and Ebert, H. G. Nijhoff Medical Division, The Hague, 1978, 168–178.
149. Radford, E. P., Doll, R., and Smith, P. G., Mortality among patients with ankylosing spondylitis not given x-ray therapy, *N Engl J Med* 297 (11), 572–576, 1977.
150. Mancuso, T. F., Stewart, A., and Kneale, G., Radiation exposures of Hanford workers dying from cancer and other causes, *Health Phys* 33, 369–385, 1977.
151. Gilbert, E. S., and Marks, S., An analysis of the mortality of workers in a nuclear facility, *Radiat Res* 79 (1), 122–148, 1979.
152. Hutchison, G. B., MacMahon, B., Jablon, S., and Land, C. E., Review of report by Mancuso, Stewart, and Kneale of radiation exposure of Hanford workers, *Health Phys* 37 (2), 207–220, 1979.

153. Dickson, R. J., Late results of radium treatment of carcinoma of the cervix, *Clin Radiol* 23 (4), 528–535, 1972.
154. Jochimsen, P. R., Pearlman, N. W., and Lawton, R. L., Pancreatic carcinoma as a sequel to therapy of lymphoma, *J Surg Oncol* 8 (6), 461–464, 1976.
155. Demidchik, Y. E., Saenko, V. A., and Yamashita, S., Childhood thyroid cancer in Belarus, Russia, and Ukraine after Chernobyl and at present, *Arq Bras Endocrinol Metabol* 51 (5), 748–762, 2007.
156. Schneider, A. B., Favus, M. J., Stachura, M. E., Arnold, J., Arnold, M. J., and Frohman, L. A., Incidence, prevalence and characteristics of radiation-induced thyroid tumors, *Am J Med* 64 (2), 243–252, 1978.
157. Greenspan, F. S., Radiation exposure and thyroid cancer, *JAMA* 237 (19), 2089–2091, 1977.
158. Hempelmann, L. H., Hall, W. J., Phillips, M., Cooper, R. A., and Ames, W. R., Neoplasms in persons treated with x-rays in infancy: Fourth survey in 20 years, *J Natl Cancer Inst* 55 (3), 519–530, 1975.
159. Saenger, E. L., Silverman, F. N., Sterling, T. D., and Turner, M. E., Neoplasia following therapeutic irradiation for benign conditions in childhood, *Radiology* 74, 889–904, 1960.
160. Janower, M. L., and Miettinen, O. S., Neoplasms after childhood irradiation of the thymus gland, *JAMA* 215 (5), 753–756, 1971.
161. Shenoy, S. N., Munish, K. G., and Raja, A., High dose radiation induced meningioma, *Br J Neurosurg* 18 (6), 617–621, 2004.
162. Neglia, J. P., Robison, L. L., Stovall, M., Liu, Y., Packer, R. J., Hammond, S., Yasui, Y., Kasper, C. E., Mertens, A. C., Donaldson, S. S., Meadows, A. T., and Inskip, P. D., New primary neoplasms of the central nervous system in survivors of childhood cancer: A report from the Childhood Cancer Survivor Study, *J Natl Cancer Inst* 98 (21), 1528–1537, 2006.
163. Henderson, T. O., Whitton, J., Stovall, M., Mertens, A. C., Mitby, P., Friedman, D., Strong, L. C., Hammond, S., Neglia, J. P., Meadows, A. T., Robison, L., and Diller, L., Secondary sarcomas in childhood cancer survivors: A report from the Childhood Cancer Survivor Study, *J Natl Cancer Inst* 99 (4), 300–308, 2007.
164. Meadows, A. T., Friedman, D. L., Neglia, J. P., Mertens, A. C., Donaldson, S. S., Stovall, M., Hammond, S., Yasui, Y., and Inskip, P. D., Second neoplasms in survivors of childhood cancer: Findings from the Childhood Cancer Survivor Study cohort, *J Clin Oncol* 27 (14), 2356–2362, 2009.
165. Bithell, J. F., and Stewart, A. M., Pre-natal irradiation and childhood malignancy: A review of British data from the Oxford Survey, *Br J Cancer* 31 (3), 271–287, 1975.
166. Macmahon, B., Prenatal x-ray exposure and childhood cancer, *J Natl Cancer Inst* 28, 1173–1191, 1962.
167. Shore, R. E., Albert, R. E., and Pasternack, B. S., Follow-up study of patients treated by x-ray epilation for Tinea capitis: Resurvey of post-treatment illness and mortality experience, *Arch Environ Health* 31 (1), 21–28, 1976.
168. Stillman, R. J., Schinfeld, J. S., Schiff, I., Gelber, R. D., Greenberger, J., Larson, M., Jaffe, N., and Li, F. P., Ovarian failure in long-term survivors of childhood malignancy, *Am J Obstet Gynecol* 139 (1), 62–66, 1981.
169. Sherins, R. J., Olweny, C. L., and Ziegler, J. L., Gynecomastia and gonadal dysfunction in adolescent boys treated with combination chemotherapy for Hodgkin's disease, *N Engl J Med* 299 (1), 12–16, 1978.
170. Rubin, P., and Casarett, G. W., *Clinical Radiation Pathology*, W. B. Saunders, Philadelphia, 1968.
171. Kusunoki, Y., and Hayashi, T., Long-lasting alterations of the immune system by ionizing radiation exposure: Implications for disease development among atomic bomb survivors, *Int J Radiat Biol* 84 (1), 1–14, 2008.

172. Akiyama, M., Late effects of radiation on the human immune system: An overview of immune response among the atomic-bomb survivors, *Int J Radiat Biol* 68 (5), 497–508, 1995.
173. Kusunoki, Y., Yamaoka, M., Kubo, Y., Hayashi, T., Kasagi, F., Douple, E. B., and Nakachi, K., T-cell immunosenescence and inflammatory response in atomic bomb survivors, *Radiat Res* 174 (6), 870–876, 2010.

5 Prevention and Mitigation of Acute Radiation Sickness (ARS)

INTRODUCTION

Acute damage refers to lethality as well as to injury to individual organs after irradiation with high doses of ionizing radiation. Acute radiation damage can also be referred to as acute radiation sickness (ARS), which can cause mortality ranging from a few to 100% of irradiated individuals. The survivors of ARS show increased risk of developing cancer and non-neoplastic diseases long after irradiation. The search for a nontoxic agent that can prevent acute radiation damage when administered before irradiation, or that can mitigate acute radiation injury when administered after irradiation, began soon after the explosion of atom bombs in Hiroshima and Nagasaki. Several chemicals that can prevent radiation damage and biological agents that can mitigate radiation injury were identified, but most of them were considered toxic in humans. When the Cold War ended, the research on radiation protection markedly slowed down. During the last decade, the threat of an explosion of a dirty bomb by terrorists, an unintentional nuclear conflict, and accidents at the nuclear power plants have increased. This has provided a new stimulus for the search for novel compounds or biologics that can prevent and/or mitigate ARS.

Prevention generally refers to devices, biologics, or chemicals that can reduce radiation damage when administered before irradiation. This can be accomplished by reducing the radiation dose and/or tissue damage. At present, prevention of radiation damage is primarily based on reducing dose levels by adopting three physical principles: increasing the distance from the radiation source, reducing the exposure time, and the shielding. Adopting one or more of these principles will no doubt reduce radiation doses to individuals, such as first responders, radiation workers, and patients receiving diagnostic radiation procedures; however, they are not applicable under all conditions of radiation exposure even among these groups of individuals. In addition, these physical strategies of prevention of radiation injury do not protect tissue damage during and after irradiation with high doses. These physical protection recommendations are of no value for personnel who are likely to be exposed to high doses of radiation in the event of a nuclear power plant accident or those who actually have been exposed to high doses of radiation, and troops or civilians who are likely to be exposed to radiation in the event of an explosion of a dirty bomb or nuclear bomb.

Therefore, it is essential to develop a biological radiation protection strategy that would complement the existing physical radiation protection recommendations. In order to develop a biological strategy for protecting tissue damage during and after irradiation, it is essential to identify nontoxic and cost-effective agents that can protect tissue against radiation injury when administered orally before and/or after irradiation. Although some agents, such as antioxidants, that can prevent and mitigate ARS are available, they are not being considered for either prevention or mitigation of radiation damage at this time. This chapter will provide scientific data on recently developed radioprotective and radiomitigating agents with a rationale for recommending antioxidants in combination with standard therapy for prevention and mitigation of ARS in humans.

DEFINITION OF RADIOPROTECTIVE AND RADIATION MITIGATING AGENTS

Radioprotective agents: Agents that can reduce radiation damage when administered before irradiation are referred to as *radioprotective agents* and can also be called *radiation preventive agents*. In this chapter, we will refer to them as radioprotective agents. These radioprotective agents are ineffective when administered soon after irradiation. This definition of radioprotective agent is based on the early concept that only free radicals generated during irradiation initiate and promote radiation damage; therefore, any potential radioprotective agent must be administered before irradiation. However, during the last decade, the existence of long-lived free radicals after irradiation has been demonstrated. These long-lived free radicals must also be quenched in order to provide an optimal reduction in radiation damage. Therefore, any potential radioprotective agent must be administered before and after irradiation for the entire observation period in order to determine its efficacy. Most previous studies on radiation protection have administered radioprotective agents only once, either before or after irradiation. It is possible that the same radioprotective agents may also be beneficial in reducing the rate of progression of radiation injury when administered after irradiation. Thus, some radioprotective agents may prevent as well as mitigate radiation injury.

The efficacy of a radioprotective agent is commonly expressed in terms of a dose-reduction factor (DRF).

$$DRF = \frac{\text{A dose to produce an effect in the presence of a radioprotective agent}}{\text{A dose to produce the same effect in the absence of a radioprotective agent}}$$

The efficacy of radioprotective agents has been determined primarily on animal models and varies from 1.2 to 1.6. The efficacy of a radioprotective agent can also be expressed as a rate of increase in survival of irradiated animals, especially in those cases in which the effective dose range of radioprotective agents is very narrow. Most identified radioprotective agents cannot be used in humans on a short-term or a long-term basis because of their toxicity. A group of dietary and endogenous antioxidants that are nontoxic to humans have exhibited strong radioprotective effects.

Radiation mitigating agents: Agents (synthetic and/or natural chemicals) or biologics that can reduce acute radiation damage when administered after irradiation can be called *radiation mitigating agents* or *radiomitigators*. The efficacy of radiation mitigating agents can be expressed as the rate of increase in survival percent or survival time or both. Since acute inflammation and free radicals contribute to the progression of radiation damage, any useful radiation mitigating agents must attenuate both free radicals and inflammation. At present, some agents (chemical and biological) are available that can be used to mitigate some of the symptoms of ARS, but most of them cannot be used in humans because of their toxicity and because some of them require Food and Drug Administration (FDA) approval. During the last five years some promising radiation mitigating agents have been identified, primarily in rodent models; however, their relevance and safety in humans remains unknown. A group of dietary and endogenous antioxidants that are nontoxic in humans have been shown to be useful in mitigating ARS in animal models by reducing oxidative damage and inflammation.

CONCEPT OF A BIOSHIELD AGAINST RADIATION DAMAGE

In contrast to the physical principles of radiation protection, which may include lead shielding, the term *bioshield* refers to chemicals or biologics that can provide tissue protection against low and high doses of ionizing radiation when administered before and/or after irradiation. Thus, a bioshield can include radioprotective as well as radiomitigating agents. At present, there is no effective bioshield that can protect against ARS. Ideally, an effective bioshield should increase the survival rate and/or survival time when administered orally before and/or after irradiation for a short or long period of time without any toxicity. Such an agent should also reduce the adverse health effects among survivors of high doses of radiation exposure. An effective bioshield must be developed on the basis of biochemical changes that initiate and contribute to the progression of acute and late radiation injury. Among various biochemical changes, free radicals and inflammation are the most important factors that initiate and promote radiation damage. Free radicals initiate as well as contribute to the progression of ARS, whereas inflammation contributes only to the progression of this damage. Both free radicals and chronic inflammation also contribute to the late adverse health effects of radiation among survivors of high doses of radiation. Therefore, agents such as antioxidants that are nontoxic to humans and that can attenuate both free radicals and inflammation would be considered an ideal bioshield. Another mechanistic basis for developing an effective bioshield includes suppression of radiation-induced apoptosis, stimulation of the proliferation of residual surviving cells after irradiation, or replacement of radiation-induced dead cells by the progenitor cells against radiation damage.

PREVENTION OF ARS BY RADIOPROTECTIVE AGENTS

Efforts to protect normal tissue were started soon after the discovery of x-rays by Dr. Roentgen in 1895. However, it was not until the World War II when the search for radioprotective and radiation mitigating agents began in earnest. The basic purposes of the research on prevention and mitigation of radiation injury have been twofold: (1) to identify agents that can reduce acute and chronic radiation damage in humans when administered shortly before or after irradiation and (2) to obtain a greater knowledge of the mechanisms of radiation injury. In spite of extensive radiobiological studies, the first objective has achieved only limited success. There are hundreds of radioprotective agents that exhibit varying degrees of radiation protection in animals, but at radioprotective doses, most of them are toxic to humans. They have been extensively discussed in a book[1] and a recent review[2] and are described briefly here. In addition, recently identified radioprotective agents and their potential use in humans are described in this chapter.

FACTORS TO BE CONSIDERED IN THE STUDY OF RADIOPROTECTIVE AGENTS

When testing the radioprotective efficacy of a new compound, the following factors must be considered in order to produce an optimal effect: (1) nontoxic doses of the compound, (2) choice of radiation dose, (3) choice of experimental models, (4) route of administration, (5) pre- and post-irradiation time of administration, and (6) criterion of the efficacy.

Nontoxic doses of the compound: The toxicity of most compounds depends upon the doses and the experimental models. For example, in a tissue culture model, the toxic doses of a compound may vary from one cell type to another and from normal cells to cancer cells. Similarly, the toxic doses in animals may vary from one species to another. The doses that are considered safe in animal models are frequently toxic in humans. Before testing the efficacy of a compound in radiation protection, it is essential to select a dose that is nontoxic for a particular experimental model.

Choice of radiation dose: Since we do not know the effect of the new compound in radiation protection, selection of an appropriate dose of radiation is very important. If the efficacy of a compound in reducing radiation damage is small, this can be masked when high doses of radiation are used. It is best to select a radiation dose that causes 50% mortality in cell culture or LD_{50} (50% lethality) in animals. The radiation dose–response curve is constructed utilizing data obtained in the presence or absence of the compound. From this curve, the DRF of the compound is calculated.

Choice of experimental models: Most studies on radiation protection have been performed on cell culture or animal models. Initially, it is useful to test the efficacy of a new compound in a cell culture model because it is time- and cost-effective. If the results are positive, then its efficacy in reducing radiation damage can be tested in animal models. Among species, mice and rats have been commonly used for radiation protection studies. However, rodents are easily infected with the pseudomonas that can increase radiation-induced mortality. Therefore, the animals must be free from infection before using them for radiation protection studies. The DRF of the compound is then determined as described previously. The radioprotective doses determined in rodents may be toxic in other animals and humans.

Route of administration: In the cell culture model, the route of administration is not important, because the compound to be tested is directly added to the culture medium from which the compound may be removed shortly after irradiation or allowed to remain in the medium for the entire observation period. However, in animal models, the route of administration markedly influences the radioprotective efficacy of a compound, because the rate of absorption, cellular and organ distribution, and excretion of the compound vary depending upon the route of injection. Generally, intraperitoneal (IP) administration is preferred when using small animals such as rats and mice, because this technique is easiest to use especially in small animals, and because the rates of absorption are relatively rapid. The rates of absorption, distribution, and excretion are most rapid when the compound is administered intravenously (IV), but this method is not easy in small animals. Occasionally, the subcutaneous (SC) route is selected, which may be less effective than that produced by IP injection. The oral administration is the least effective. In addition to affecting the radioprotective efficacy of the compound, the route of administration also influences the toxicity of the compound. A compound that is not toxic when administered IP could become toxic when injected intravenously.

Pre- and post-irradiation time of administration: To test the radioprotective efficacy of a compound, only pre-irradiation time of varying lengths have been utilized in most radiation protection studies in the past. This is due to the assumption that free radicals generated during irradiation are primarily responsible for inducing radiation damage. The pre-irradiation time of administration of the compound has varied from a few minutes to a few hours. Most previously identified radioprotective agents were ineffective when injected shortly after irradiation. In view of the demonstration of long-lived free radicals after irradiation[3,4] and acute inflammation[5-8] that contribute to the progression of radiation damage, it is essential that an appropriate pre- and post-irradiation time of injection are used in determining the efficacy of a compound in reducing ARS.

Criterion of the efficacy: The following criteria are commonly used to evaluate the radioprotective efficacy of a compound:

1. Dose reduction factor (DRF), using the $^{30}LD_{50}$ for rodents as end point
2. Percent survivors as a function of radiation dose when the dose of a compound (often optimal) is constant
3. Percent survivors as a function of dose of a compound when the radiation dose (often $^{30}LD_{50}$) is constant
4. Percent survivors as a function of radiation dose delivered at high dose rate or delivered at any dose rate of high linear energy transfer (LET) radiation when the dose of a compound (often optimal) is constant
5. Incidence of cancer and nonneoplastic diseases among survivors of high doses of radiation

RADIOPROTECTIVE PHARMACOLOGICAL AGENTS

Extensive radiobiological research has identified numerous radioprotective pharmacological agents, ranging from the component of snake venom to alcohol. These agents, when administered IP shortly before irradiation with x- radiation or gamma radiation, protected animals (primarily rodents) against ARS. The radioprotective efficacy of previously identified agents was none or minimal in animals irradiated with high LET radiation.[1,2,9,10] Previously identified radioprotective pharmacological agents can be divided into four major groups: (1) radioprotective pharmacological agents not approved by the FDA, (2) radioprotective drugs approved by the FDA for other indications in humans, (3) radioprotective herbal extracts not requiring FDA approval, and (4) radioprotective dietary and endogenous antioxidants not requiring FDA approval. The radioprotective efficacy of some of them and their potential mechanisms of action in animals are briefly described in the following text.

RADIOPROTECTIVE PHARMACOLOGICAL AGENTS NOT APPROVED BY THE FDA

Thiols: Thiols exhibited very strong radioprotective effects in rodents when administered IP before whole-body irradiation with a dose that produced bone marrow syndrome. They were ineffective when administered after irradiation. These agents did not provide any protection when injected before or after irradiation against doses that produced gastrointestinal (GI) syndrome. The radioprotective thiols include cysteine, 2-mercaptoethylamine (MEA, also called cysteamine), cystamine, aminoethylisothiourea dihydrobromide (AET), 2-mercaptoethylguanidine (MEG), and amifostine, an analogue of cysteamine. The DRF value of these agents varied from 1.4 to 2.0. Although they were powerful radioprotective agents against radiation doses that produced bone marrow syndrome in rodents without producing any adverse health effects, all were toxic to humans at radioprotective doses.[1,9,11–14] Therefore, these thiols were of no value for radiation protection in humans. Some features of the above thiols are described in the following text.

Cysteine: One of the first radioprotective agents discovered was cysteine. The DRF values of cysteine (1200 mg/kg of body weight) on the criterion of the $^{30}LD_{50}$ for rats and mice were 1.5 and 1.7, respectively. Both the L- and D-isomers of cysteine were equally effective in providing radiation protection in rodents.[15] In order to reduce the toxicity of cysteine, prodrugs of cysteine, ribose-cysteine (RibCys), and glucose-cysteine (GlcCys) were developed. Both RibCys and GlcCys provided levels of radiation protection similar to those produced by cysteine.[16] Because of high toxicity of cysteamine in humans, the US Army synthesized large numbers of compounds at the Walter Reed Army Hospital (about 3,000 were tried), some of which are described in the following text.

Cysteamine: Cysteamine, at a dose of 150 mg/kg of body weight, produced the same degree of radiation protection as that produced by a dose of cysteine at 1200 mg/kg in rodents. In humans, cysteamine at very low doses (3–4 mg/kg of body weight) administered IV did not provide radiation protection against ARS.[17]

Aminoethylisothiourea dihydrobromide (AET): The DRF values of AET when administered IP before whole-body irradiation varied from 1.45 to 2.10 in rodents. It was also effective in reducing radiation damage in monkeys,[18] but not in dogs.[19] AET was found to be highly toxic (nausea, vomiting, and circulatory disturbances) even at doses of only 10–20 mg when administered orally or intravenously.[20]

2-Mercaptoethylguanidine (MEG): The neutralization of AET to pH 7 quantitatively converts this compound to MEG. The oxidation product of MEG is bis-(2-guanidinoethyl) disulfide (GED). The DRF value of MEG at a dose of 275 mg/kg of body weight was about 1.7, whereas it was 1.4 for GED at a dose of 200 mg/kg of body weight. [21] This compound is also considered toxic to humans.

Amifostine: Amifostine, also known as WR-2721(S-2-(3-aminopropylamino-ethylphosphorothioic acid), when administered IP at a dose of 400 mg/kg of body weight before whole-body irradiation, increased the survival of mice irradiated with gamma radiation doses that produce bone marrow syndrome, yielding a DRF value of 1.91. DRF values as high as 2.7 have been observed in animal models.[22] Amifostine is metabolized by the enzyme alkaline phosphatase to an active sulfhydryl compound WR-1065. It is well established that ionizing radiation increases the frequency of gamma-H2AX (histone phosphorylated at serine 139) in a dose-dependent manner. Treatment of human microvascular endothelial cells (HMEC) with WR-1065 or WR-2721 was effective in reducing radiation-induced gamma-H2AX formation after irradiation with a dose of 8 Gy. They were ineffective when administered after irradiation.[23]

Mechanisms of radiation protection by thiols: Excessive levels of free radicals and products of acute inflammation are responsible for initiating and promoting radiation damage. Therefore, any agents that can inhibit one or both of these biochemical events would be of radioprotective value. Indeed, thiols provide radiation protection by scavenging free radicals. They also produce hypoxia that makes cells resistant to radiation. The thiols also reversibly inhibit DNA synthesis, which may delay DNA replication. This, then, allows time for the DNA molecule to repair its damage.[10] They also reduced radiation-induced double-strand DNA breaks.[23] The effects of thiols on inflammation have not been investigated.

Other Sulfur-Containing Compounds: Other sulfur-containing compounds include thiourea, thiouracil, sufoxide, and sulfones. The DRF value for dimethyl sulfoxide (DMSO) in mice was 1.33. This value compares with the DRF value of AET (1.45) for the same strain of mice.[9]

Alcohol: Ethyl alcohol, when administered IP at a dose of 6–7.5 ml/Kg of body weight shortly before whole-body irradiation with x-ray doses that produced bone marrow syndrome significantly improved the survival of irradiated mice. Respiratory depression after ingestion of large amounts of alcohol can lead to tissue hypoxia, which may be responsible for radiation protection in mice. The high dose of alcohol used in this study, if translated from mouse to man, would correspond to about a quart of 100-proof whiskey. The ingestion of this amount of alcohol all at once would be toxic in humans.

Dopamine: Dopamine at a dose of 400 mg/kg of body weight, when administered immediately before whole-body x-irradiation, protected 80% of irradiated mice exposed to a dose of 700 R (about 7 Gy), which produced 100% lethality.[24] The radioprotective efficacy of dopamine did not decrease in spleenectomized mice. L-dopa, a precursor of dopamine, was ineffective. Dopamine also protected rats against radiation-induced lethality.[25] The DRF value of dopamine was about 1.3. The exact mechanisms of dopamine-induced radiation protection are unknown; however, hypoxia did not appear to be a major factor. The following possibilities have been suggested:

1. Dopamine protected radiation-induced DNA damage in vitro,[26] probably by scavenging free radicals. The fact that the concentration of dopamine in the spleen and small intestine is high suggests that dopamine protected these radiosensitive organs by scavenging free radicals.
2. Dopamine caused reversible inhibition of DNA synthesis in the mouse spleen.[27] It is presumed that a transient inhibition of DNA synthesis may delay DNA replication and thus enhances the probability of the repair of radiation damage.
3. Unlike epinephrine, dopamine protected spleenectomized mice.
4. The effectiveness of dopamine in protecting irradiated animals exists for a very short time; therefore, it is unlikely that the radioprotective effect of dopamine is mediated via epinephrine.

Histamine: The DRF value of histamine at a dose of 500 mg/kg of body weight in CBA mice was about 1.5, while in the C57BL strain of mice it was only 1.1. The oxygen tension in the spleen after histamine treatment decreased by about 77–93%, suggesting that hypoxia was a major factor in radiation protection.

Serotonin: Administration of serotonin at a dose of 25 mg/kg of body weight IV or 95 mg/kg of body weight IP shortly before whole-body irradiation with doses that produced bone marrow syndrome increased the survival of irradiated mice to the same degree as that produced by cysteamine. The DRF value of serotonin at a dose of 50 mg/kg of body weight was about 1.84. The production of hypoxia was the major factor in radiation protection.

Hormones: The DRF value of hormones, such as adrenal hormones and thyroid hormones, was about 1.1. The melanocytes-stimulating hormone (MSH) at a dose of 1 mg/mouse, when injected IP before whole-body x-irradiation with doses that produced bone marrow syndrome, increased the survival of irradiated animals.[28] In addition, certain hormone derivatives exhibited radioprotective activity in rodents.[29]

Sodium cyanide: Sodium cyanide at a dose of 5 mg/kg of body weight was of some radioprotective value. Tissue hypoxia appears to be a major factor in radiation protection.

Nucleic acid derivatives: Nucleic acid derivatives, such as uracil, 5-hydroxy-4-methyluracil, 5-amino-4-methyluracil, and 5-amino-4-methylcytosine, are of some radioprotective value in rodents. They increase the survival at the $^{30}LD_{50}$ by 20–40%. Adenosine triphosphate (ATP) treatment also increased survival at the $^{30}LD_{50}$ by about 30%. When ATP and pyridoxal-5-phosphate were administered together, the survival at the $^{30}LD_{50}$ was increased by 60%. The DRF value of ATP was about 1.55, whereas the DRF value of the combined treatment with ATP and pyridoxal-5-phosphate was about 1.7.

Sodium fluoroacetate: This compound, at a dose of 7.5 mg/kg of body weight, when injected IP in mice 3 hours before whole-body irradiation increased the average $^{30}LD_{50}$ value from 648 to 998 R (about 6.48 to 9.98 Gy).

Para-aminopropiophenone: This compound, at a dose of 40 mg/kg of body weight, when injected IP 15 minutes before whole-body irradiation produced a DRF value of 1.7 for bone marrow syndrome. The mechanism of radiation protection was primarily through production of tissue hypoxia.[30]

Melittin: The chief component of free venom is melittin, a strong basic polypeptide of molecular weight 2080. A dose of 5 mg/kg of body weight injected SC 24 hours before whole-body irradiation increased the survival of irradiated animals from 50% to 100%. The mechanisms of radiation protection are not known.[31]

Imidazole: This compound, at a dose of 350 mg/kg of body weight, when administered 5 minutes before whole-body irradiation, increased the survival of irradiated animals from 14% to about 80%.

Fullerene compound DF-1: This fullerene compound is known to posses antioxidant properties. Treatment of mice with DF-1 at a dose of 300 mg/kg before whole body gamma-ray irradiation increased the value of $^{30}LD_{50}$ from 8.29 Gy to 10.09 Gy. [32] The DRF value in mammalian cells in culture at a LD_{50} was 2.0.[33]

Adenosine 3′, 5′-cyclic monophosphate (cyclic AMP): We demonstrated for the first time that cyclic AMP stimulating agents, such as inhibitors of cyclic AMP phosphodiesterase activity and prostaglandin E1, a stimulator of adenylate cyclase activity when administered IP before x-irradiation, protected mammalian cells in culture.[34] This observation on radiation protection was confirmed on intestine and hair follicles of mice.[35,36] The efficacy of cyclic AMP–stimulating agents in radiation protection has not been tested in higher animal species.

Alpha2-macroglobulin: Serum alpha-macroglobulin, when administered before or shortly after irradiation, enhanced the recovery of hematopoietic tissues and increased the survival of irradiated rodents.[37,38] Alpha2-macroglobulin, when administered IP before whole-body irradiation with a dose of 6.7 Gy, increased 30-day survival from 50% to 100%, a value similar to that produced by amifostine.[39]

Substance P: Substance P (SP) is a neuropeptide and acts as a neurotransmitter, neuromodulator, and inflammation mediator. Several studies suggest that SP mediates multiple cellular activities, including cell proliferation, anti-apoptotic responses, and inflammatory reactions. It has been reported that administration of SP reduced radiation-induced apoptosis in mouse bone marrow stem cells and human mesenchymal stem cells in culture and stimulated cell proliferation in gamma-irradiated mouse bone marrow cells.[40]

Tumor suppressor gene p53: The role of p53 in modifying radiation-induced bone marrow syndrome is different from that of radiation-induced gastrointestinal (GI) syndrome. For example, pifithrin-alpha, an inhibitor of p53, protected mice from radiation-induced bone marrow syndrome, but it failed to protect against radiation-induced GI syndrome.[41] Overexpression of p53 may be involved in radiation-induced bone marrow syndrome. On the other hand, transgenic mice overexpressing p53 in all tissues were protected from the radiation-induced GI syndrome, and selective deletion of p53 from the GI epithelial cells, but not from endothelial cells, sensitized irradiated mice to GI syndrome.[44] The role of p53 in bone marrow radiation injury is further confirmed by the fact that a transient induction of p53. Short hairpin RNA (shRNA) (inhibits p53 expression) protected radiation-induced apoptosis in the thymus and protected mice against bone marrow syndrome produced by whole-body irradiation with a dose of 7.5 Gy.[42]

The mechanisms of radiation-induced elevation of p53 are not well understood. It has been proposed that the increased level of p53 after irradiation may be due to the binding of p53 with a heat shock protein (Hsp90).[43] If this is the case, then an inhibitor of Hsp90 should decrease the levels of p53 after irradiation. Indeed, 17-DMAG, an inhibitor of Hsp90, protected human T cells in culture.[43] Normally, the oncoproteins Mdm2 and Mdm4 are required to restrain p53 function in order to allow normal cell proliferation and/or cell viability. It is unknown if lethal doses of ionizing radiation impair the binding of p53 with Mdm4 and Mdm4. Increased activity of p53 occurs upon the loss of Mdm2 or Mdm4, which causes apoptosis. In addition, certain anti-apoptotic peptides exhibited radioprotective activity in rodents.[44]

Kinase inhibitor: A novel small-molecule kinase inhibitor, Ex-Rad, when administered orally or subcutaneously 15 minutes or 24 hours before or 24 hours and 36 hours after whole-body gamma irradiation with a dose of 7.5 Gy, markedly enhanced the survival of irradiated mice.[45] It was also demonstrated that an oral dose of 500 mg/kg of body weight of Ex-Rad, when administered 24 hours and 36 hours after irradiation, increased the survival of irradiated mice from 50% to 90%. A lower oral dose of 200 mg/kg of body weight also provided significant protection against radiation damage. Preclinical pharmacokinetic studies in rats, dogs, rabbits, and monkeys showed that Ex-Rad is absorbed well and its bioavailability ranged from 47% to 91%. It also has a very good safety profile. Ex-Rad, when administered SC 15 min or 24 hours before whole-body irradiation with a dose of 7 Gy, protected against bone marrow syndrome in mice, yielding a DRF value of 1.16. The mechanisms of protection of hematopoietic tissue may involve downregulation of phosphorylated p53. It also protected against radiation-induced damage to intestinal crypt cells.[46] The mechanism of radiation protection and/or radiation mitigation may involve activation of an AKT signaling pathway.[47] An inhibitor of p38 MAPK signal pathway, SB203580(SB), when injected IP before and after irradiation, protected mice against injury to bone marrow and intestinal crypts after whole-body irradiation with lethal doses.[48] The combination of SB203580 and granulocyte colony-stimulating factor (G-CSF) when administered IP after whole-body irradiation was more effective to some degree than the individual agents in reducing radiation-induced bone marrow injury.[48]

Inhibitors of inflammation: Minozac, a selective inhibitor of pro-inflammatory microglia cytokines, when administered IP to rats at a dose of 5 mg/kg of body weight per day 24 hours after whole-body irradiation with a gamma radiation dose of 10 Gy, and continued daily for 1 or 4 weeks after irradiation, reduced brain injury.[49] Anti-inflammatory drugs, when administered before irradiation, can also partially prevent radiation-induced brain damage following lethal doses of irradiation. Ethyl pyruvate (EP), a simple aliphatic ester pyruvic acid, acts as an anti-inflammatory agent. Ethyl pyruvate, when administered IP 1 hour before whole-body irradiation with a dose of 9.75 Gy and continued daily for 5 days after irradiation, increased the survival of irradiated mice, and protected cells against radiation-induced apoptosis without affecting the levels of oxidative stress.[50,51]

Mitochondrial targeting agents: A small-molecule mitochondrial-targeted nitroxide (JP4-039), when administered IP at a dose of 10 mg/kg of body weight

10 minutes before or at several time points after whole-body irradiation with a dose of 9.5 Gy, increased the survival of irradiated mice. Significant protection was obtained even when JP4-039 was injected 4 hours after irradiation. A dose as low as 0.5 mg/kg of body weight was effective in providing protection against radiation-induced bone marrow syndrome.[52] Another mitochondrial-targeted compound, cytochrome/cardiolipin peroxidase inhibitor, exhibited radioprotective and radiomitigator properties in rodents.[53] Treatment with triphenyl-phosphonium (TPEY-Tempo), a conjugate of nitroxide with a hydrophobic cation, significantly reduced gamma irradiation–induced apoptosis in brain endothelial cells, while enhancing the radiosensitivity of brain tumor cells.[54] Treatment of mouse embryonic cells with hemigramicidin S-conjugated 4-amoni-2,2,6,6-tetramethyl-piperidine-N-oxyl (hemi-GS-TEMPO) 5-125, 10 minutes before or 1 hour after gamma irradiation, markedly reduced radiation-induced superoxide generation, cardiolipin oxidation, and apoptosis. In addition, it increased the survival of irradiated animals. The mechanisms of protection may involve prevention of release of pro-apoptotic factors from mitochondria and cell-cycle arrest at the G2M phase.[55]

Agonists of Toll-like receptor (TLR): CBLB502, a recombinant protein derivative of Salmonella enterica flagellin, exhibited strong radioprotective and radiation mitigating effects in mice with a DRF value of 1.5–1.8. It acts via interaction with Toll-like receptor 5 and subsequent NF-kappaB-mediated induction of multiple protective mechanisms. Flagellin CBLB502, when administered intramuscularly at a dose of 0.04 mg/kg of body weight before whole-body irradiation with a dose of 6.5 Gy, increased the survival of irradiated animals from 25% to 64% in nonhuman primates (rhesus monkey, *Macaca mulatta*).[56] Administration of CBLB502 at 1, 16, 25, or 48 hours after whole-body irradiation with doses of 6.5–6.75 Gy increased the survival of irradiated animals from 20–40% to 70–100%. Intestinal epithelial cells in irradiated animals were also protected by CBLB502. Salmonella flagellin is an agonist of Toll-like receptor (TLR5), which acted as a radioprotective agent against radiation-induced bone marrow as well as gastrointestinal syndromes by activating anti-apoptotic and pro-survival NF-kappaB pathway in mice.[57] Lipoproteins from different bacteria stimulate cell surface TLR2-containing complexes. A synthetic lipopeptide, CBLB600, when administered 24 hours before whole-body irradiation with a dose of 10 Gy increased the survival of irradiated mice from 0% to 100%. Significant radiation protection was obtained when the drug was administered between 30 minutes to 48 hours before irradiation. Injection of CBLB600 15 minutes to 9 hours after irradiation also increased the survival of irradiated mice from 0% to 70%.[58]

Agonist of thrombopoietin: A thrombopoietin agonist, Alxn4100TPO (4100TPO), when administered SC at a dose of 1 mg/kg of body weight 24 hours before or 12 hours after whole-body irradiation with a dose of 7 Gy, protected mice against hematopoietic damage by stimulating hematopoietic recovery.[59]

Cytokines and growth factors: Interleukin-12 (IL-12), when injected 24 hours before or 2 hours after whole-body irradiation with a dose of 6.25 Gy, provided protection against hematopoietic tissue damage in mice. Administration of IL-12 up to 48 hours after whole-body irradiation increased the survival of irradiated mice from 0% to 70% at 30 days after irradiation. Injection of IL-12 6 hours after whole-body irradiation with a dose that produced GI syndrome also increased the survival of irradiated mice from 0% to 100%.[60] Synthokine SC-55494, a synthetic cytokine, is a high-affinity interleukin-3 (IL-3) receptor ligand that stimulates multi-lineage hematopoietic activity more than that produced by native IL-3. Administration of Synthokine SC at a dose of 100 mcg/kg/day 1 day after whole-body irradiation with gamma radiation with a dose of 7 Gy and continued daily for a period of 23 days enhanced bone marrow recovery in rhesus monkeys.[61] A combination of Synthokine SC-55494 and recombinant human granulocyte colony-stimulating factor (rhG-CSF) was more effective in bone marrow recovery in rhesus monkeys than Synthokine alone.[62] Administration of IL-1 alpha and tumor necrosis factor-alpha (TNF-alpha) exhibited radioprotective activity in mice; the combination of the two produced an additive effect on radiation protection.[63] Treatment of rat intestinal epithelial cells in culture with a recombinant human IL-11 reduced radiation-induced apoptosis.[64] When IL-4 or IL-11 was injected as a single dose for 5 consecutive days 2 hours after whole-body irradiation with a dose of 8.5 Gy, the 30-day survival increased without significant recovery in hematopoietic cells.[65,66] IL-11 in combination with thrombopoietin increased 30-day survival after irradiation with a dose of 10 Gy more than that produced by either agent alone. In addition, IL-11 in combination with bone marrow transplant increased 30-day survival of irradiated mice with a dose of 15 Gy more than that produced by individual agents.[66] Treatment of adult human umbilical vein endothelial cells in culture with hepatocyte growth factor (HGF) at a dose of 20 or 40 ng/ml 3 hours before irradiation with a dose of 20 Gy increases the proliferation rate and inhibited radiation-induced apoptosis, when determined 48 hours after irradiation.[67]

Analogs of somatostatin: Administration of somatostatin analog SOM230 (pasireotide) SC twice a day for 14 days at a dose of 1 or 4 mg/kg of body weight 24 hours after whole-body irradiation with a dose of 9–10 Gy increased the survival of irradiated mice. Injection of SOM230-LAR, a long-acting release form of SOM230, once at a dose of 8–32 mg/kg of body weight after irradiation increased the survival of irradiated mice at 30 days similar to that produced by SOM230 treatment.[68]

BIO 300: The radioprotective effect of BIO 300 was demonstrated by the investigators at the Armed Forces Radiobiology Research Institute, Bethesda, Maryland. This is a nonsynthetic single molecular agent and is effective when administered SC or orally before whole-body irradiation with lethal doses of ionizing radiation that produce bone marrow syndrome in rodents. A review has discussed in detail the radioprotective efficacy of this agent.[69]

Vaccine: When the typhoid-paratyphoid vaccine containing 1.5×10^9 killed organisms was injected SC in mice 24 hours before irradiation, the survival of irradiated animals was markedly increased.[70] The pretreatment period was very important in producing an optimal radiation protection. The stimulation of bone marrow by the vaccine treatment was considered a major factor in radiation protection.

Isolation of molecules from the lymph of lethally irradiated animals that can produce the same signs and symptoms of damage when administered intramuscularly to unirradiated animals as that produced by exposure to ionizing radiation was a major discovery in radiation biology by Russian investigators. These molecules were referred to as specific radiation determinant (SRD) toxins and consisted of protein (about 50%), lipid (38%), and carbohydrate (10%). The molecular weight of the SRD toxin ranges from 200–250 kDa. It was then possible to prepare small amounts of vaccine against the SRD toxin. This vaccine, when injected intramuscularly into animals (mice, rats, cattle, pigs, sheep, dogs, rabbits, and horses) before whole-body irradiation with lethal doses of radiation, increased the survival rates and survival time of irradiated animals. Therefore, this vaccine is referred to as *anti-radiation vaccine*. The protective effects of the immunization with anti-radiation vaccine begin to manifest 15–35 days after an injection.[71] The use of antiradiation vaccine in humans will require FDA approval, which is a very costly and time-consuming process.

RADIOPROTECTIVE DRUGS APPROVED BY THE FDA FOR OTHER INDICATIONS IN HUMANS

Some drugs, which have been approved by the FDA for the treatment of other indications in humans, exhibited radioprotective effects. They are described in the following text.

Amifostine: Among sulfhydryl compounds, amifostine, an analog of cysteamine, has been approved by the FDA to be used under certain conditions of radiation therapy of some tumors. In patients undergoing radiation therapy for head and neck cancer, SC administration of amifostine provided similar degrees of protection as that produced by IV injection.[72] Topical application of amifostine failed to protect skin against acute or late adverse effects of irradiation.[73] Amifostine selectively protected some normal tissues during radiation therapy of cancer,[11,74] but it also induced bone marrow hypoxia and hypotension.[75] Other toxicities of amifostine include Stevens-Johnson syndrome and epidermal necrolysis,[76,77] neurotoxicity,[78] impairment of learned performance tasks in rats,[75] hypocalcemia, and inhibition of parathyroid secretion.[79] Because of potential toxicity, amifostine may not be suitable for use in normal human population as a radioprotective agent. In addition, it cannot be used daily for a long period of time in order to reduce the adverse late effects of radiation among survivors of high doses of radiation.

Diltiazem (DTZ): This drug, at a dose of 100 mg/kg of body weight administered IP before whole-body irradiation with a dose of 7.5 Gy, improved radiation-induced decline in hematological and biochemical parameters.[80] It also increased the serum levels of erythropoietin and glutathione in comparison with the irradiated control animals.

Angiotensin converting enzyme (ACE) inhibitors: Three structurally different ACE inhibitors (captopril, enalpril, and fosinopril), when administered through drinking water 4 hours to 49 days after whole-body irradiation with a dose of 11 Gy followed by a syngeneic bone marrow transplant, protected against radiation-induced pneumonitis in rats.[81] Administration of captopril or losartan orally through drinking water at a dose of 1 g/m2/day 10 days after whole-body irradiation mitigated renal failure in rodents.[82] Treatment with captopril (200 mg/m2/day), or enalpril (25 mg/m2/day), but not fosinopril (25 mg/m2/day), 1 week after whole thorax irradiation with a dose of 13 Gy reduced morbidity during the pneumonitis phase. However, all three ACE inhibitors inhibited the onset of radiation-induced lung fibrosis.[83] Among four ACE inhibitors, captopril, enalapril, ramipril, and fosinopril, only captopril contains a reducing sulfhydryl (-SH) group). Studies on the ACE inhibitors showed that the mechanism of protection against radiation-induced nephropathy in rats was due to ACE inhibition, rather than due to reduction in oxidative stress by the SH group of captopril. The longer-acting enalapril may be more effective than captopril, which has only 2 hours of biological half-life.[84] Administration of captopril (100 mg/m2/day) and EUK-207, a SOD/catalase mimetic, (1.8 mg/m2/day) alone or in combination when administered one week after whole-body irradiation with a dose of 11 Gy and continued daily for the entire observation period, provided protection against radiation-induced injury to the lungs in rats; however, a combination of EUK-207 and enalapril (10 mg/m2/day) was ineffective.[85] These studies suggest that an elevation of angiotensin may be useful in prevention and mitigation of radiation damage. Indeed, it has been reported that administration of angiotensin 1-7 accelerated the recovery of hematopoietic tissues after whole-body irradiation.[86]

Statins: Statins such as Lovastatin, which inhibits HMG-CoA-reductase when administered orally (gavage) before and after a single or fractionated doses, reduced radiation-induced oral mucositis.[87] Lovastatin was administered at 8 mg/kg of body weight or 16 mg/kg of body weight 3 days before and at various time intervals after irradiation. Lovastatin, when administered 4 to 24 hours after whole-body irradiation with a single dose of 6 Gy or fractionated dose of 2.5 Gy twice, reduced normal tissue injury by inhibiting radiation-induced activation of pro-inflammatory markers (NF-kappaB and cell adhesion molecules) and pro-fibrotic marker genes (TNF-alpha, IL-6, and type I and type III collagen).[88] Statins are commonly used to lower cholesterol levels in humans. They exhibit both antioxidant and anti-inflammatory activities; therefore, it is expected that they will be of some radioprotective value. Indeed, pravastatin administered through drinking water daily 3 days before and 14 days after irradiation of surgically exteriorized small intestine with a dose of 19 Gy protected intestinal injury, but it did not protect tumor cells in culture against radiation injury.[89] Pravastatin also reduced radiation-induced skin lesions in mice when administered daily for 28 days.[90] Administration of simvastatin through food 2 weeks before and continued daily after local irradiation to the intestine for the entire observation period reduced intestinal injury in rats.[91] The efficacy of statins in treating radiation-exposed individuals remains uncertain. It is very interesting to note that they mimic the effects of antioxidants in reducing radiation injury. Although statins are FDA-approved drugs, it is unknown whether or not they would be toxic to humans at radioprotective doses.

Analgesics: Administration of morphine (60 mg/kg of body weight) or heroin (60 mg/kg of body weight) IP shortly before whole-body irradiation increased the value of $^{30}LD_{50}$ in mice from 609 to 830 R (about 6.09 to 8.30 Gy). Administration of sodium salicylate (600 mg/kg of body weight) IP shortly before whole-body irradiation increased the survival of animals irradiated with a dose of 700 R (about 7 Gy) from 0% to 50%.

Tranquilizers: Injection of reserpine (4 mg/kg of body weight) IP 12 hours before whole-body irradiation increased the value of $^{30}LD_{50}$ in male mice from 605 to 825 R (about 6.05 to 8.25 Gy), and from 635 to 727 R (about 6.35 to 7.27 Gy) in female mice. Reserpine did not provide radiation protection in rats under similar experimental conditions. This observation suggests that radioprotective efficacy of a given compound may not be observed in all species.

Epinephrine and norepinephrine: Administration of epinephrine IP at a dose of 5 mg/kg of body weight shortly before whole-body irradiation with doses that produced bone marrow syndrome increased the survival of irradiated mice; however, norepinephrine was ineffective. This may be due to the fact that tissue hypoxia is not achieved by a comparable dose of norepinephrine. For example, an equal dose (1.5 mg/kg of body weight) of both drugs raised the arterial blood pressure to approximately the same extent, but norepinephrine decreased oxygen tension of the spleen by about 48%, whereas epinephrine decreased it by about 90%. Norepinephrine treatment produced only 5% survival at a dose that produced 100% mortality without any treatment; however, epinephrine treatment produced 95% survival. Tyramine was less effective than epinephrine. Ephedrine at a dose of 80 mg/kg of body weight did not provide any radiation protection in rodents. Administration of methoxamine IP at a dose of 80 mg/kg of body weight shortly before whole-body irradiation with doses that produce bone marrow syndrome increased the value of $^{30}LD_{50}$ from 825 to 1100 R (about 8.25 to 11 Gy), yielding a DRF value of about 1.3.[92]

Inhibitors of histone deacetylase: The inhibitor of histone deacetylase, valproic acid, has been shown to suppress cutaneous radiation syndrome and stimulate hematopoiesis in clinical studies.[93] Another inhibitor of histone deacetylase, phenylbutyrate (PB), when administered IP at doses of 10–500 mg/kg of body weight 24 hours before whole-body irradiation with doses of 8–8.5 Gy of gamma radiation, increased the survival at 30 days after irradiation.[94] When PB was administered 0 and 12 hours after irradiation, significant protection was achieved at a radiation dose of 8 Gy, but not at 8.5 Gy.

RADIOPROTECTIVE HERBAL EXTRACTS
NOT REQUIRING FDA APPROVAL

Herbal extracts individually or in combination have been used in Asia to treat chronic human diseases in which increased oxidative stress and inflammation play a central role in the initiation and progression of the disease. Since increased oxidative stress and inflammation are also involved in the initiation and progression of radiation injury, the study of the radioprotective efficacy of herbal extracts began about a decade ago. Several herbal ethanol-, methanol-, water-extracts, or semipurified herbal extracts have exhibited varying degrees of radioprotective activities in both cell culture and rodent models, when administered before irradiation with lethal doses of x-rays or gamma rays. These herbal extracts consistently have exhibited antioxidant activity, anti-inflammation activity, hematopoietic stimulation, and immunostimulation; therefore, the radioprotective effects of these extracts were not unexpected. An excellent review on this issue has been published.[95] The efficacy of herbal extracts in larger animal species has not been tested. The efficacy and toxicity of these agents at radioprotective doses in humans remains uncertain. In addition, certain herbs have exhibited adverse drug interactions; therefore, these issues must be resolved before their efficacy in humans can be tested. The radioprotective effectiveness of selected herbal extracts is described in the following text.

Nigella sativa (**EE-NS**): An ethanol extract of *Nigella sativa* (EE-NS) exhibited antioxidant properties and provided significant protection against radiation-induced oxidative damage, including DNA damage, in murine spleen lymphocytes in culture when administered 1 hour before irradiation.[96] Oral administration of EE-NS at doses of 0–100 mg/kg of body weight for five consecutive days before whole-body irradiation with a dose of 7.5 Gy increased the survival of irradiated Swiss albino mice.[96]

Acorus calamus (**EE-AC**): Oral administration of EE-AC at a dose of 250 mg/kg of body weight 1 hour before whole-body gamma irradiation with a dose of 10 Gy delivered to mice in a single dose increased the activities of antioxidant enzymes superoxide dismutase (SOD), catalase and glutathione peroxidase, and levels of glutathione, and decreased the formation of malondialdehyde (MDA) and DNA strand breaks in comparison to irradiated control animals. The survival rate also increased.[97]

Aloe vera (**LE-AV**): Oral administration of LE-AV at a dose of 1 g/kg of body weight daily for 15 days and on the 15th day, 30 minutes before whole-body gamma irradiation with a dose of 6 Gy delivered to Swiss albino mice in a single dose increased the activities of antioxidant enzymes SOD and catalase in the skin and levels of glutathione in the liver and blood, and decreased the formation of lipid peroxidation products.[98]

Mentha piperita (**LE-MP**): Oral administration of LE-MP at a dose of 1 g/kg of body weight daily for 3 days and on the 3rd day, 30 minutes before whole-body gamma irradiation with a dose of 8 Gy delivered to Swiss albino mice in a single dose increased the activities of antioxidant enzymes glutathione peroxidase, glutathione reductase, glutathione-S-transferase, SOD, and catalase, and decreased the formation of MDA in the liver.[99] The radioprotective effect of LE-MP may be due to an increase of erythropoietin level in the serum.[100]

Tinospora cordifolia (**TC**): Administration of extract of TC IP at a dose of 200 mg/kg of body weight 1 hour before whole-body irradiation with gamma radiation doses of 5, 7.5, and 10 Gy increased the survival of irradiated mice from 0% to about 76%. Pretreatment with the extract of TC also increased the formation of spleen colony-forming units (CFU), and restored lymphocyte counts in irradiated mice.[101]

Podophyllum hexandrum (**Himalayan may apple**): Administration of ethanol extract of *Podophyllum hexandrum* (EE-PH), a high-altitude Himalayan plant, IP at a dose of 200 mg/kg of body weight 2 hours before whole-body irradiation with a gamma radiation dose of 10 Gy increased the survival of irradiated mice from 0% to about 80%. This level of protection was related to the modulation of proteins associated with cell death.[102]

The maximum tolerated dose (MTD) of a purified EE-PH was 60 mg/kg of body weight, whereas a dose of 90 mg/kg of body weight when injected IP produced 50% mortality within 72 hours. Administration of semipurified EE-PH IP at a dose of 15–20 mg/kg of body weight 2 hours before whole-body irradiation with gamma radiation with a dose of 10 Gy increased the survival of irradiated mice from 0%

to about 66%, while at doses 10–15 mg/kg of body weight administered 1 hour before irradiation, it increased the survival from 0% to 90%, yielding a DRF value of 1.625.[103] Although the radioprotective efficacy of semipurified EE-PH is very impressive in mice; its efficacy in larger animal species remains to be determined.

Vernonia cinerea (**VC**): Administration of methanolic extract of VC (ME-VC) at a dose of 20 mg/kg of body weight before whole-body gamma irradiation with a dose of 6 Gy delivered to mice in a single dose increased the activities of antioxidant enzymes glutathione peroxidase, SOD and catalase, and glutathione level in the irradiated animals. In addition, this treatment decreased radiation-induced elevation of pro-inflammatory cytokines, such as IL-1B, TNF-alpha, and c-reactive proteins (CRP), and stimulated the production of granulocyte-monocyte colony-stimulating factor (GM-CSF).[104]

Centella asiatica (**CA**): Indian traditional medicine recommends this herb for the treatment of various chronic illnesses in humans. Administration of aqueous extract of CA (AE-CA) IP at a dose of 100 mg/kg of body weight before whole-body gamma radiation with a dose of 2 Gy provided protection against radiation-induced body weight loss and conditioned taste aversion in rats.[105] Oral administration of AE-CA at the same dose before whole-body gamma radiation with a dose of 8 Gy increased the survival and decreased the body weight loss in irradiated mice.[106]

Ginkgo biloba (**GB**): This herb is commonly found in China, Japan, and Korea and has been used to treat human chronic diseases. Clastogenic factors composed of lipid peroxidation products, cytokines, and other oxidants are commonly present in the plasma of irradiated individuals. A 30% ethanol extract of GB (EE-GB) of dried leaf at a dose of 100 µg/ml protected cells in culture grown in the plasma from irradiated individuals against clastogenic factors.[107] The leaf extract of GB at a concentration of 100 µg/ml protected rat neurons in culture against hydroxyl radical-induced apoptosis.[108] An oral dose of 40 mg/day administered 3 times a day for a period of 2 months was effective in treating recovery workers from the Chernobyl nuclear power plant accident site.[109] An ethanol leaf extract of GB when administered IV at a dose of 100 mg/person was effective in reducing vasogenic edema in patients after irradiation of the brain.[110]

Hippophae rhamnoides (**HR**): This plant has been used in traditional Indian and Tibetan systems of medicine to treat several chronic diseases for centuries. The berries of HR contain polyphenolic compounds that are known to exhibit antioxidant activity. The radioprotective effect of a 50% ethanol extract of berry at a dose of 30 mg/kg of body weight when administered IP in mice 30 minutes before whole-body gamma irradiation at a dose of 10 Gy was evaluated. The results showed that berry extract increased the survival from 0% to 82%.[111,112] The extract also protected the gastrointestinal system against a lethal dose of whole-body irradiation when administered before irradiation.[113]

Ocimum sanctum (**OC**): This is an Indian herb, commonly known as Tulasi, which is widely distributed throughout the India subcontinent. Every part of this plant has been used in traditional medicine systems like Ayurveda for thousands of years to treat several human chronic diseases. The aqueous and ethanol extracts were used in radioprotective studies.[114] The aqueous extract of dried leaves was more effective than the ethanol extract. Administration of aqueous extract IP at a dose of 10 mg/kg of bodyweight daily for 5 days before whole-body gamma irradiation yielded a DRF of 1.28. The IP route of injection was more effective than the oral route. This extract was found to be more effective in comparison to amifostine.[115]

***Panax ginseng* (PG):** Ginseng is widely used in traditional medicine in oriental countries to treat various human chronic diseases. The radioprotective effects of the aqueous or ethanol extract of whole ginseng or the aqueous extract of roots of ginseng have been demonstrated in mice.[116–120] Ginsan, a purified polysaccharide (molecular weight 2000 kD) isolated from the ethanol-soluble fraction of PG aqueous extract, when administered IP at a dose of 100 mg/kg of body weight 24 hours before whole-body irradiation with a dose of 8 Gy significantly improved the survival of irradiated mice. It was ineffective when injected after irradiation.[121]

Tea polyphenols and epigallocatechin: Administration of tea polyphenols (TP) and its major constituent epigallocatechin gallate (EGCG) at doses 10 and 50 mg/kg of body weight after whole-body gamma irradiation of mice reduced radiation-induced damage to spleen and hematological parameters. In addition, they decreased radiation-induced elevation of MDA (melondialdehyde). Oral administration of TP at a dose of 50 mg/kg of body weight showed the best radioprotective effects.[122]

***Ecklonia cava* (EC):** EC is an edible brown algae widely found in the subtidal regions of Korea. Triphlorethol-A, a purified fraction isolated from EC, protected radiation-induced cellular damage by quenching free radicals.[123] The radioprotective efficacy of another compound, phloroglucinol (1, 3, 5-trihydroxybenzene) isolated from EC was evaluated in cell culture and in mice. The results showed that this compound also exhibited radioprotective activity in cell culture and in animals. The DRF on the criterion of intestinal crypt damage was 1.24.[124]

***Psidium guajava* (common guava):** Morin (2, 3, 4, 5, 7-pentahydroflavone), a flavonoid compound that can be isolated from *Maclura pomifera* and from the leaves of *Psidium guajava*, reduced radiation-induced cellular injuries by quenching free radicals.[125]

Other herbs: The other ethanol alcohol or water extracts of herbs that exhibited radioprotective effects in mice included *Biophytum sensitivum*, *Alstonia scholaris* (bark extract), *Emblica officinalis* (fruit extract), *Boerhaavia diffusa*, and *Phyllanthus amarus*.[126–130]

Quercetin: Administration of quercetin IP daily for 3 days at a dose of 100 mg/kg before whole-body irradiation with a dose of 4 Gy gamma radiation increased the survival of irradiated mice similar to that produced by AET.[131]

Genistein: Genistein, a soy isoflavon, is a nonspecific protein kinase inhibitor and a scavenger of free radicals. It also blocked the oxidative stress–induced activation of NF-kappaB, a transcriptional factor that regulates the expression of many cytokines, chemokines, immune receptors, and cell adhesion molecules.[132,133] Dietary supplementation of genistein (750 mg/kg of diet) immediately after irradiation with a dose of 18 Gy gamma rays to whole lungs of rats caused a delay in mortality by 50–80 days, and decreased the levels of the inflammatory cytokines TNF-alpha, IL-1β, and TGF-β, and DNA damage in the lung.[134] In addition, genistein treatment also reduced collagen content and 8-hydroxy deoxyguanosine (8-OHdG). These results suggest that dietary supplementation with genistein immediately after irradiation can provide partial protection against radiation-induced early pneumonitis and reduce the extent of fibrosis, although not sufficient to prevent lethality at this high dose in rats. Genistein, when administered orally or SC before lethal doses of whole-body irradiation, protected against bone marrow syndrome in mice.[135,136] The radioprotective effect of genistein appears to be mediated via granulocyte colony-stimulating factor (G-CSF) and IL-6.[136] Subcutaneous injection of genistein 24 hours before whole-body irradiation with a dose of 8.75 Gy increased the 30-day survival of irradiated mice from 31% (vehicle treated group) or 0% (irradiated control group) to 97%.[137] Genistein (50 mg/kg of body weight) or EUK-207, a SOD/catalase mimetic (8 mg/kg of body weight), when administered SC daily for 2 weeks after irradiation of whole lung with a dose of 12 Gy, mitigated lung injury in rats.[138]

Curcumin: Pretreatment of mice with curcumin before irradiation accelerated healing of radiation-induced skin wounds.[139] Supplementation with diet containing 5% curcumin significantly improved the survival of irradiated mice and reduced radiation-induced lung fibrosis.[140]

Resveratrol: Pretreatment of rats with resveratrol administered orally at a dose of 10 mg/kg of body weight daily for 10 days before and 10 days after whole-body irradiation with a dose of 8 Gy reduced damage to the liver and ileum tissues by reducing oxidative damage.[141]

Flaxseed: Treatment of mice with a diet containing 10% flaxseed daily for a period of 3 weeks before thoracic-irradiation with a dose of 13.5 Gy reduced lung injury by reducing the biomarkers of oxidative damage and pro-inflammatory cytokines.[142]

LIMITATIONS OF USING MOST IDENTIFIED RADIOPROTECTIVE AGENTS FOR PREVENTION OF ARS IN HUMANS

The efficacy of most radioprotective agents has been tested primarily on rodents (mice and rats), but most were toxic to humans; a few of them that appear nontoxic in animal models may require FDA approval before they can be recommended as radioprotective agents for humans. Some FDA-approved drugs for use in specific human diseases have exhibited varying degrees of radiation protection primarily in rodents; however, it is unknown whether these drugs at radioprotective doses would be safe in humans. Additional studies will be required before these drugs can be approved by the FDA for use as radioprotective and/or radiation mitigating agents. Many herbs not requiring FDA approval for human consumption exhibited varying degrees of radioprotective activities against ARS, primarily in rodents. Since some of them interact with prescription or nonprescription drugs in an adverse manner, their use in humans cannot be recommended for the purpose of radiation protection. It is also unknown whether herbs or herbal extracts at radioprotective doses would be safe in humans. Additional studies are needed before they can be recommended for radiation protection in humans. Until then, the use of multiple dietary and endogenous antioxidants remains one of the best choices for protection against ARS in humans.

RADIOPROTECTIVE DIETARY AND ENDOGENOUS ANTIOXIDANTS NOT REQUIRING FDA APPROVAL

Several studies have shown that dietary antioxidants, such as beta-carotene, vitamin A, vitamin C, vitamin E, and the mineral selenium, and endogenous antioxidants, such as N-acetylcysteine, R-alpha-lipoic acid, coenzyme Q10, when used individually IP shortly before and/or after x-irradiation or gamma irradiation exhibit varying degrees of radioprotective activity in cell culture and animal models. Limited studies on humans have revealed that some of these antioxidants when administered orally before and/or after irradiation provided some protection against radiation damage.

Based on the results of our studies and those published by others, and based on the fact that antioxidants neutralize free radicals and reduce inflammation that contributes to the initiation and progression of radiation damage, it is possible to develop a formulation of micronutrients containing dietary and endogenous antioxidants that have been consumed by the public for decades without any reported toxicity for reducing ARS in humans. Indeed, a commercial preparation referred to as Bioshield-R2 containing multiple micronutrients including dietary and endogenous antioxidants for reducing ARS in humans is available. All ingredients in Bioshield-R2 are considered safe in humans. Several laboratory and limited human studies that support the concept of using antioxidants as radioprotective and radiation mitigating agents in reducing ARS are described in the following text.

Radioprotective effect of antioxidant enzymes: The radioprotective effect of antioxidant enzymes cannot be studied directly because they will be hydrolyzed quickly. Therefore, the efficacy of antioxidants in radiation protection has been investigated by overexpressing a specific antioxidant enzyme in the cells, preparing a liposomal encapsulated antioxidant enzyme or a mimetic of a specific antioxidant enzyme. Indeed, it has been shown that overexpression of manganese-superoxide dismutase (MnSOD) protects hematopoietic progenitor cells against radiation damage.[143] Overexpression of CuZnSOD or MnSOD in human primary lung fibroblasts increased the survival of irradiated cells exposed to doses of 1–6 Gy. More recently, manganese superoxide dismutase mimetic, M40403, a stable non-peptidyl mimetic of MnSOD, when administered at a dose of 40 mg/kg of body weight 30 minutes before whole-body irradiation with a dose of 8.5 Gy, increased the survival from 0% to 100%. The DRF of M40403 when administered at a dose of 30 mg/kg of body weight SC 30 minutes before irradiation was 1.41.[144] Overexpression of extracellular SOD in transgenic mice protected against radiation-induced lung injury.[145] Administration of manganese superoxide dismutase-plasmid liposome (MnSOD-PL) provided local radiation protection to the lung, esophagus, oral cavity, urinary bladder, and intestine when administered before gamma irradiation.[146] It is not known whether MnSOD-PL can reduce the risk of late adverse health effects among survivors of high doses of radiation. The relevance of this observation in humans remains uncertain. Administration of polyethylene glycol (PEGylated) antioxidant enzymes (1:1 mixture of PEG-catalase and PEG-SOD) at a dose of 100 μg IV before whole-thorax irradiation with a dose of 13.5 Gy reduced radiation-induced pulmonary fibrosis in mice.[147]

A metalloporphyrin-based SOD mimic, MnTnHex-2-PyP^{5+} (hexyl), when administered at a dose of 0.05 mg/kg of body weight 2 hours after irradiation reduced radiation injury to the lung in rodents. Administration of hexyl SC at a dose of 0.05 mg/kg of body weight daily for the first two months after thorax irradiation with 10 Gy also mitigated radiation-induced lung injury in Rhesus monkeys. There was no demonstrable toxicity of this agent. Supportive care was given as needed to these animals.[148] Hexyl, a powerful antioxidant, when administered SC at a dose of 6 mg/kg of body weight 6 hours after whole-body irradiation with a sublethal dose of 6.5 Gy, reduced bone marrow injury by inhibiting oxidative stress and senescence.[149]

Radioprotective effect of individual antioxidants in cell culture: In 1982, we discovered that alpha-tocopheryl succinate (alpha-TS) is the most effective form of vitamin E in inducing selective cell death in cancer cells.[150,151] It has been shown that alpha-TS and selenium, but not alpha-tocopheryl acetate, reduced radiation-induced transformation in mammalian cells in culture; the combination of alpha-TS and selenium was more effective than the individual agents.[152,153] Alpha-TS was effective in decreasing radiation-induced chromosomal damage in normal human fibroblasts in culture.[154] Vitamin E (alpha tocopherol and alpha-tocopheryl acetate), vitamin C, and beta-carotene inhibited radiation-induced mutations, chromosomal damage, and lethality in mammalian cells in culture.[155–160] Gamma-tocotrienol, a well-known antioxidant, and pentoxifylline that improved local blood flow to the intestine, protected radiation-induced oxidative stress in the intestine when administered before whole-body irradiation with a dose of 12 Gy, which produces GI syndrome. Gamma-tocotrienol was more effective than that produced by the combination of gamma-tocotrienol and pentoxifylline.[161]

Natural beta-carotene was more effective than the synthetic form in reducing radiation-induced transformation in mammalian cells in culture.[162] Treatment of lymphocytes in culture with lycopene at a dose of 1, 5, and 10 μg/ml, before irradiation with 1, 2, and 4 Gy significantly decreased the frequency of micronuclei formation, dicentric and translocation types of chromosomal aberrations compared to irradiation control cells. In addition, lycopene treatment of lymphocytes before irradiation decreased the levels of thiobarbituric acid reactive substances (TBARS) and hydroperoxides, and increased the activities of antioxidant enzymes, such as superoxide dismutase, catalase, and glutathione peroxidase, and the level of glutathione.[163] Pretreatment of lymphocytes in culture with other antioxidants, such as n-acetylcysteine, glutathione, and thioproline before irradiation with doses of 2–4 Gy also produced similar protective effects.[164] A dose of 10 Gy caused apoptosis and increased the amounts and activation of matrix metalloproteinase 1 (MMP1) and MMP2 in bovine adrenal capillary endothelial cells in culture. Treatment of cells with all-trans-retinol or all-trans-retinoic acid 6 days before irradiation inhibited apoptosis. The fact that a tissue inhibitor of metalloproteinases (TIMP) blocked radiation-induced apoptosis in these cells suggests that the mechanism of protection by retinoids is mediated through the MMP. Retinoic acid at a dose of 100 mcg/day when administered IP daily for 5 days before whole-body irradiation with a dose of 9–16 Gy increased the survival of intestinal crypt cells. The dose was increased from 1.30 Gy to 1.85 Gy.[165] Vitamin C (ascorbic acid) at a dose of 250 mg/kg of body weight administered IP shortly before whole-body irradiation with a dose of 10 Gy increased the 30-day survival from 0% to 33%. Treatment of mice with ascorbic acid also improved the healing of radiation-induced wounds.[166]

In addition to dietary antioxidants, endogenous antioxidants such as n-acetyl-cysteine (NAC), a glutathione-elevating agent, attenuated radiation-induced toxicity in mammalian cells in culture.[167] Calcitriol, the hormonally active vitamin D metabolite, protected keratinocytes against radiation-induced caspase-dependent and -independent programmed cell death in mammalian cells in culture.[168]

Radioprotective effect of a mixture of antioxidants in cell culture: It was thought earlier that the radiation damage produced by high LET radiation cannot be modified by any pharmacological agents, because most of the damage is produced by direct ionization. However, recent studies have evaluated the radioprotective efficacy of antioxidants on space radiation (proton radiation and HZE-particle radiation, which are high-LET radiation) irradiated cells revealed that antioxidants can protect against space radiation–induced injury. For example, irradiation with space radiation caused cytotoxicity in human breast epithelial cells in culture and transformation in human thyroid cells in culture. Pretreatment of these cells with a mixture of soy-derived Bowman-Birk inhibitor (BBI), ascorbic acid, and coenzyme Q10, selenomethionine and alpha-tocopheryl succinate protected against space radiation–induced cytotoxicity and transformation in mammalian cells in culture.[169] It has been reported that the treatment of human thyroid epithelial cells in culture with selenomethionine alone also protected against space radiation–induced increase in oxidative stress, cytotoxicity, and cell transformation, possibly by enhancing DNA repair machinery in irradiated cells.[170] These studies suggest that the early radiobiology concept that the effect of high LET radiation cannot be modified by chemical agents is no longer valid.

Radioprotective effect of individual antioxidants in animals: Using animal models, several studies revealed that a single IP or subcutaneous SC administration of individual dietary or endogenous antioxidants before whole-body gamma irradiation with high doses enhanced the survival rate in varying degrees in rodents.[171–183]

In 1982, we identified alpha-tocopheryl succinate (alpha-TS) as the most effective form of vitamin E in reducing the growth of tumors without affecting the growth of normal cells.[150,151] In 2002, we discovered that alpha-TS protected human fibroblasts in culture against radiation-induced chromosomal damage.[172] Later in the same year, it was shown that alpha-TS, when administered before irradiation, also protected mice against radiation-induced bone marrow syndrome.[173] The radioprotective effect of alpha-tocopheryl succinate is mediated through the release of granulocyte colony-stimulating factor (G-CSF).[184] Intraperitoneal administration of vitamin E and L-carnitine individually before irradiation markedly reduced radiation-induced cataract formation in rats.[185,186] Pre-irradiation treatment of rats with vitamin E or L-carnitine alone significantly reduced severity of brain and retinal damage in rats; however, the combination of the two did not provide an additive protective effect,[187] suggesting that they were protecting radiation damage by a similar mechanism. It has been reported that IP administration of L-carnitine reduced gamma radiation–

induced cochlear damage in guinea pigs.[188] Vitamin E administered IP before irradiation reduced radiation-induced damage to salivary glands.[189]

Radiation-induced heart disease is a severe side effect of thoracic radiation therapy. The efficacy of a combined treatment with vitamin E (alpha-tocopherol) and pentoxifylline (PTX) in reducing radiation-induced heart disease was evaluated in three groups of rats, irradiated control animals, animals receiving a combination of vitamin E (20 IU/day), and PTX (100 mg/day) 1 week before and 6 months after irradiation, and animals receiving the same treatment but treatment starting at 3 months after irradiation with a fractionated daily dose of 9 Gy to the heart. The results showed that treatment with vitamin E and PTX produced beneficial effects on radiation-induced myocardial fibrosis and left ventricular function in both groups of animals in comparison to irradiated control animals.[190]

Delta-tocotrienol, when administered 24 hours before or 6 hours after whole-body irradiation with a dose of 8.75 Gy, provided significant protection against radiation-induced damage to hematopoietic tissue.[191] The DRF value for tocotrienol at a dose of 200 mg/kg of body weight was 1.29.[192] In another study, administration of gamma-tocotrienol at a dose of 200 mg/kg of body weight 24 hours before and pentoxifylline at a dose of 200 mg/kg of body weight 15 minutes before whole-body irradiation with a dose of 8 Gy was more effective in providing protection against bone marrow syndrome than gamma-tocotrienol alone. The DRF value for the combined treatment with gamma-tocotrienol and PTX was 1.45, whereas it was only 1.3 for gamma-tocotrienol treatment alone.[193]

N-acetylcysteine (NAC) increases the intracellular level of glutathione. Daily administration of NAC at a dose of 300 mg/kg of body weight SC for 7 days starting 4 hours before or 2 hours after abdominal irradiation with a dose of 20 Gy improved 10- and 30-day survival from 5% (irradiated control group) to greater than 50% (NAC-treated group). It also reduced the loss of villi from the small intestine and suppressed oxidative stress in nonirradiated bone marrow.[194] Pretreatment of mice with NAC attenuated radiation-induced liver injury by reducing free radical–mediated oxidative damage.[195] Ionizing radiation is known to activate downstream pro-inflammatory responses through the activation of NF-kappaB. It has been reported that this effect of radiation may be mediated thorough the inhibition of 26S proteasome activity rather than through the radiation-induced generated free radicals. This was confirmed by the results showing that NAC at a dose that inhibited proteasome activity also reduced activation of NF-kappaB.[196] Alpha-lipoic acid is also known to increase the intracellular level of glutathione, in addition to performing other biological functions. Whole-body x-irradiation of mice with a dose of 6 Gy increases the levels of markers of oxidative stress and impairs cognitive function. Injection of alpha-lipoic acid IP before irradiation markedly reduced cognitive dysfunction and oxidative stress. In addition, irradiated mice treated with alpha-lipoic acid showed intact structure of the cerebellum, higher counts of intact Purkinje cells and granular cells in comparison to irradiated animals that did not receive alpha-lipoic acid.[197] Administration of natural beta-carotene through diet (50 mg/kg of diet) daily for 1 week before whole-body irradiation with a dose of 4 Gy provided significant protection against radiation damage.[198] Pycnogenol, when administered

orally to rats before irradiation with a dose of 15 Gy of x-rays significantly pre-
served the height and number of villi, suggesting that this antioxidant can protect
intestinal mucosa.[171]

Melatonin, the chief secretory product of the pineal gland, exhibited strong
radioprotective effects both in cell culture and in animal models.[199] Melatonin and
vitamin E administered IP before irradiation reduced radiation-induced damage to
brain.[200] Melatonin significantly reduced radiation-induced edema, necrosis, and
neuronal degeneration, whereas vitamin E reduced only necrosis. The reasons for
these differential effects of melatonin and vitamin E are unknown, but their mechanisms of radiation protection may in part be different.

Radioprotective effect of a mixture of antioxidants in animals: The radioprotective effects of an antioxidant mixture were confirmed on mammalian cells in
culture as well as on animal models irradiated with high doses of high-LET radiation. For example, irradiation of mice with space radiation, a high-LET radiation,
increased oxidative stress, and this effect of space irradiation was reduced by pretreatment with dietary supplementation containing Bowman-Birk inhibitor concentrate
(BBIC), L-selenomethionine, or a mixture of n-acetylcysteine, sodium ascorbate,
coenzyme Q10, alpha-lipoic acid, L-selenomethionine, and alpha-tocopheryl succinate.[201] Alpha-lipoic acid administered IP before whole-body irradiation significantly attenuated high-LET radiation (^{56}Fe-beams)-induced radiation damage, such
as impairment in the reference memory, apoptotic damage in the cerebellum, and
increase in DNA and markers of oxidative damage.[202]

A few studies have shown that a mixture of dietary antioxidants administered
IP before irradiation reduced radiation-induced myelosuppression and oxidative
stress in rodents.[201,203] They were ineffective when injected after irradiation. An oral
administration of antioxidants before or after whole body x-irradiation was ineffective. It was thought that only free radicals generated during radiation exposure contribute to radiation injury; therefore, antioxidants were used only before irradiation.

Radioprotective effect of individual or a mixture of antioxidants in humans:
Direct radiation protection studies with antioxidants cannot be performed in humans
for obvious reasons. The limited data on this issue primarily come from experiences
of patients and radiation oncologists during radiation therapy of cancer or individuals exposed to high doses of radiation following an accident in nuclear power plants.
Administration of beta-carotene orally reduced the severity of radiation-induced
mucositis during radiation therapy of the head and neck cancer without affecting
the efficacy of therapy.[204] A combination of dietary antioxidants was more effective
in protecting normal tissue during radiation therapy than the individual agents.[205,206]
Vitamin A and NAC may be effective against radiation-induced cancer.[207] Alpha-
lipoic acid treatment alone for 28 days lowered lipid peroxidation among children
chronically exposed to low doses of radiation daily in the area of radioactive contamination by the Chernobyl nuclear power accident.[208] In another study involving
709 children (324 boys and 385 girls) who had been exposed to low levels of radiation during and after the Chernobyl nuclear power plant accident and moved to Israel
between 1990 and 1994, the effect of daily natural beta-carotene supplementation
(40 mg/day) for a period of 3 months on the blood level of oxidative damage was
evaluated. The results showed that the blood levels of oxidized conjugated dienes in

262 children were increased compared to those who did not receive radiation. After a 3-month supplementation with beta-carotene alone, the serum levels of this biomarker of oxidative damage was decreased without any significant changes in the level of total carotenoids, retinol, or alpha-tocopherol in these children.[209] A combination of vitamin E and alpha-lipoic acid was more effective than the individual agents.[208] In a randomized study involving 91 patients with lung cancer, the efficacy of a combined treatment with vitamin E and PTX in reducing lung toxicity was evaluated. Forty-four patients received vitamin E 300 mg twice a day orally and PTX 400 mg three times a day orally during the entire period of radiation therapy. Vitamin E at a dose of 300 mg once a day and PTX at a dose of 400 mg once a day were administered for an additional 3 months after irradiation. The results showed that the treatment with vitamin E and PTX markedly reduced radiation-induced lung toxicity.[210]

Radiation dermatitis occurs in about 95% of cancer patients receiving radiation therapy. At present, there are no effective treatment methods for these skin reactions. A well-designed clinical study has reported that daily oral supplementation with 6 g of curcumin during the entire course of treatment significantly reduced the severity of radiation dermatitis in breast cancer patients.[211] These studies strongly suggest that dietary and endogenous antioxidants can be of radioprotective value in humans.

SCIENTIFIC RATIONALE FOR USING MULTIPLE ANTIOXIDANTS

Although administration of a single antioxidant once shortly before irradiation produced consistent radioprotective effects in rodents, they were ineffective when administered only once after irradiation. In addition, the use of a single antioxidant, such as synthetic beta-carotene, increased the incidence of lung cancer among male heavy cigarette smokers.[212,213] In view of the fact that the level of oxidative stress is high among heavy tobacco smokers due to production of excessive amounts of free radicals and depletion of antioxidants, supplemented individual antioxidant may be oxidized, and thus may act as a pro-oxidant rather than as an antioxidant. Therefore, the above adverse effect of administration of beta-carotene alone could have been predicted. In order to avoid the risk of using individual antioxidants, the use of multiple antioxidants is proposed. Other rationales for using multiple antioxidants in radiation protection studies are described in the following text. The mechanisms of action of antioxidants and their distribution at the cellular and organ levels differ, their internal cellular and organ environments (oxygenation, aqueous, and lipid components) differ, and their affinity for various types of free radicals differ. For example, beta-carotene (BC) is more effective in quenching oxygen radicals than most other antioxidants.[214] BC can perform certain biological functions that cannot be produced by its metabolite, vitamin A, and vice versa.[215,216] It has been reported that BC treatment enhances the expression of the connexin gene, which codes for a gap junction protein in mammalian fibroblasts in culture, whereas vitamin A treatment does not produce such an effect.[216] Vitamin A can induce differentiation in certain normal and cancer cells, whereas BC and other carotenoids do not.[217,218] Thus, BC and vitamin A have, in part, different biological functions. Therefore, both BC and vitamin A should be added to a multiple micronutrient preparation.

The gradient of oxygen pressure varies within the cells. Some antioxidants, such as vitamin E, are more effective as quenchers of free radicals in reduced oxygen pressure, whereas BC and vitamin A are more effective in higher atmospheric pressures.[219] Vitamin C is necessary to protect cellular components in aqueous environments, whereas carotenoids, vitamin A, and vitamin E protect cellular components in lipid environments. Vitamin C also plays an important role in maintaining cellular levels of vitamin E by recycling vitamin E radical (oxidized) to the reduced (antioxidant) form.[220] Also, oxidative DNA damage produced by high levels of vitamin C could be protected by vitamin E. Oxidized forms of vitamin C and vitamin E can also act as radicals; therefore, excessive amounts of any one of these forms, when used as a single agent, could be harmful over a long period of time.

The form of vitamin E used in any multiple micronutrient preparation is important. It has been established that d-alpha-tocopheryl succinate (alpha-TS) is the most effective form of vitamin E both in vitro and in vivo.[151,221] This form of vitamin E is more soluble than alpha-tocopherol and enters cells more readily. We have reported that oral ingestion of alpha-TS (800 I.U. /day) in humans increased plasma levels of not only alpha-tocopherol, but also alpha-TS, suggesting that a portion of alpha-TS can be absorbed from the intestinal tract before hydrolysis.[222] This observation is important because the conventional assumption based on the rodent studies has been that esterified forms of vitamin E, such as alpha-TS, alpha-tocopheryl nicotinate, or alpha-tocopheryl acetate, can be absorbed from the intestinal tract only after they are hydrolyzed to alpha-tocopherol. Our preliminary data showed that this assumption may not be true for the absorption of alpha-TS in humans. Therefore, for an optimal effect of vitamin E, both alpha-TS and alpha-tocopherol or alpha-tocopheryl acetate should be added to a micronutrient preparation.

Glutathione is effective in catabolizing H_2O_2 and anions. However, an oral supplementation with glutathione failed to significantly increase the plasma levels of glutathione in human subjects,[223] suggesting that this tripeptide is completely hydrolyzed in the GI tract. Therefore, both N-acetylcysteine and alpha-lipoic acid, which increase the cellular levels of glutathione by different mechanisms, should be added to multiple micronutrient preparations.

Another endogenous antioxidant, coenzyme Q10, may also have some potential value in radiation protection. Since mitochondrial dysfunction may be associated with acute and late effects of radiation, and since coenzyme Q10 is needed for the generation of ATP by mitochondria, it is essential to add this antioxidant in a multiple micronutrient preparation in order to improve the function of mitochondria. A study has shown that ubiquinol (coenzyme Q10) scavenges peroxy radicals faster than alpha-tocopherol,[224] and like vitamin C, can regenerate vitamin E in a redox cycle.[225] However, it is a weaker antioxidant than alpha-tocopherol.

Selenium is a co-factor of glutathione peroxidase, and Se-glutathione peroxidase increases the intracellular level of glutathione, which is a powerful antioxidant. There may be some other mechanisms of action of selenium. Therefore, selenium and co-enzyme Q10 should be added to a multiple micronutrient preparation for an optimal radiation protection.

SCIENTIFIC RATIONALE FOR UTILIZING
ORAL ROUTE OF ADMINISTRATION

Individual antioxidants, when administered orally only once before or after irradiation, failed to provide any significant degree of protection. This could be due to the fact that absorption of a single antioxidant through the intestine was not sufficient to increase the tissue levels of antioxidants high enough to provide any significant radiation protection. In addition, a single antioxidant could not quench different types of free radicals that are produced during irradiation. Therefore, it is possible that a preparation of multiple micronutrients containing dietary and endogenous antioxidants administered orally once shortly before irradiation may be effective in reducing radiation damage. In addition, the oral route of administration is more relevant to humans for short- and long-term daily administration in order to reduce radiation damage.

SCIENTIFIC RATIONALE FOR ADMINISTERING MULTIPLE
ANTIOXIDANTS BEFORE AND AFTER IRRADIATION

Most radiation protection studies have been performed by administering antioxidants before irradiation, because it was thought that free radicals generated during irradiation are responsible for initiating and promoting radiation damage. However, long-lived free radicals exist after irradiation. In addition, reactive oxygen species (ROS) are released from inflammatory reactions that occur after irradiation. The inflammatory reactions also release toxic chemicals such as pro-inflammatory cytokines, adhesion molecules, and complement proteins, all of which are toxic to cells. Therefore, these post-irradiation biological events should be attenuated in order to maximize the efficacy of potential radioprotective agents. Therefore, pre- and post-irradiation treatment with a radioprotective agent is essential. Antioxidants not only neutralize free radicals, but also reduce inflammatory reactions;[5,226] therefore, both pre- and post-irradiation treatments with antioxidants become essential in order to maximize their effectiveness in reducing radiation damage. The pre-irradiation treatment period may include daily for 3–7 days both for animals and humans in order to increase the tissue levels of antioxidants optimally before irradiation, and post-irradiation treatment may not start until 24 hours after irradiation, and then continue daily for the entire observation period. In spite of the fact that antioxidants are fairly nontoxic in humans and that they exhibit radioprotective potential that has a mechanistic basis, no significant research was conducted to explore the capacity of multiple antioxidants in radiation protection, using the above dose schedules for pre- and post-irradiation periods.

ANIMAL STUDIES TO DEMONSTRATE THE RADIOPROTECTIVE
EFFICACY OF A MIXTURE OF MULTIPLE DIETARY
AND ENDOGENOUS ANTIOXIDANTS ADMINISTERED
ORALLY BEFORE AND/OR AFTER IRRADIATION

Sheep: The radiation-induced GI syndrome in sheep appears to be more sensitive than in rodents or humans. A pilot study was performed to evaluate the radioprotective

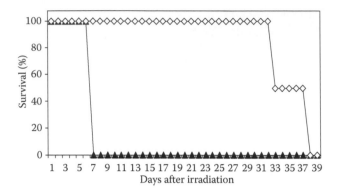

FIGURE 5.1 Antioxidant mixture increased survival time of sheep exposed to a whole-body gamma-radiation dose of 4.11 Gy that produced GI syndrome associated with a mild CNS syndrome with a 100% mortality in 7 days. Sheep were pretreated with placebo (black triangle) or a mixture of dietary and endogenous antioxidants (open diamond) 7 days before and 7 days after irradiation. A fivefold increase in survival time was observed in antioxidant-treated sheep. The number of irradiated animals receiving placebo or no treatment was 4, whereas the number of animals receiving antioxidant treatment was 2 (unpublished observation, in collaboration with Dr. Jones of NASA, Houston).

efficacy of a mixture of multiple dietary antioxidants, such as vitamin A (retinyl palmitate), vitamin C (calcium ascorbate), vitamin E (d-alpha tocopheryl succinate and d-alpha tocopheryl acetate), and selenomethionine, and endogenous antioxidants, such as n-acetylcysteine (NAC), R-alpha-lipoic acid, and coenzyme Q10 in sheep exposed to a dose of gamma rays that produce GI syndrome (in collaboration with Dr. Jones of NASA, Houston). The results showed that a dose of 4.41 Gy produced GI syndrome associated with a mild CNS syndrome in sheep, causing 100% lethality in 7 days. This radiation dose is significantly lower than that needed to produce GI syndrome in rodents (12 Gy) and in humans (about 6 Gy). An oral administration of the antioxidant mixture daily for 7 days before and daily for 7 days after irradiation increased the survival time of irradiated sheep from 7 days to 38 days. To our knowledge, this is a first demonstration in which antioxidant treatment orally before and after irradiation increased the survival time of irradiated animals exhibiting GI syndrome by about 5 times (Figure 5.1). It should be noted that these animals received no supportive care such as antibiotic, fluid replacement, or blood transfusion. The exact mechanisms of this high level of radiation protection in irradiated sheep exhibiting GI syndrome remain unknown. However, I suggest that in addition to scavenging free radicals during radiation exposure, the antioxidant mixture administered after irradiation also neutralized radiation-induced long-lived free radicals and reduced inflammation. It remains uncertain whether the continuation of antioxidant treatment after irradiation for a period longer than 7 days would have provided better radiation protection. It also remains unknown whether the concentration of antioxidants used in this study represented an optimal dose. It is possible that the use of other treatment modalities, such as restoration of electrolyte imbalance, blood transfusions, and antibiotic treatment in a combination of multiple antioxidants may have

(A)   (B)

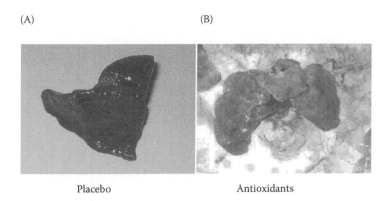

Placebo   Antioxidants

FIGURE 5.2   Protection of lung in rabbits exposed to CNS syndrome dose (9.011Gy) by the mixture of dietary and endogenous antioxidants. The autopsy of irradiated rabbits receiving placebo (and died in 4 h exhibiting central nervous system syndrome) revealed that lung was necrotic and without a lobular architecture (A), whereas antioxidant treated irradiated rabbits showed minimal pulmonary hemorrhage while maintaining the lobular architecture of the lung (B). This level of protection has never been achieved by any pharmacological or biological agents (unpublished observation, in collaboration with Dr. Jones of NASA, Houston).

increased the survival rate in irradiated sheep. At present, except for bone marrow transplant, no other effective FDA-approved treatment is available for treating irradiated individuals exhibiting GI syndromes, but nearly all survivors of bone marrow transplants eventually die of host versus graft rejection.

Rabbits: A pilot study was performed to evaluate the radioprotective efficacy of a mixture of multiple antioxidants (same as used in sheep) in rabbits irradiated with a dose of gamma rays that produce GI syndrome associated with CNS syndrome (in collaboration with Dr. Jones of NASA, Houston). The results showed that a dose of 9.011 Gy produced GI syndrome associated with the CNS syndrome. About 25% of irradiated rabbits died of CNS syndrome within 4 hours, the remaining died within 7 days. An oral administration of the mixture of antioxidants before irradiation did not prolong the survival time of irradiated rabbits. This was expected because the dose used to produce GI syndrome was closer to a dose that produced CNS syndrome in a significant number of irradiated animals. However, the necropsy of those irradiated rabbits dying of CNS syndrome showed that damage to the lungs was markedly reduced in the antioxidant-treated group in comparison to that observed in the placebo-treated group (Figure 5.2).

Mice: A pilot study was performed at the Armed Forces Radiobiology Research Institute (AFRRI) to evaluate the radioprotective efficacy of a mixture of multiple antioxidants (same as used in sheep) in mice. The results showed that a dose of 8.5 Gy produced bone marrow syndrome with a 100% lethality in 30 days. Oral administration of the antioxidant mixture daily for 7 days or 24 hours before whole-body gamma irradiation increased the survival rate of irradiated animals from 0% to 40% (Table 5.1). Placebo treatment of irradiated mice was ineffective. To our knowledge, this level of protection has not been achieved by an oral administration of a single antioxidant or its derivatives before whole-body gamma irradiation with a

TABLE 5.1

Effect of an Antioxidant Mixture or Placebo Administered Orally at a Dose of 222.5 mg/kg of Body Weight 7 Days before Various Doses of Whole-Body Gamma Irradiation on Survival in Mice

Agents	Dose (Gy)	% Survival
Placebo	7.50	90
Antioxidant mixture	7.50	80
Placebo	8.00	30
Antioxidant mixture	8.00	60
Placebo	8.50	0
Antioxidant mixture	8.50	40
Placebo	9.00	0
Antioxidant mixture	9.00	0
Placebo	9.25	0
Antioxidant mixture	9.25	0

Note: Antioxidant mixture contained neither micro-crystalline cellulose nor dextrose, but it contained about 11% silica by weight. Doses administered orally to animals 7 days before irradiation were corrected for the presence of 11% silica. Each group contained 10 animals.

dose that produced bone marrow syndrome with a 100% mortality. The results also revealed that a placebo containing cellulose and dextrose administered orally exhibited some radioprotective effect in mice at lower doses of radiation similar to that produced by the antioxidant mixture (data not shown). Thus, our initial assumption that the placebo used in this study was inert was found to be incorrect. It is interesting to point out that certain polysaccharides when administered IP shortly before whole-body irradiation have been reported to be of some radioprotective value in irradiated mice.[227–229] Therefore, it is not surprising that a mixture of cellulose and dextrose provided some degree of radiation protection at lower doses; however, the protective effect of the placebo was not observed at higher radiation doses in mice. In contrast to the observation made in mice, placebo treatment was totally ineffective in irradiated sheep and rabbits.

Drosophila melanogaster: In collaboration with Dr. Sharmila Bhattacharya of NASA, Moffat Field, California, a pilot study was performed to evaluate the effects of a mixture of multiple antioxidants (same as used in sheep) on proton radiation–induced cancer in *Drosophila melanogaster* (fruit fly). The female flies carrying mutant HOP

(TUM-1) become sensitive to developing leukemia-like cancer. The results showed that proton radiation markedly enhanced the incidence of tumor in these flies, and dietary supplementation with the antioxidant mixture 7 days daily before and for the entire observation period after irradiation, prevented proton radiation–induced cancer in female fruit flies carrying mutant HOP (TUM-1). It is impossible to extrapolate data obtained from fruit flies to humans. However, the possibility exists that daily supplementation with a mixture of antioxidants may reduce the risk of cancer among individuals with a family history of cancer. To our knowledge, this is a first demonstration in which the genetic basis of a disease can be prevented by antioxidants.

The published studies on radioprotective efficacy of antioxidants and our results of pilot studies suggest that the use of multiple dietary and endogenous antioxidants that are nontoxic to humans may reduce ARS when consumed orally before and daily 24 hours after irradiation for the entire observation period. The recommendation of antioxidants is also based on the fact that they can neutralize excessive amounts of free radicals generated during and after irradiation and reduce production of pro-inflammatory cytokines. Thus, it is possible to develop a commercial preparation of a micronutrient containing dietary and endogenous antioxidants that can reduce ARS in humans. Indeed, a commercial preparation of micronutrients containing multiple dietary and endogenous antioxidants referred to as BioShield-R2 is available.

PROPOSED RECOMMENDATION FOR USING BIOSHIELD-R2 IN COMBINATION WITH STANDARD THERAPY FOR PREVENTION OF ARS IN HUMANS

BioShield-R2 contains vitamin A (retinyl palmitate), vitamin E (both d-alpha-tocopheryl acetate and d-alpha-TS), natural mixed carotenoids, vitamin C (calcium ascorbate), coenzyme Q10, R-alpha-lipoic acid, n-acetylcysteine, L-carnitine, vitamin D, B vitamins, selenium, zinc, and chromium. No iron, copper, or manganese is included because these trace minerals are known to interact with vitamin C to produce free radicals. These trace minerals are absorbed from the intestinal tract more in the presence of antioxidants than in their absence, which could result in increased body stores of these minerals. Increased iron stores have been linked to increased risk of several chronic diseases. Consuming BioShield-R2 orally twice a day before irradiation would maintain high tissue levels of antioxidants in all individuals. In the event these individuals are exposed to lethal doses of radiation that produce bone marrow syndrome, the symptoms of ARS may be reduced, and the likelihood of survival from such doses may be enhanced. Consuming BioShield-R2 orally twice a day daily beginning 24 hours after irradiation for a period of 60 days may further enhance the survival rates of irradiated individuals. Combining Bioshield-R2 treatment with standard therapy may increase the survival rate of irradiated individuals more than that produced by the individual therapeutic modality. It is possible that the combined treatment may save some irradiated individuals who are exposed to radiation doses that produce GI syndrome.

MITIGATION OF ARS

The studies on radiation mitigating agents are described iin the following text. They can be grouped in three broad categories: (1) FDA-approved drugs and biologics, (2) drugs and biologics not approved by the FDA, and (3) multiple antioxidants that do not require approval by the FDA.

DRUGS AND BIOLOGICS APPROVED BY THE FDA

Replacement Therapy: Replacement therapy includes biologics, such as antibiotics, electrolytes, platelets and whole blood, and is administered after irradiation with lethal doses of radiation. However, each of these biologics should be administered when indicated by the laboratory tests. The effectiveness of antibiotics in protecting against high doses of radiation that produce bone marrow syndrome is observed only when infection is present after irradiation. This is due to the fact that infection is the major cause of death in irradiated individuals dying from bone marrow syndrome. In addition, loss of electrolytes and depletion of platelets and leukocytes occur after irradiation; therefore, transfusion with electrolytes and blood when indicated becomes necessary to prevent the irradiated individuals from dying of bone marrow syndrome. Replacement therapy is mostly effective against radiation doses that produce $^{60}LD_{50}$ in humans. This therapy can safely be applied to humans after irradiation. It should be mentioned that the replacement therapy requires large amounts of each of the biologics for the treatment of irradiated individuals; therefore, it is only suitable for the management of a small number of lethally irradiated individuals. This therapy may not be able to provide adequate amounts of biologics in case of managing mass casualties.

This therapy is ineffective in reducing mortality in irradiated humans exposed to radiation doses that cause GI syndrome. The best result one can hope for from the replacement therapy is an increase in survival time by about twofold (from 14 days to 28 days). Therefore, additional treatment modalities must be added in order to improve the efficacy of replacement therapy in management of irradiated individuals exhibiting GI syndrome.

Erythropoietin: The survival of mice irradiated whole-body with a dose of 10 Gy increased from 0% to 40% when administered after irradiation.[230] The DRF value of erythropoietin dose (10 units) injected 1 hour after whole-body irradiation with a dose of 6.51 Gy is 1.12.[231] The efficacy of erythropoietin in irradiated individuals has not been evaluated, but it is an FDA-approved drug that is used in certain human diseases to boost hemoglobin levels; therefore, it can be used as a radiation mitigating agent in combination with other therapeutic modalities.

Bone marrow and newborn liver cell transplants: Bone marrow transplant is often administered intravenously to irradiated individuals exposed to doses of radiation that produce GI syndrome, because replacement therapy is not effective in increasing the survival of irradiated individuals. On the other hand, bone marrow transplantation can save many irradiated individuals dying from GI syndrome; however, the survivors die within a few years because of host versus graft rejection events. Therefore, bone marrow transplant cannot be considered a suitable treatment option for those individuals exhibiting radiation-induced GI syndrome. Isologous bone marrow cells can be used without the fear of rejection, but they are not available.

A recent study has revealed that transplantation of newborn liver cells (as a source of hematopoietic stem cells) plus newborn thymus markedly improved the survival rate of mice irradiated with a dose that produced GI syndrome.[232] The relevance of this observation in the treatment of irradiated individuals remains uncertain.

The Chernobyl experience in treating irradiated individuals: On April 26, 1986, the world's most serious nuclear accident occurred at the Chernobyl nuclear power station in the former Soviet Union, releasing excessive amounts of radioactive substances into the atmosphere.[233] The care and treatment of individuals receiving high doses of radiation has been discussed in a book.[234] The medical response to this emergency involved five phases: assessment, containment, reduction of radiation exposure, estimation of dose to individuals, and treatment of irradiated individuals.

Assessments were made on important parameters such as amount and type of radioactive material released. Measures to contain the spread of radioactivity were carried out. To limit the radiation exposure to individuals located near the reactor, massive evacuations were conducted. Within 36 hours of the accident, 45,000 persons were evacuated, and two weeks later, 90,000 additional individuals were evacuated.

In order to reduce the dose to exposed individuals, exposed persons were thoroughly washed. In the event of exposure to radioactive iodine (^{131}I), a dose of potassium iodide is recommended as soon as possible in order to block the uptake of ^{131}I by the thyroid gland. An oral administration of potassium iodide 2 hours after exposure to ^{131}I will reduce the uptake by the gland by 80%, and 6 hours after, by about 30%.[235]

Physical dosimeters, such as individual radiation meters or film badges, were of limited value at Chernobyl, because the monitoring devices were either unrecoverable or destroyed by the high radiation doses. Biological dosimeters proved more effective at providing information on exposure levels, but demanded enormous amounts of medical and technical expertise and resources. Furthermore, there were limitations to the accuracy of the doses because of thermal and chemical injuries that have an impact on biological damage.[236]

About 200 people were exposed to large doses of whole-body radiation. One hundred and five received about 1–2 Gy or more, 33 received less than 6 Gy, and 10 received 6 Gy or more. Thirteen persons received doses of 5.6–13.4 Gy that produced GI syndrome and received homologous bone marrow transplants.[236] Most died within a few months. Two transplanted patients who received an estimated dose of 5.6 and 8.7 Gy were surviving 3 years after the accident; they all later died due to host versus graft rejection events. Six irradiated patients received fetal liver transplants. Most died in about 3 months, and two patients who survived 3 years also soon died primarily due to host versus graft rejection events. The studies discussed previously suggest that the current treatment of irradiated patients exposed to doses that produce GI syndrome is not effective; therefore, additional approaches should be developed.

DRUGS AND BIOLOGICS NOT APPROVED BY THE FDA

Cytokines and growth factors: Interleukin-4 (IL-4) injected once or for 5 consecutive days and 2 hours after whole-body irradiation of mice with doses of 7–10 Gy increased the survival of irradiated animals in the presence of poor recovery of the hematopoietic system.[65] The cytokines IL-4 and IL-13 appear to have pleiotropic effects affecting both pathological changes and tissue remodeling. Comparing the effects of IL-13 Ralpha2 (IL-13 receptor alpha2) in irradiated wild-type mice and IL-4 receptor alpha gene-deficient mice revealed that IL-13Ralpha2 plays a major role in regeneration of epithelial cells after irradiation.[237]

Pre- or post-irradiation of mice treated with recombinant murine (rM) granulocyte-macrophage colony-stimulating factor (GM-CSF) or recombinant human (rh) granulocyte-CSF (G-CSF) reduced radiation damage. These growth factors, when injected 20 hours before irradiation or soon after irradiation, increased the survival of irradiated animals.[238] The efficacy of recombinant human interleukin-3 (rhIL-3) and GM-CSF on the peripheral lymphocytes of rhesus monkeys after whole-body irradiation with a 3 Gy dose of gamma rays was evaluated. The results showed that GM-CSF alone or GM-CSF + rhIL-3 reduced radiation-induced apoptosis in peripheral lymphocytes. The combination was more effective than the individual agents.[239] Administration of G-CSF 7 at 14 and 21 days after local liver irradiation initiated endogenous hepatic cell regeneration and reduced hepatic fibrosis in mice.[240] Acidic or basic fibroblast factor (FGF-1 or FGF-2) injected intravenously 24 hours before or 1 hour after whole-body irradiation with a dose of 7–18 Gy reduced radiation-induced apoptosis in crypt stem cells of intestine of mice.[241] These growth factors, such as GM-CSF, can be used in the management of radiation injury associated with GI syndrome in humans.

Nucleic acid derivatives: Post-irradiation treatment with DNA or RNA (3–6 hours after irradiation) increased the survival of irradiated mice from 50% to 85%, an increase of 35%.[1,242] Guanosine or inosine administered IP 15 minutes after whole-body irradiation with a dose of 7 Gy delivered in a single dose produced a DRF value of 1.23 for guanosine and 1.15 for inosine. This effect is mediated by preventing oxidative damage and reducing the generation of reactive oxygen species.[243,244] Thus, these two compounds act as an antioxidant and promoted complete and rapid repair of DNA in mouse leukocytes in vitro.[244] The relevance of these observations in treating radiation exposed humans remains unknown.

Imidazole: Intraperitoneal administration of imidazole immediately after whole-body irradiation increased the survival of irradiated mice from 14% to 42%.[231] It increased the survival of irradiated mice from 14% to 80% when administered IP 5 min before whole-body irradiation.[245] The toxicity of this drug is unknown; therefore, it is not recommended in the management of radiation injury in humans at this time.

RADIOPROTECTIVE DIETARY AND ENDOGENOUS ANTIOXIDANTS NOT REQUIRING APPROVAL BY THE FDA

Since antioxidants neutralize free radicals and inhibit inflammation, and since these two biological events play a central role in the progression of radiation damage after irradiation, the use of multiple antioxidants after irradiation appears to be one of the rational choices for mitigating ARS. Indeed, a recent study suggests that dietary supplementation with a mixture of antioxidants (sodium ascorbate, N-acetylcysteine, alpha-lipoic acid, alpha-tocopheryl succinate, coenzyme Q10, and L-selenomethionine) can mitigate ARS in mice when administered after irradiation; however, the time of initiation of post-irradiation treatment was critical in obtaining optimal protection.[246] Animals were fed diets rich in antioxidants immediately, and at 12, 24, and 48 hours after whole-body irradiation with gamma-ray doses of 8 Gy, and the percent of survival was determined 30 days after irradiation. The results showed that the survival was 0% in irradiated control animals or in those animals receiving an antioxidant-rich diet immediately after irradiation. However, about 29%, 78%, and 20% of irradiated animals receiving antioxidant-rich diets 12, 24, and 48 hours after irradiation survived. These results suggest that initiation of an antioxidant-rich diet 24 hours after irradiation provided optimal radiation protection in mice. The exact reasons for this phenomenon are unknown. However, the authors have suggested that delaying the start of an antioxidant-rich diet allows for the most efficient repair of radiation injury and the highest increase in survival of bone marrow cells. I suggest that immediately after irradiation, pro- and anti-inflammatory cytokines are released in response to cellular injuries. The proportion of anti-inflammatory cytokines responsible for repair may be higher than those of pro-inflammatory cytokines that further damage the irradiated cells. Inhibition of inflammation soon after irradiation may prevent the release of both anti- and pro-inflammatory cytokines, and thereby can prevent the repair of radiation injury. On the other hand, by about 24 hours after irradiation, pro-inflammatory cytokines dominate; therefore, inhibition of inflammation at this time may help to prevent progression of radiation damage. As discussed earlier, a mixture of multiple dietary and endogenous antioxidants administered orally before irradiation increased the survival of irradiated mice from 0% to 40%. It increased the survival time of irradiated sheep, which received a radiation dose that produced GI syndrome, by about fivefold when administered orally daily for 7 days before and daily for 7 days after irradiation.

Based on our preliminary data and published studies on the radioprotective efficacy of individual antioxidants, it is possible to develop a micronutrient formulation containing antioxidants that can reduce ARS in humans when administered after irradiation. Indeed, a commercial preparation of multiple micronutrients containing dietary and endogenous antioxidants referred to as BioShield-R3 is available. The micronutrients present in BioShield-R3 are similar to those present in BioShield-R2, except that the levels of antioxidants are higher than those in BioShield-R2. The additional amounts of antioxidants were added to BioShield-R3 in order to quickly increase the tissue levels of antioxidants.

POTENTIAL CANDIDATES FOR AN IDEAL BIOSHIELD AGAINST ARS

An ideal bioshield must satisfy the following criteria: (1) agents should exhibit radioprotective and radiation mitigating properties against ARS in laboratory experiments (cell culture and animal models), (2) they should show at least some evidence that they can protect human normal tissues against radiation damage, and (3) they must be safe for humans when consumed orally on a short- and long-term basis.

Although many pharmacological agents not approved by the FDA and some that have been approved by the FDA prevented or mitigated ARS in animal models, only a few of them prevented as well as mitigated ARS in humans. They are described here.

Pharmacological agents not approved by the FDA: Agents that satisfy only the first criterion of an ideal bioshield for preventing and mitigating ARS include the following:

1. Kinase inhibitors: Ex-Rad and SB203580 (SB)
2. Mitochondria targeting agents
3. Agonists of Toll-like receptor (TLR): CBLB 502 and CBLB 600
4. Agonists of thrombopoietin: Alxn4100TPO (4100TPO)
5. Cytokines

The toxicity of these agents at radioprotective and radiation mitigating doses in humans is unknown; therefore, they cannot be recommended at this time. In addition, these agents will require FDA approval before any one of them can be recommended for human use; therefore, they cannot be recommended for radiation prevention or mitigation in humans at this time.

Drugs approved by the FDA for other indications: The FDA drugs approved for specific indications in humans that meet only the first criterion of an ideal bioshield for preventing and mitigating ARS include the following:

1. Angiotensin converting enzyme (ACE) inhibitors: captopril, enlapril, and fosinopril
2. Statins: lovastatin and pravastatin
3. Inhibitors of histone deacetylase: valporic acid

The doses used for the treatment of human diseases are well defined and relatively safe. It is not known whether they, at similar doses, can protect against radiation damage in humans. It is also unknown if at radioprotective doses they would remain safe on a short- and long-term basis. Therefore, these drugs cannot be recommended for radiation prevention or mitigation in humans at this time.

Herbs not requiring approval by the FDA: Several herbal extracts described in this chapter have been used as a traditional medicine for centuries in the treatment of varieties of human diseases in the Asian subcontinent. These herbs exhibit antioxidant and anti-inflammation properties. They satisfy only the first criterion of an ideal bioshield in preventing ARS in animal models. It is not known whether they can protect against ARS in humans. It is also unknown if at radioprotective doses they would remain safe on a short- and long-term basis. Therefore, herbal extracts cannot be recommended for radiation prevention or mitigation of ARS in humans at this time.

Antioxidants not requiring approval of the FDA: The dietary supplements BioShield-R2 and BioShield-R3 contain multiple dietary and endogenous antioxidants that satisfy all three criteria of an ideal bioshield against ARS. Therefore, they can be considered as an ideal bioshield that can be recommended for prevention and/or mitigation of ARS in humans. The adoption of this recommendation must be done in consultation with physicians or knowledgeable health professionals.

PROPOSED RECOMMENDATIONS FOR USING A MICRONUTRIENT PREPARATION (BIOSHIELD-R3) IN COMBINATION WITH STANDARD THERAPY FOR MITIGATING ARS IN HUMANS

Standard therapy includes replacement therapy (antibiotics, and blood and electrolyte transfusion when indicated) and should be provided to all individuals exposed to lethal doses of radiation causing bone marrow syndrome. Those receiving doses that produce GI syndrome should receive replacement therapy as well as certain FDA-approved growth factors, such as G-CSF. However, it is recommended to start consuming BioShield-R3 twice a day 24 hours after irradiation and continue daily twice a day for 60 days in order to mitigate ARS associated with bone marrow syndrome. The post-irradiation period of 24 hours for the start of BioShield-R3 treatment was selected because immediately after irradiation, pro- and anti-inflammatory cytokines are released in response to cellular injuries. During this period, the proportion of anti-inflammatory cytokines responsible for repair of radiation injury may be higher than those of pro-inflammatory cytokines that further damage the cells. Inhibition of inflammation by high-dose antioxidants soon after irradiation may prevent the release of both anti- and pro-inflammatory cytokines, and thereby prevent the repair of radiation injury. On the other hand, by about 24 hours after irradiation, pro-inflammatory cytokines may dominate; therefore, inhibition of inflammation at that time may help to prevent progression of radiation damage. The antioxidant treatment period of 60 days was selected because humans exposed to doses that produce bone marrow syndrome will die within this period, and those who are exposed to doses that produce GI syndrome will die within 14 days. The study using BioShield-R3 in combination with standard therapy should be initiated to determine the efficacy of this strategy on the criteria of survival rate and survival time in animals irradiated with doses of radiation that produce bone marrow syndrome and GI syndrome. It is possible that the combination of BioShield-R3, replacement therapy, and (G-CSF) when indicated after irradiation may mitigate ARS in irradiated animals or humans exhibiting bone marrow syndrome as well as GI syndrome.

Since survivors of high doses of radiation will have increased risk of neoplastic and nonneoplastic diseases, it is recommended that after treatment with BioShield-R3 for a period of 60 days, BioShield-R2 can be consumed twice daily for the remainder of lifespan in order to reduce the risk of late adverse health effects of radiation. The advantage of combining BioShield with the replacement therapy and growth factor is that continued treatment with micronutrients may also reduce the risk of late adverse health effects (neoplastic and nonneoplastic diseases) among survivors of ARS. Meanwhile, those individuals who are at risk of receiving high doses of radiation, and those who are already exposed to these doses of radiation, may like to adopt the proposed recommendations in consultation with their physicians in order to prevent or mitigate ARS as well as to reduce the risk of late adverse health effects of radiation.

SUMMARY

Acute radiation damage, also referred to as acute radiation sickness (ARS), is produced by high doses of ionizing radiation, which can cause mortality ranging from a few to 100% of irradiated individuals. ARS can be associated with bone marrow syndrome as well as GI syndrome. Prevention generally refers to devices, biologics, or chemicals that can reduce ARS when administered before irradiation, and agents that can accomplish it are referred to as radioprotective agents. Mitigation refers to chemicals or biologics that can reduce ARS when administered after irradiation, and agents that can accomplish it are referred to as radiation mitigating agents. At present, prevention of radiation damage is primarily based on reducing the dose levels by adopting three physical principles: increasing the distance from the radiation source, reducing the exposure time, and shielding. Because of several limitations of the physical principles, the concept of a bioshield, which refers to tissue protection and includes both radioprotective and radiation mitigating agents, is proposed.

Since World War II, many radioprotective and radiation mitigating agents have been identified using cell culture and animal models (primarily rats and mice). Radioprotective agents were divided into four major groups (1) radioprotective pharmacological agents not approved by the FDA, (2) radioprotective drugs approved by the FDA for other indications in humans, (3) radioprotective herbal extracts not requiring FDA approval, and (4) radioprotective dietary and endogenous antioxidants not requiring FDA approval.

Radioprotective pharmacological agents not approved by the FDA include thiols (cysteine, cysteamine, aminoethylisothiourea dihydrobromide, 2-mercaptoethylguanidine, and amifostine), other sulfur-containing compounds, alcohol, dopamine, histamine, serotonin, hormones, sodium cyanide, nucleic acid derivatives, sodium fluoroacetate, para-aminopropiophenone, melittin, imidazole, adenosine 3',5'-cyclic monophosphate inhibitors of cyclic nucleotide phosphodiesterase, and prostaglandin E1, a stimulator of adenylate cyclase, tumor suppressor gene p53, kinase inhibitors (Ex-Rad and SB203580), inhibitors of inflammation (Minozac and ethyl pyruvate), mitochondrial targeting agents (JP4-309), Toll-like receptor agonists (CBLB 502 and CBLB 600), agonists of thrombopoietin (Alnx41100TPO), cytokines and growth factors (IL-1, IL-3, Il-4, IL-11, IL-12), TNF-alpha, hepatocyte growth factor and recombination human granulocyte colony-stimulating factor (rhG-CSF), and analogs of somatostatin (SOM230 and SOM230-LAR). In addition to these radioprotective pharmacological agents, administration of typhoid-paratyphoid vaccine and anti-radiation vaccine also protected animals against ARS. These studies have been performed primarily on rodents.

Radioprotective drugs approved by the FDA for other indications in humans include amifostine, diltiazem, ACE inhibitors (captopril, enalpril, and fosinopril), statins (lovastatin and pravastatin), analgesics (morphine and sodium salicylate), tranquilizers (reserpine), epinephrine, norepinephrine, and inhibitors of histone deacetylase (valporic acid and phenylbutyrate). These studies have been performed primarily on rodents.

Radioprotective herbal extracts not requiring FDA approval include *Nigella sativa* (EE-NS), *Acorus calamus* (EE-AC), *Aloe vera, Mentha piperita* (LE-MP),

Tinospora cordifolia (TC), *Podophyllum hexandrum* (Himalayan mayapple), *Vernonia cinerea* (VC), *Centella asiatica* (CA), *Ginkgo biloba* (GB), *Hippophae rhamnoides* (HR), *Ocimum sanctum* (OC), *Panax ginseng* (PG), tea polyphenols and epigallocatechin, *Ecklonia cava* (EC), *Psidium guajava* (common guava), other herbs, curcumin, resveratrol, flaxseed, quercetin, and genistein. These studies have been performed primarily on rodents.

Radioprotective antioxidants include dietary antioxidants such as vitamin A, vitamin C, beta-carotene, lycopene, alpha-tocopherol and alpha-tocopheryl succinate, tocotrienol, selenium, and endogenous antioxidants such as antioxidant enzymes, primarily superoxide dismutase (SOD), glutathione and glutathione-elevating agents (n-acetylcysteine, alpha-lipoic acid), co-enzyme Q10, and L-carnitine, and reduce ARS. Most studies have been performed in rodents, but a few have also been performed in humans. The rationale for using multiple antioxidants rather than the individual antioxidant in reducing ARS has been presented. In addition, the rationale for using the oral route of administration before and after irradiation has been discussed.

The radiation mitigating agents are grouped into two broad categories: (1) FDA-approved drugs and biologics approved by the FDA, and (2) drugs and biologics not approved by the FDA.

FDA-approved drugs and biologics include replacement therapy (antibiotics and replacement of fluids, including blood transfusion when indicated), erythropoietin, cytokines, and growth factors, and bone marrow transplant when needed. Drugs and biologics not approved by the FDA include nucleic acid derivatives, imidazole, and antioxidants not requiring approval by the FDA.

Agents that satisfy only the first criterion of an ideal bioshield for radiation protection include kinase inhibitors (Ex-Rad and SB203580 [SB]), mitochondria targeting agents, Toll-like receptor agonists (CBLB 502 and CBLB 600), agonists of thrombopoietin (Alxn4100TPO (4100TPO), and cytokines. The toxicity of these agents at radioprotective and radiation mitigating doses in humans is unknown; therefore, they cannot be recommended at this time.

The FDA drugs approved for specific indications in humans that meet only the first criterion of an ideal bioshield for radiation protection include ACE inhibitors (captopril, enlapril, and fosinopril), statins (lovastatin and pravastatin), and inhibitors of histone deacetylase (valporic acid and phenylbutyrate). It is not known whether they, at doses similar to those recommended for human use for other specific indications, can protect against ARS in humans. It is also unknown if at radioprotective or radiation mitigating doses they would remain safe on a short- and long-term basis. Therefore, these drugs cannot be recommended for radiation prevention or mitigation in humans at this time.

Radioprotective herbal extracts satisfy only the first criterion of an ideal bioshield against radiation damage. They are considered relatively safe in humans at least for a short period of time. It is not known whether they can protect against ARS in humans. It is also unknown if at radioprotective doses they would remain safe on a short- and long-term basis. Therefore, herbal extracts cannot be recommended for radiation prevention or mitigation of ARS in humans at this time.

Dietary (vitamin A, beta-carotene, vitamin C, vitamin E, vitamin D, and mineral selenium) and endogenous (glutathione elevating agents, such as n-acetylcysteine and alpha-lipoic acid, L-carnitine, and coenzyme Q10) antioxidants satisfy all three criteria of an ideal bioshield against radiation injury. Therefore, they can be considered as an ideal bioshield that can be recommended for prevention and mitigation of ARS in humans. A micronutrient preparation containing high doses of dietary and endogenous antioxidants and their derivatives, vitamin D, all B vitamins, mineral selenium, and zinc, but no trace minerals (iron, copper, or manganese) or heavy metals (zirconium and molybdenum), meets the criteria of an ideal bioshield. Trace minerals, such as iron, copper, and manganese interact with vitamin C to generate an excessive amount of free radicals. The excretion of these trace minerals or heavy metals is negligible; therefore, these minerals and heavy metals may accumulate in excessive amounts in the body after long-term consumption. Excessive levels of heavy metals could be neurotoxic. A separate micronutrient preparation referred to as BioShield-R2 and BioShield-R3, which can act as a bioshield for prevention and mitigation of ARS, is available commercially. BioShield-R2 is recommended to consume orally twice a day before irradiation and BioShield-R3 is recommended to start consuming 24 hours after irradiation and continue daily for 60 days in combination with standard therapy (replacement therapy with antibiotics, electrolytes, blood transfusion when indicated, and G-CSF when needed in order to prevent and mitigate ARS associated with bone marrow syndrome and GI syndrome.

After 60 days of treatment with BioShield-R3, BioShield-R2 should be consumed orally twice a day for the remainder of the patient's lifespan. The continuation of BioShield-R2 is recommended in order to reduce the long-term adverse health effects of radiation among survivors of ARS.

REFERENCES

1. Thomson, J., *Radiation Protection in Mammals*, Reinhold, New York, 1962.
2. Weiss, J. F., and Landauer, M. R., History and development of radiation-protective agents, *Int J Radiat Biol* 85 (7), 539–573, 2009.
3. Kumagai, J., Masui, K., Itagaki, Y., Shiotani, M., Kodama, S., Watanabe, M., and Miyazaki, T., Long-lived mutagenic radicals induced in mammalian cells by ionizing radiation are mainly localized to proteins, *Radiat Res* 160 (1), 95–102, 2003.
4. Waldren, C. A., Vannais, D. B., and Ueno, A. M., A role for long-lived radicals (LLR) in radiation-induced mutation and persistent chromosomal instability: Counteraction by ascorbate and RibCys but not DMSO, *Mutat Res* 551 (1–2), 255–265, 2004.
5. Abate, A., Yang, G., Dennery, P. A., Oberle, S., and Schroder, H., Synergistic inhibition of cyclooxygenase-2 expression by vitamin E and aspirin, *Free Radic Biol Med* 29 (11), 1135–1142, 2000.
6. Fedorocko, P., Egyed, A., and Vacek, A., Irradiation induces increased production of haemopoietic and proinflammatory cytokines in the mouse lung, *Int J Radiat Biol* 78 (4), 305–313, 2002.
7. Mizutani, N., Fujikura, Y., Wang, Y. H., Tamechika, M., Tokuda, N., Sawada, T., and Fukumoto, T., Inflammatory and anti-inflammatory cytokines regulate the recovery from sublethal X irradiation in rat thymus, *Radiat Res* 157 (3), 281–289, 2002.
8. Popp, W., Plappert, U., Muller, W. U., Rehn, B., Schneider, J., Braun, A., Bauer, P. C., Vahrenholz, C., Presek, P., Brauksiepe, A., Enderle, G., Wust, T., Bruch, J., Fliedner, T. M., Konietzko, N., Streffer, C., Woitowitz, H. J., and Norpoth, K., Biomarkers of genetic damage and inflammation in blood and bronchoalveolar lavage fluid among former German uranium miners: A pilot study, *Radiat Environ Biophys* 39 (4), 275–282, 2000.

9. Prasad, K., *Handbook of Radiobiology,* 2nd Ed. CRC Press, Boca Raton, FL, 1995.

10. Bacq, Z. M., and Alexander, P., *Fundamentals of Radiobiology*, Pergamon Press, Elmsford, NY, 1961.

11. Capizzi, R. L., and Oster, W., Chemoprotective and radioprotective effects of amifostine: An update of clinical trials, *Int J Hematol* 72 (4), 425–435, 2000.

12. Allalunis-Turner, M. J., Walden, T. L. Jr., and Sawich, C., Induction of marrow hypoxia by radioprotective agents, *Radiat Res* 118 (3), 581–586, 1989.

13. Anne, P. R., Phase II trial of subcutaneous amifostine in patients undergoing radiation therapy for head and neck cancer, *Semin Oncol* 29 (6 Suppl 19), 80–83, 2002.

14. Weiss, J. F., and Landauer, M. R., Protection against ionizing radiation by antioxidant nutrients and phytochemicals, *Toxicology* 189 (1–2), 1–20, 2003.

15. Devik, F., and Lothe, F., The effect of cysteamine, cystamine and hypoxia on mortality and bone marrow chromosome aberrations in mice after total body roentgen irradiation, *Acta Radiol* 44 (3), 243–248, 1955.

16. Roberts, J. C., Koch, K. E., Detrick, S. R., Warters, R. L., and Lubec, G., Thiazolidine prodrugs of cysteamine and cysteine as radioprotective agents, *Radiat Res* 143 (2), 203–213, 1995.

17. Court Brown, W. M., A clinical trial of cysteamine (beta-mercaptoethyamine) in radiation sickness, *Br J Radiol* 28, 325–326, 1955.

18. Crouch, B. G., and Overman, R. R., Chemical protection against x-radiation death in primates: A preliminary report, *Science* 125 (3257), 1092, 1957.

19. Benson, R. E., Michaelson, S. M., Downs, W. L., Maynard, E. A., Scott, J. K., Hodge, H. C., and Howland, J. W., Toxicological and radioprotection studies on S,beta-aminoethylisothiuronium bromide (AET), *Radiat Res* 15, 561–572, 1961.

20. Condit, P. T., Levy, A. H., Van Scott, E. J., and Andrews, J. R., Some effects of beta- aminoethylisothiourea bromide (AET) in man, *J Pharmacol Exp Ther* 122, 13–23, 1958.

21. Cronkite, E. P., and Bond, V. P., *Radiation Injury in Man*, Charles C. Thomas, Springfield, IL, 1960.

22. Vasilescu, D., Broch, H., and Hamza, A., Quantum molecular simulation of the radioprotection by the aminothiol WR-1065, active metabolite of amifostine (WR-2721) modeling the hydroxyl scavenging process, *J. Mol. Struc. (Thoechem)* 538, 133–144, 2001.

23. Kataoka, Y., Murley, J. S., Baker, K. L., and Grdina, D. J., Relationship between phosphorylated histone H2AX formation and cell survival in human microvascular endothelial cells (HMEC) as a function of ionizing radiation exposure in the presence or absence of thiol-containing drugs, *Radiat Res* 168 (1), 106–114, 2007.

24. Prasad, K. N., and Van Woert, M. H., Dopamine protects mice against whole-body irradiation, *Science* 155 (761), 470–472, 1967.

25. Prasad, K. N., and Van Woert, M. H., Radioprotective action of dihydroxyphenylethylamine (Dopamine) on whole-body x-irradiated rats, *Radiat Res* 37 (2), 305–315, 1969.

26. Prasad, K. N., and Vanwoert, M. H., Effect of dopamine on DNA x-irradiated in vitro, *Int J Radiat Biol* 14, 79–82, 1968.

27. Prasad, K. N., and Kollmorgen, G. M., Effect of dihydroxyphenylethylamine (dopamine) on mammalian cells in vivo and in vitro, *Proc Soc Exp Biol Med* 132 (2), 426–430, 1969.

28. Sakamoto, A., and Prasad, K. N., Radioprotective effect of beta-melanocyte-stimulating hormone (MSH) in rodents exposed to whole-body x-irradiation, *Int J Radiat Biol* 12, 97–100, 1967.

29. Stickney, D. R., Dowding, C., Garsd, A., Ahlem, C., Whitnall, M., McKeon, M., Reading, C., and Frincke, J., 5-androstenediol stimulates multilineage hematopoiesis in rhesus monkeys with radiation-induced myelosuppression, *Int Immunopharmacol* 6 (11), 1706–13, 2006.

30. Yuhas, J. M., and Storer, J. B., Chemoprotection against three modes of radiation death in the mouse, *Int J. Radiat Biol* 15, 233–237, 1969.

31. Ginsberg, N. J., Dauer, M., and Slotta, K. H., Melittin used as a protective agent against x-irradiation, *Nature* 220 (5174), 1334, 1968.

32. Brown, A. P., Chung, E. J., Urick, M. E., Shield, W. P. III, Sowers, A. L., Thetford, A., Shankavaram, U. T., Mitchell, J. B., and Citrin, D. E., Evaluation of the fullerene compound DF-1 as a radiation protector, *Radiat Oncol* 5, 34, 2010.

33. Theriot, C. A., Casey, R., Coyers, J., and Wu, H., Effectiveness of Df-1, a nontoxic carbon fullerene based antioxidant, as a biomedical countermeasure against radiation, *Radiat Res* 56th Annual Meeting Radiation Research Society, Maui, Hawaii, September 25–29, 2010.

34. Prasad, K. N., Radioprotective effect of prostaglandin and an inhibitor of cyclic nucleotide phosphodiesterase on mammalian cells in culture, *Int J Radiat Biol* 22, 187–189, 1972.

35. Lehnert, S., Radioprotection of mice intestine by inhibitors of cyclic AMP phosphodiesterase, *Int J Radiat Oncol Biol Biophys* 5, 825–833, 1979.

36. Dubravsky, N. B., Hunter, N., Mason, K., and Withers, H. R., Dibutryl cyclic adenosine monophosphate: effect on radiosensitivity of tumors and normal tissues in mice, *Radiology* 126 (3), 799–802, 1978.

37. Nettesheim, P., Hanna, M. G. Jr., and Fisher, W. D., Further studies on the effect of serum alpha macroglobulin on regeneration of hemopoietic tissue after x-irradiation, *Radiat Res* 35 (2), 379–389, 1968.

38. Bogojevic, D., Poznanovic, G., Grdovic, N., Grigorov, I., Vidakovic, M., Dinic, S., and Mihailovic, M., Administration of rat acute-phase protein alpha(2)-macroglobulin before total-body irradiation initiates cytoprotective mechanisms in the liver, *Radiat Environ Biophys* 50 (1), 167–179, 2011.

39. Mihailovic, M., Dobric, S., Poznanovic, G., Petrovic, M., Uskokovic, A., Arambasic, J., and Bogojevic, B., The acute-phase protein alpha2-macroglobulin plays an important role in radioprotection in the rat, *Shock* 31 (6), 607–614, 2009.

40. An, Y. S., Lee, E., Kang, M. H., Hong, H. S., Kim, M. R., Jang, W. S., Son, Y., and Yi, J. Y., Substance P stimulates the recovery of bone marrow after the irradiation, *J Cell Physiol* 226 (5), 1204–1213, 2011.

41. Komarova, E. A., Kondratov, R. V., Wang, K., Christov, K., Golovkina, T. V., Goldblum, J. R., and Gudkov, A. V., Dual effect of p53 on radiation sensitivity in vivo: p53 promotes hematopoietic injury, but protects from gastro-intestinal syndrome in mice, *Oncogene* 23 (19), 3265–3271, 2004.

42. Lee, C.-L., Kim, Y., Sullivan, J. M., Jeffords, L., Lowe, S. W., and Kirsch, D. G., Gentic dissection of the temporal role of p53 in regulating radiation-induced carcinogenesis, *Radiat Res* 56th Annual Meeting Radiation Research Society, Maui, Hawaii, September 25–29, 2010.

43. Fukomoto, R., Kiang, J. G., Inhibition of 53 activation and its interaction with Hsp90 limits apoptosis after ionizing irradiation in human peripheral blood cells, *Radiat Res,* 56th Annual Meeting Radiation Research Society, Maui, Hawaii, September 25–29, 2010.

44. McConnell, K. W., Muenzer, J. T., Chang, K. C., Davis, C. G., McDunn, J. E., Coopersmith, C. M., Hilliard, C. A., Hotchkiss, R. S., Grigsby, P. W., and Hunt, C. R., Anti-apoptotic peptides protect against radiation-induced cell death, *Biochem Biophys Res Commun* 355 (2), 501–507, 2007.

45. Kumar, R., Radioprotection and radiomitigation properties of Ex-RAD, *Radiat Res* 56th Annual Meeting Radiation Research Society, Maui, Hawaii, September 25–29, 2010.

46. Ghosh, S. P., Kulkarni, S., Perkins, M. W., Gambles, K., Heiber, K., Maniar, M., Seed, T. W., and Kumar, K. C., Recovery from radiation-induced hematopoietic and gastrointestinal sub-syndromes by Ex-RAD in murine model, *Radiat Res* 56th Annual Meeting Radiation Research Society, Maui, Hawaii, September 25–29, 2010.

47. Kang, A. D., Coseza, S. C., Reddy, M. V. R., and Reddy, E. P., Radioprotection of human bone marrow by ON01210.Na (ExRADTM) through AKT mediated signaling pathway, *Radiat Res* 56th Annual Meeting Radiation Research Society, Maui, Hawaii, September 25–29, 2010.

48. Li, D., Wang, Y., Wu, H., Lu, L., Zhang, H., Chang, J., Zhai, Z., Zhang, J., Wang, Y., Zhou, D., and Meng, A., Protection of mice against irradiation injuries by the post-irradiation combined administration of P38 inhibitor and G-CSF, *Radiat Res* 56th Annual Meeting Radiation Research Society, Maui, Hawaii, September 25–29, 2010.

49. Jenrow, K. A., Brown, S. L., Lapanowski, K., and Kim. J. H., Minozac mitigates rat brain injury following whole brain irradiation, *Radiat Res* 56th Annual Meeting Radiation Research Society, Maui, Hawaii, September 25–29, 2010.

50. Epperly, M., Jin, S., Nie, S., Cao, S., Zhang, X., Franicola, D., Wang, H., Fink, M. P., and Greenberger, J. S., Ethyl pyruvate, a potentially effective mitigator of damage after total-body irradiation, *Radiat Res* 168 (5), 552–529, 2007.

51. Pecaut, M. J., Bayeta, E. Pan, C., Perez, C. P., and Gridley, D. S., Does ethyl pyruvate act as a radioprotectant in macrophage cell lines?, *Radiat Res* 56th Annual Meeting Radiation Research Society, Maui, Hawaii, September 25–29, 2010.

52. Zhang, X., Greenberger, J., Dixon, T., Franicola, D., Wipf, P., and Epperly, M., Mitochondrial targeting peptide isostere bound 4-amino-tempo (JP4-039) mitigates against the ionizing irradiation-induced hematopoietic syndrome, *Radiat Res* 56th Annual Meeting Radiation Research Society, Maui, Hawaii, September 25–29, 2010.

53. Atkinson, J., Kapralov, A., Huang, Z., Belikova, N. A., Yanamala, N., Jiang, J., Klein-Seetharaman, J., Epperly, M., Stoyanovsky, D. A., and Greenberger, J. S., Mitochondria-targeted ligands of hene-iron in cytochrome c as novel radioprotectors/radiomitigators, *Radiat Res* 56th Annual Meeting Radiation Research Society, Maui, Hawaii, September 25–29, 2010.

54. Huang, Z., Jiang, J., Belikova, N. A., Stoyanovsky, D. A., Kagan, V. E., and Mintz, A. H., Protection of normal brain cells from gamma-irradiation-induced apoptosis by a mitochondria-targeted triphenyl-phosphonium-nitroxide: A possible utility in glioblastoma therapy, *J Neurooncol* 100 (1), 1–8, 2010.

55. Jiang, J., Belikova, N. A., Hoye, A. T., Zhao, Q., Epperly, M. W., Greenberger, J. S., Wipf, P., and Kagan, V. E., A mitochondria-targeted nitroxide/hemigramicidin S conjugate protects mouse embryonic cells against gamma irradiation, *Int J Radiat Oncol Biol Phys* 70 (3), 816–825, 2008.

56. Krivokrysenko, V., Toshkov, I., Gleiberman, A., Gudkov, A., and Feinstein, E., Single injection of novel medical radiation countermeasure CBLB502 rescues non-human primates within broad time window after lethal irradiation, *Radiat Res* 56th Annual Meeting Radiation Research Society, Maui, Hawaii, September 25–29, 2010.

57. Burdelya, L., Tallant, T., Aygun-Sunar, S., Kojouharov, B., Haderski, G., DiDonato, J., and Gudkov, A., The role of NF-KB-activation in radioprotection by flagellin in mice, *Radiat Res* 56th Annual Meeting Radiation Research Society, Maui, Hawaii, September 25–29, 2010.

58. Shakhov, A., Bone, F., Kononov, E., Cheney, A., Krasnov, P., Toshkova, T., Shakhova, V., Singh, V., and Feinstein, E. E., CBLB600: A novel class of radioprotective drugs acting via activation of TLR2 receptor complexes, *Radiat Res* 56th Annual Meeting Radiation Research Society, Maui, Hawaii, September 25–29, 2010.

59. Satayamitra, M. M., Lombardini, E., Mullaney, C., Graves, J. III, Harrison, L., Johnson, K., Hunter, J., Tamburini, P., Wang, Y., Springhorn, J. P., and Srinivasan, V., Alxn4100TPO, a tpo agonist, ameliorates radiation-induced injury by stimulating proliferation and differentiation of multi-lineage hematopoietic progenitors, *Radiat Res* 56th Annual Meeting Radiation Research Society, Maui, Hawaii, September 25–29, 2010.

60. Ellefson, D. D., Gallaher, T., Miller, J., and Basile, L., Mitigation of acute radiation injury by interleukin-12, *Radiat Res* 56th Annual Meeting Radiation Research Society, Maui, Hawaii, September 25–29, 2010.

61. Farese, A. M., Herodin, F., McKearn, J. P., Baum, C., Burton, E., and MacVittie, T. J., Acceleration of hematopoietic reconstitution with a synthetic cytokine (SC-55494) after radiation-induced bone marrow aplasia, *Blood* 87 (2), 581–591, 1996.

62. MacVittie, T. J., Farese, A. M., Herodin, F., Grab, L. B., Baum, C. M., and McKearn, J. P., Combination therapy for radiation-induced bone marrow aplasia in nonhuman primates using synthokine SC-55494 and recombinant human granulocyte colony-stimulating factor, *Blood* 87 (10), 4129–4135, 1996.

63. Neta, R., Oppenheim, J. J., and Douches, S. D., Interdependence of the radioprotective effects of human recombinant interleukin 1 alpha, tumor necrosis factor alpha, granulocyte colony-stimulating factor, and murine recombinant granulocyte-macrophage colony-stimulating factor, *J Immunol* 140 (1), 108–111, 1988.

64. Uemura, T., Nakayama, T., Kusaba, T., Yakata, Y., Yamazumi, K., Matsuu-Matsuyama, M., Shichijo, K., and Sekine, I., The protective effect of interleukin-11 on the cell death induced by x-ray irradiation in cultured intestinal epithelial cell, *J Radiat Res (Tokyo)* 48 (2), 171–177, 2007.

65. Van der Meeren, A., Gaugler, M. H., Mouthon, M. A., Squiban, C., and Gourmelon, P., Interleukin 4 promotes survival of lethally irradiated mice in the absence of hematopoietic efficacy, *Radiat Res* 152 (6), 629–636, 1999.

66. Van der Meeren, A., Mouthon, M. A., Gaugler, M. H., Vandamme, M., and Gourmelon, P., Administration of recombinant human IL11 after supralethal radiation exposure promotes survival in mice: Interactive effect with thrombopoietin, *Radiat Res* 157 (6), 642–649, 2002.

67. Hu, S. Y., Duan, H. F., Li, Q. F., Yang, Y. F., Chen, J. L., Wang, L. S., and Wang, H., Hepatocyte growth factor protects endothelial cells against gamma ray irradiation-induced damage, *Acta Pharmacol Sin* 30 (10), 1415–1420, 2009.

68. Wang, W., Fu, Q., Biju, P., Garg, S., Schmid, H., and Hauer-Jensen, M., Efficacy of "single-shot" SOM230-LAR as a mitigator of total-body irradiation-induced lethality: Comparison with SOM230, *Radiat Res* 56th Annual Meeting Radiation Research Society, Maui, Hawaii, September 25–29, 2010.

69. Zenk, J. L., New therapy for the prevention and prophylactic treatment of acute radiation syndrome, *Expert Opin Investig Drugs* 16 (6), 767–770, 2007.

70. Smith, W. W., Alderman, I. M., and Gillespie, R. E., Increased survival in irradiated animals treated with bacterial endotoxins, *Am J Physiol* 191 (1), 124–130, 1957.

71. Maliev, V. P., Jones, J. A., and Casey, R. C., Mechanisms of action of anti-radiation vaccine in reducing the biological impact of high-dose gamma-irradiation, *J Adv Space Res* 40, 586–590, 2007.

72. Bardet, E., Martin, L., Calais, G., Alfonsi, M., Feham, N. E., Tuchais, C., Boisselier, P., Dessard-Diana, B., Seng, S. H., Garaud, P., Auperin, A., and Bourhis, J., Subcutaneous compared with intravenous administration of amifostine in patients with head and neck cancer receiving radiotherapy: Final results of the GORTEC, 2000–02 phase III randomized trial, *J Clin Oncol* 29 (2), 127–133, 2011.

73. Verhey, L. J., and Sedlacek, R., Determination of the radioprotective effects of topical applications of MEA, WR-2721, and N-acetylcysteine on murine skin, *Radiat Res* 93 (1), 175–183, 1983.

74. Movsas, B., Scott, C., Langer, C., Werner-Wasik, M., Nicolaou, N., Komaki, R., Machtay, M., Smith, C., Axelrod, R., Sarna, L., Wasserman, T., and Byhardt, R., Randomized trial of amifostine in locally advanced non-small-cell lung cancer patients receiving chemotherapy and hyperfractionated radiation: Radiation therapy oncology group trial 98-01, *J Clin Oncol* 23 (10), 2145–2154, 2005.

75. Bogo, V., Jacobs, A. J., and Weiss, J. F., Behavioral toxicity and efficacy of WR-2721 as a radioprotectant, *Radiat Res* 104 (2 Pt 1), 182–190, 1985.
76. Valeyrie-Allanore, L., Poulalhon, N., Fagot, J. P., Sekula, P., Davidovici, B., Sidoroff, A., and Mockenhaupt, M., Stevens-Johnson syndrome and toxic epidermal necrolysis induced by amifostine during head and neck radiotherapy, *Radiother Oncol* 87 (2), 300–303, 2008.
77. Demiral, A. N., Yerebakan, O., Simsir, V., and Alpsoy, E., Amifostine-induced toxic epidermal necrolysis during radiotherapy: A case report, *Jpn J Clin Oncol* 32 (11), 477–479, 2002.
78. Church, M. W., Blakley, B. W., Burgio, D. L., and Gupta, A. K., WR-2721 (Amifostine) ameliorates cisplatin-induced hearing loss but causes neurotoxicity in hamsters: Dose-dependent effects, *J Assoc Res Otolaryngol* 5 (3), 227–237, 2004.
79. Glover, D., Riley, L., Carmichael, K., Spar, B., Glick, J., Kligerman, M. M., Agus, Z. S., Slatopolsky, E., Attie, M., and Goldfarb, S., Hypocalcemia and inhibition of parathyroid hormone secretion after administration of WR-2721 (a radioprotective and chemoprotective agent), *N Engl J Med* 309 (19), 1137–1141, 1983.
80. Nunia, V., Sancheti, G., and Goyal, P. K., Protection of Swiss albino mice against whole-body gamma irradiation by diltiazem, *Br J Radiol* 80 (950), 77–84, 2007.
81. Medhora, M., Ghosh, S. N., Fish, B. L., Gao, F., Gao, Y., Kma, L., Gruenloh, S., Narayanan, J., Jacobs, E. R., and Moulder, J. E., Efficacy of structurally-different angiotensin converting enzyme (ACE) inhibitors as mitigators of radiation pneumonitis, *Radiat Res* 56th Annual Meeting Radiation Research Society, Maui, Hawaii, September 25–29, 2010.
82. Moulder, J. E., Cohen, E. P., and Fish, B. L., Captopril and losartan for mitigation of renal injury caused by single-dose total-body irradiation, *Radiat Res* 175 (1), 29–36, 2011.
83. Jacobs, E. R., Kma L., Gao, F., Ghosh, S. N., Fish, B. L., Narayanan, J., Moulder, J. E., and Medhora, M., Angiotensin converting enzyme (ACE) inhibitors mitigate pulmonary fibrosis in rats, *Radiat Res* 56th Annual Meeting Radiation Research Society, Maui, Hawaii, September 25–29, 2010.
84. Fish, B. L., Moulder, J. E., Mader, M. M., Schock, A. M., and Medhora, M. M., Understanding the mechanisms of mitigation of radiation-induced neuropathy by ACE inhibitors, *Radiat Res* 56th Annual Meeting Radiation Research Society, Maui, Hawaii, September 25–29, 2010.
85. Molthen, R. C., Wu, Q. P., Fish, B. L., Jacobs, E. R., Moulder, J. E., Doctrow, S. R., and Medhora, M., Mitigation of radiation-induced pneumonitis and pulmonary vascular injury using combined therapy: Angiotensin converting enzyme inhibitors and super-oxide dismutase (SOD) mimetics, *Radiat Res* 56th Annual Meeting Radiation Research Society, Maui, Hawaii, September 25–29, 2010.
86. Meeks, C. M., Hill, C. K., Espinoza, T., Roda, N., Louie, S. L., diZerga, G. S., and Rodgers, K. E., Recovery from radiation-induced thrombocytopenia by angiotensin 1-7, *Radiat Res* 56th Annual Meeting Radiation Research Society, Maui, Hawaii, September 25–29, 2010.
87. Doerr, W., Buettner, S., Wildermuth, C., and Schmidt, M., Effect of Lovastatin on oral mucositis (mouse) after single dose or during daily fractionated irradiation, *Radiat Res* 56th Annual Meeting Radiation Research Society, Maui, Hawaii, September 25–29, 2010.
88. Ostrau, C., Hulsenbeck, J., Herzog, M., Schad, A., Torzewski, M., Lackner, K. J., and Fritz, G., Lovastatin attenuates ionizing radiation-induced normal tissue damage in vivo, *Radiother Oncol* 92 (3), 492–499, 2009.
89. Haydont, V., Gilliot, O., Rivera, S., Bourgier, C., Francois, A., Aigueperse, J., Bourhis, J., and Vozenin-Brotons, M. C., Successful mitigation of delayed intestinal radiation injury using pravastatin is not associated with acute injury improvement or tumor protection, *Int J Radiat Oncol Biol Phys* 68 (5), 1471–1482, 2007.

90. Holler, V., Buard, V., Gaugler, M. H., Guipaud, O., Baudelin, C., Sache, A., Perez Mdel, R., Squiban, C., Tamarat, R., Milliat, F., and Benderitter, M., Pravastatin limits radiation-induced vascular dysfunction in the skin, *J Invest Dermatol* 129 (5), 1280–1291, 2009.

91. Wang, J., Boerma, M., Fu, Q., Kulkarni, A., Fink, L. M., and Hauer-Jensen, M., Simvastatin ameliorates radiation enteropathy development after localized, fractionated irradiation by a protein C-independent mechanism, *Int J Radiat Oncol Biol Phys* 68 (5), 1483–1490, 2007.

92. Smith, A. D., Ashwood-Smith, M. J., and Lowman, D., Radioprotective action of methoxamine, *Nature* 184 (Suppl. 22), 1729–1730, 1959.

93. Chung, Y. L., Wang, A. J., and Yao, L. F., Antitumor histone deacetylase inhibitors suppress cutaneous radiation syndrome: Implications for increasing therapeutic gain in cancer radiotherapy, *Mol Cancer Ther* 3 (3), 317–325, 2004.

94. Miller, A., Radioprotection by the histone deacetylase inhibitor phenylbutyrate, *Radiat Res* 56th Annual Meeting Radiation Research Society, Maui, Hawaii, September 25–29, 2010.

95. Arora, R., Gupta, D., Chawla, R., Sagar, R., Sharma, A., Kumar, R., Prasad, J., Singh, S., Samanta, N., and Sharma, R. K., Radioprotection by plant products: Present status and future prospects, *Phytother Res* 19 (1), 1–22, 2005.

96. Rastogi, L., Feroz, S., Pandey, B. N., Jagtap, A., and Mishra, K. P., Protection against radiation-induced oxidative damage by an ethanolic extract of Nigella sativa L, *Int J Radiat Biol* 86 (9), 719–731, 2010.

97. Sandeep, D., and Nair, C. K., Protection from lethal and sub-lethal whole body exposures of mice to gamma-radiation by *Acorus calamus* L.: Studies on tissue antioxidant status and cellular DNA damage, *Exp Toxicol Pathol*, July 3, 2010.

98. Goyal, P. K., and Gehlot, P., Radioprotective effects of Aloe vera leaf extract on Swiss albino mice against whole-body gamma irradiation, *J Environ Pathol Toxicol Oncol* 28 (1), 53–61, 2009.

99. Samarth, R. M., Panwar, M., Kumar, M., and Kumar, A., Radioprotective influence of *Mentha piperita* (Linn) against gamma irradiation in mice: Antioxidant and radical scavenging activity, *Int J Radiat Biol* 82 (5), 331–337, 2006.

100. Samarth, R. M., Protection against radiation induced hematopoietic damage in bone marrow of Swiss albino mice by *Mentha piperita* (Linn), *J Radiat Res (Tokyo)* 48 (6), 523–528, 2007.

101. Goel, H. C., Prasad, J., Singh, S., Sagar, R. K., Agrawala, P. K., Bala, M., Sinha, A. K., and Dogra, R., Radioprotective potential of an herbal extract of Tinospora cordifolia, *J Radiat Res (Tokyo)* 45 (1), 61–68, 2004.

102. Kumar, R., Singh, P. K., Sharma, A., Prasad, J., Sagar, R., Singh, S., Arora, R., and Sharma, R. K., *Podophyllum hexandrum* (Himalayan mayapple) extract provides radioprotection by modulating the expression of proteins associated with apoptosis, *Biotechnol Appl Biochem* 42 (Pt 1), 81–92, 2005.

103. Gupta, M. L., Agrawala, P. K., Kumar, P., Devi, M., Soni, N. L., and Tripathi, R. P., Modulation of gamma radiation-inflicted damage in Swiss albino mice by an alcoholic fraction of *Podophyllum hexandrum* rhizome, *J Med Food* 11 (3), 486–492, 2008.

104. Pratheeshkumar, P., and Kuttan, G., Protective role of *Vernonia cinerea* L. against gamma radiation-induced immunosupression and oxidative stress in mice, *Hum Exp Toxicol*, doi: 10.1177/0960327110385959, 2010.

105. Shobi, V., and Goel, H. C., Protection against radiation-induced conditioned taste aversion by *Centella asiatica*, *Physiol. Behavior* 73, 19–23, 2001.

106. Sharma, J., and Sharma, R., Radioprotection of Swiss albino mouse by *Centella asiatica* extract, *Phytother Res* 16 (8), 785–786, 2002.

107. Emerit, I., Arutyunyan, R., Oganesian, N., Levy, A., Cernjavsky, L., Sarkisian, T., Pogossian, A., and Asrian, K., Radiation-induced clastogenic factors: Anticlastogenic effect of *Ginkgo biloba* extract, *Free Radic Biol Med* 18 (6), 985–991, 1995.

108. Ni, Y. B., Preventive effect of *Ginkgo biloba* extract on apoptosis in rat cerebellar neuronal cells induced by hydroxy radicals, *Neurosci. Lett.* 214, 115–118, 1996.

109. Emerit, I., Oganesian, N., Sarkisian, T., Arutyunyan, R., Pogosian, A., Asrian, K., Levy, A., and Cernjavski, L., Clastogenic factors in the plasma of Chernobyl accident recovery workers: Anticlastogenic effect of *Ginkgo biloba* extract, *Radiat Res* 144 (2), 198–205, 1995.

110. Hannequin, D., Thibert, A., and Vaschalde, Y., Development of a model to study the anti-edema properties of *Ginkgo biloba* extract [in French], *Presse Med* 15 (31), 1575–1576, 1986.

111. Goel, H. C., Prasad, J., Singh, S., Sagar, R. K., Kumar, I. P., and Sinha, A. K., Radioprotection by a herbal preparation of *Hippophae rhamnoides*, RH-3, against whole body lethal irradiation in mice, *Phytomedicine* 9 (1), 15–25, 2002.

112. Sharma, R. K. A., R.; and Prasad, J. et al., Radioprotective efficacy of *Hippophae rhamnoides*, Economy., I. P. o. t. S. o. S. B. A. R. f. E. H. a. Directorate of Life Sciences, Defense Research Development Organization., New Delhi, 2004, pp. 21–22.

113. Goel, H. C., Salin, C. A., and Prakash, H., Protection of jejunal crypts by RH-3 (a preparation of *Hippophae rhamnoides*) against lethal whole body gamma irradiation, *Phytother Res* 17 (3), 222–6, 2003.

114. Devi, P. U., and Ganasoundari, A., Radioprotective effect of leaf extract of Indian medicinal plant *Ocimum sanctum*, *Indian J Exp Biol* 33 (3), 205–208, 1995.

115. Ganasoundari, A., Devi, P. U., and Rao, M. N., Protection against radiation-induced chromosome damage in mouse bone marrow by Ocimum sanctum, *Mutat Res* 373 (2), 271–276, 1997.

116. Kim, S. H., Cho, C. K., Yoo, S. Y., Koh, K. H., Yun, H. G., and Kim, T. H., In vivo radio-protective activity of Panax ginseng and diethyldithiocarbamate, *In Vivo* 7 (5), 467–470, 1993.

117. Kim, S. H., Son, C. H., Nah, S. Y., Jo, S. K., Jang, J. S., and Shin, D. H., Modification of radiation response in mice by *Panax ginseng* and diethyldithiocarbamate, *In Vivo* 15 (5), 407–411, 2001.

118. Kumar, M., Sharma, M. K., Saxena, P. S., and Kumar, A., Radioprotective effect of *Panax ginseng* on the phosphatases and lipid peroxidation level in testes of Swiss albino mice, *Biol Pharm Bull* 26 (3), 308–312, 2003.

119. Takeda, A., Katoh, N., and Yonezawa, M., Restoration of radiation injury by ginseng. III. Radioprotective effect of thermostable fraction of ginseng extract on mice, rats and guinea pigs, *J Radiat Res (Tokyo)* 23 (2), 150–167, 1982.

120. Yonezawa, M., Katoh, N., and Takeda, A., Restoration of radiation injury by ginseng. IV. Stimulation of recoveries in CFUs and megakaryocyte counts related to the prevention of occult blood appearance in x-irradiated mice, *J Radiat Res (Tokyo)* 26 (4), 436–442, 1985.

121. Lee, Y. S., Chung, I. S., Lee, I. R., Kim, K. H., Hong, W. S., and Yun, Y. S., Activation of multiple effector pathways of immune system by the antineoplastic immunostimulator acidic polysaccharide ginsan isolated from *Panax ginseng*, *Anticancer Res* 17 (1A), 323–331, 1997.

122. Guo, S., Hu, Y., Liu, P., Wang, Y., Guo, D., Wang, D., and Liao, H., Protective activity of different concentration of tea polyphenols and its major compound EGCG against whole body irradiation-induced injury in mice [in Chinese], *Zhongguo Zhong Yao Za Zhi* 35 (10), 1328–1331, 2010.

123. Kang, K. A., Zhang, R., Lee, K. H., Chae, S., Kim, B. J., Kwak, Y. S., Park, J. W., Lee, N. H., and Hyun, J. W., Protective effect of triphlorethol-A from Ecklonia cava against ionizing radiation in vitro, *J Radiat Res (Tokyo)* 47 (1), 61–68, 2006.

124. Kang, K. A., Zhang, R., Chae, S., Lee, S. J., Kim, J., Jeong, J., Lee, J., Shin, T., Lee, N. H., and Hyun, J. W., Phloroglucinol (1,3,5-trihydroxybenzene) protects against ionizing radiation-induced cell damage through inhibition of oxidative stress in vitro and in vivo, *Chem Biol Interact* 185 (3), 215–226, 2010.

125. Zhang, R., Kang, K. A., Kang, S. S., Park, J. W., and Hyun, J. W., Morin (2′,3,4′,5,7-pentahydroxyflavone) protected cells against gamma-radiation-induced oxidative stress, *Basic Clin Pharmacol Toxicol* 108 (1), 63–72, 2011.

126. Guruvayoorappan, C., and Kuttan, G., Protective effect of *Biophytum sensitivum* (L.) DC on radiation-induced damage in mice, *Immunopharmacol Immunotoxicol* 30 (4), 815–835, 2008.

127. Singh, I., Soyal, D., and Goyal, P. K., *Emblica officinalis* (Linn.) fruit extract provides protection against radiation-induced hematological and biochemical alterations in mice, *J Environ Pathol Toxicol Oncol* 25 (4), 643–654, 2006.

128. Jahan, S., and Goyal, P. K., Protective effect of *Alstonia scholaris* against radiation-induced clastogenic and biochemical alterations in mice, *J Environ Pathol Toxicol Oncol* 29 (2), 101–111, 2010.

129. Manu, K. A., Leyon, P. V., and Kuttan, G., Studies on the protective effects of *Boerhaavia diffusa* L. against gamma radiation induced damage in mice, *Integr Cancer Ther* 6 (4), 381–388, 2007.

130. Harikumar, K. B., and Kuttan, R., An extract of *Phyllanthus amarus* protects mouse chromosomes and intestine from radiation induced damages, *J Radiat Res (Tokyo)* 48 (6), 469–476, 2007.

131. Benkovic, V., Knezevic, A. H., Dikic, D., Lisicic, D., Orsolic, N., Basic, I., and Kopjar, N., Radioprotective effects of quercetin and ethanolic extract of propolis in gamma-irradiated mice, *Arh Hig Rada Toksikol* 60 (2), 129–138, 2009.

132. Li, X., and Stark, G. R., NFkappaB-dependent signaling pathways, *Exp Hematol* 30 (4), 285–296, 2002.

133. Kang, J. L., Lee, H. W., Lee, H. S., Pack, I. S., Chong, Y., Castranova, V., and Koh, Y., Genistein prevents nuclear factor-kappa B activation and acute lung injury induced by lipopolysaccharide, *Am J Respir Crit Care Med* 164 (12), 2206–2212, 2001.

134. Calveley, V. L., Jelveh, S., Langan, A., Mahmood, J., Yeung, I. W., Van Dyk, J., and Hill, R. P., Genistein can mitigate the effect of radiation on rat lung tissue, *Radiat Res* 173 (5), 602–611, 2010.

135. Landauer, M. R., Lowery, L. J., and Davis, T. A., Time course of radiation protection in mice following oral administration of genistein, *Radiat Res* 56th Annual Meeting Radiation Research Society, Maui, Hawaii, September 25–29, 2010.

136. Singh, V. K., Grace, M. B., Parekh, V. I., Whitnall, M. H., and Landauer, M. R., Effects of genistein administration on cytokine induction in whole-body gamma irradiated mice, *Int Immunopharmacol* 9 (12), 1401–1410, 2009.

137. Davis, T. A., Clarke, T. K., Mog, S. R., and Landauer, M. R., Subcutaneous administration of genistein prior to lethal irradiation supports multilineage, hematopoietic progenitor cell recovery and survival, *Int J Radiat Biol* 83 (3), 141–151, 2007.

138. Mahmood, J., Jelveh, S., and Hill, R. P., Mitigation of radiation-induced lung injury with genistein or EUK-207, *Radiat Res* 56th Annual Meeting Radiation Research Society, Maui, Hawaii, September 25–29, 2010.

139. Jagetia, G. C., and Rajanikant, G. K., Curcumin treatment enhances the repair and regeneration of wounds in mice exposed to hemibody gamma-irradiation, *Plast Reconstr Surg* 115 (2), 515–528, 2005.

140. Lee, J. C., Kinniry, P. A., Arguiri, E., Serota, M., Kanterakis, S., Chatterjee, S., Solomides, C. C., Javvadi, P., Koumenis, C., Cengel, K. A., and Christofidou-Solomidou, M., Dietary curcumin increases antioxidant defenses in lung, ameliorates radiation-induced pulmonary fibrosis, and improves survival in mice, *Radiat Res* 173 (5), 590–601, 2010.

141. Velioglu-Ogunc, A., Sehirli, O., Toklu, H. Z., Ozyurt, H., Mayadagli, A., Eksioglu-Demiralp, E., Erzik, C., Cetinel, S., Yegen, B. C., and Sener, G., Resveratrol protects against irradiation-induced hepatic and ileal damage via its anti-oxidative activity, *Free Radic Res* 43 (11), 1060–1071, 2009.
142. Lee, J. C., Krochak, R., Blouin, A., Kanterakis, S., Chatterjee, S., Arguiri, E., Vachani, A., Solomides, C. C., Cengel, K. A., and Christofidou-Solomidou, M., Dietary flaxseed prevents radiation-induced oxidative lung damage, inflammation and fibrosis in a mouse model of thoracic radiation injury, *Cancer Biol Ther* 8 (1), 47–53, 2009.
143. Epperly, M. W., Epperly, L. D., Niu, Y., Wang, H., Zhang, X., Franicola, D., and Greenberger, J. S., Overexpression of the MnSOD transgene product protects cryopreserved bone marrow hematopoietic progenitor cells from ionizing radiation, *Radiat Res* 168 (5), 560–566, 2007.
144. Thompson, J. S., Chu, Y., Glass, J., Tapp, A. A., and Brown, S. A., The manganese superoxide dismutase mimetic, M40403, protects adult mice from lethal total body irradiation, *Free Radic Res* 44 (5), 529–540, 2010.
145. Kang, S. K., Rabbani, Z. N., Folz, R. J., Golson, M. L., Huang, H., Yu, D., Samulski, T. S., Dewhirst, M. W., Anscher, M. S., and Vujaskovic, Z., Overexpression of extracellular superoxide dismutase protects mice from radiation-induced lung injury, *Int J Radiat Oncol Biol Phys* 57 (4), 1056–1066, 2003.
146. Greenberger, J. S., and Epperly, M. W., Review. Antioxidant gene therapeutic approaches to normal tissue radioprotection and tumor radiosensitization, *In Vivo* 21 (2), 141–146, 2007.
147. Machtay, M., Scherpereel, A., Santiago, J., Lee, J., McDonough, J., Kinniry, P., Arguiri, E., Shuvaev, V. V., Sun, J., Cengel, K., Solomides, C. C., and Christofidou-Solomidou, M., Systemic polyethylene glycol-modified (PEGylated) superoxide dismutase and catalase mixture attenuates radiation pulmonary fibrosis in the C57/bl6 mouse, *Radiother Oncol* 81 (2), 196–205, 2006.
148. Cline, M., Dugan, G., Perry, D., Batinic-Haberie, I., and Vujaskovic, Z., MnTnHex-2-PyP5+ provides protection in non-human primate lungs after whole-thorax exposure to ionizing irradiation, *Radiat Res* 56th Annual Meeting Radiation Research Society, Maui, Hawaii, September 25–29, 2010.
149. Li, H., Wang, Y., Pazhanisamy, S. K., Shao, L., Meng, A., Batinic-Haberle, I., and Zhou, D., Treatment with MnTn-2-Py5+ mitigates total body irradiation-induced long-term bone marrow suppression, *Radiat Res* 56th Annual Meeting Radiation Research Society, Maui, Hawaii, September 25–29, 2010.
150. Prasad, K. N., and Edwards-Prasad, J., Effects of tocopherol (vitamin E) acid succinate on morphological alterations and growth inhibition in melanoma cells in culture, *Cancer Res* 42 (2), 550–555, 1982.
151. Prasad, K. N., Kumar, B., Yan, X. D., Hanson, A. J., and Cole, W. C., Alpha-tocopheryl succinate, the most effective form of vitamin E for adjuvant cancer treatment: A review, *J Am Coll Nutr* 22 (2), 108–117, 2003.
152. Borek, C., Ong, A., Mason, H., Donahue, L., and Biaglow, J. E., Selenium and vitamin E inhibit radiogenic and chemically induced transformation in vitro via different mechanisms, *Proc Natl Acad Sci U S A* 83 (5), 1490–1494, 1986.
153. Radner, B. S., and Kennedy, A. R., Suppression of x-ray induced transformation by vitamin E in mouse C3H/10T1/2 cells, *Cancer Lett* 32 (1), 25–32, 1986.
154. Chandrashekara, S., Anilkumar, T., and Jamuna, S., Complementary and alternative drug therapy in arthritis, *J Assoc Physicians India* 50, 225–227, 2002.
155. Ushakova, T., Melkonyan, H., Nikonova, L., Afanasyev, V., Gaziev, A. I., Mudrik, N., Bradbury, R., and Gogvadze, V., Modification of gene expression by dietary antioxidants in radiation-induced apoptosis of mice splenocytes, *Free Radic Biol Med* 26 (7–8), 887–891, 1999.

156. Konopacka, M., Widel, M., and Rzeszowska-Wolny, J., Modifying effect of vitamins C, E and beta-carotene against gamma-ray-induced DNA damage in mouse cells, *Mutat Res* 417 (2–3), 85–94, 1998.

157. Gaziev, A. I., Podlutsky, A., Panfilov, B. M., and Bradbury, R., Dietary supplements of antioxidants reduce hprt mutant frequency in splenocytes of aging mice, *Mutat Res* 338 (1–6), 77–86, 1995.

158. O'Connor, M. K., Malone, J. F., Moriarty, M., and Mulgrew, S., A radioprotective effect of vitamin C observed in Chinese hamster ovary cells, *Br J Radiol* 50 (596), 587–591, 1977.

159. Weiss, J. F., and Landauer, M. R., Radioprotection by antioxidants, *Ann N Y Acad Sci* 899, 44–60, 2000.

160. Konopacka, M., and Rzeszowska-Wolny, J., Antioxidant vitamins C, E and beta-carotene reduce DNA damage before as well as after gamma-ray irradiation of human lymphocytes in vitro, *Mutat Res* 491 (1–2), 1–7, 2001.

161. Cui, L., Fu, Q., Kumar, K. S., and Hauer-Jensen,M., Gamma-tocotrienol and/or pentoxifylline attenuates DNA and lipid oxidative damage in mice intestine after total body irradiation, *Radiat Res* 56th Annual Meeting Radiation Research Society, Maui, Hawaii, September 25–29, 2010.

162. Kennedy, A. R., and Krinsky, N. I., Effects of retinoids, beta-carotene, and canthaxanthin on UV- and x-ray-induced transformation of C3H10T1/2 cells in vitro, *Nutr Cancer* 22 (3), 219–232, 1994.

163. Srinivasan, M., Devipriya, N., Kalpana, K. B., and Menon, V. P., Lycopene: An antioxidant and radioprotector against gamma-radiation-induced cellular damages in cultured human lymphocytes, *Toxicology* 262 (1), 43–49, 2009.

164. Tiwari, P., Kumar, A., Balakrishnan, S., Kushwaha, H. S., and Mishra, K. P., Radiation-induced micronucleus formation and DNA damage in human lymphocytes and their prevention by antioxidant thiols, *Mutat Res* 676 (1–2), 62–68, 2009.

165. Ruifrok, A. C., Mason, K. A., and Thames, H. D., Changes in clonogen number and radiation sensitivity in mouse jejunal crypts after treatment with dimethylsulfoxide and retinoic acid, *Radiat Res* 145 (6), 740–745, 1996.

166. Jagetia, G. C., Rajanikant, G. K., Baliga, M. S., Rao, K. V., and Kumar, P., Augmentation of wound healing by ascorbic acid treatment in mice exposed to gamma-radiation, *Int J Radiat Biol* 80 (5), 347–354, 2004.

167. Wu, W., Abraham, L., Ogony, J., Matthews, R., Goldstein, G., and Ercal, N., Effects of N-acetylcysteine amide (NACA), a thiol antioxidant on radiation-induced cytotoxicity in Chinese hamster ovary cells, *Life Sci* 82 (21–22), 1122–1130, 2008.

168. Langberg, M., Rotem, C., Fenig, E., Koren, R., and Ravid, A., Vitamin D protects keratinocytes from deleterious effects of ionizing radiation, *Br J Dermatol* 160 (1), 151–161, 2009.

169. Kennedy, A. R., Zhou, Z., Donahue, J. J., and Ware, J. H., Protection against adverse biological effects induced by space radiation by the Bowman-Birk inhibitor and antioxidants, *Radiat Res* 166 (2), 327–332, 2006.

170. Kennedy, A. R., Ware, J. H., Guan, J., Donahue, J. J., Biaglow, J. E., Zhou, Z., Stewart, J., Vazquez, M., and Wan, X. S., Selenomethionine protects against adverse biological effects induced by space radiation, *Free Radic Biol Med* 36 (2), 259–266, 2004.

171. de Moraes Ramos, F. M., Schonlau, F., Novaes, P. D., Manzi, F. R., Boscolo, F. N., and de Almeida, S. M., Pycnogenol protects against ionizing radiation as shown in the intestinal mucosa of rats exposed to x-rays, *Phytother Res* 20 (8), 676–679, 2006.

172. Kumar, B., Jha, M. N., Cole, W. C., Bedford, J. S., and Prasad, K. N., D-alpha-tocopheryl succinate (vitamin E) enhances radiation-induced chromosomal damage levels in human cancer cells, but reduces it in normal cells, *J Am Coll Nutr* 21 (4), 339–343, 2002.

173. Kumar, K. S., Srinivasan, V., Toles, R., Jobe, L., and Seed, T. M., Nutritional approaches to radioprotection: Vitamin E, *Mil Med* 167 (2 Suppl), 57–59, 2002.

174. Manda, K., Ueno, M., Moritake, T., and Anzai, K., Alpha-lipoic acid attenuates x-irradiation-induced oxidative stress in mice, *Cell Biol Toxicol* 23 (2), 129–137, 2007.

175. Mansour, H. H., Hafez, H. F., Fahmy, N. M., and Hanafi, N., Protective effect of N-acetylcysteine against radiation induced DNA damage and hepatic toxicity in rats, *Biochem Pharmacol* 75 (3), 773–780, 2008.

176. Mutlu-Turkoglu, U., Erbil, Y., Oztezcan, S., Olgac, V., Toker, G., and Uysal, M., The effect of selenium and/or vitamin E treatments on radiation-induced intestinal injury in rats, *Life Sci* 66 (20), 1905–1913, 2000.

177. Satyamitra, M., Devi, P. U., Murase, H., and Kagiya, V. T., In vivo radioprotection by alpha-TMG: Preliminary studies, *Mutat Res* 479 (1–2), 53–61, 2001.

178. Shirazi, A., Ghobadi, G., and Ghazi-Khansari, M., A radiobiological review on melatonin: A novel radioprotector, *J Radiat Res (Tokyo)* 48 (4), 263–272, 2007.

179. Sridharan, S., and Shyamaladevi, C. S., Protective effect of N-acetylcysteine against gamma ray induced damages in rats: Biochemical evaluations, *Indian J Exp Biol* 40 (2), 181–186, 2002.

180. Srinivasan, M., Sudheer, A. R., Pillai, K. R., Kumar, P. R., Sudhakaran, P. R., and Menon, V. P., Lycopene as a natural protector against gamma-radiation induced DNA damage, lipid peroxidation and antioxidant status in primary culture of isolated rat hepatocytes in vitro, *Biochim Biophys Acta* 1770 (4), 659–665, 2007.

181. Harapanhalli, R. S., Yaghmai, V., Giuliani, D., Howell, R. W., and Rao, D. V., Antioxidant effects of vitamin C in mice following x-irradiation, *Res Commun Mol Pathol Pharmacol* 94 (3), 271–287, 1996.

182. Narra, V. R., Harapanhalli, R. S., Howell, R. W., Sastry, K. S., and Rao, D. V., Vitamins as radioprotectors in vivo. I. Protection by vitamin C against internal radionuclides in mouse testes: Implications to the mechanism of damage caused by the Auger effect, *Radiat Res* 137 (3), 394–399, 1994.

183. Ramakrishnan, N., Wolfe, W. W., and Catravas, G. N., Radioprotection of hematopoietic tissues in mice by lipoic acid, *Radiat Res* 130 (3), 360–365, 1992.

184. Singh, V. K., Brown, D. S., and Kao, T. C., Alpha-tocopherol succinate protects mice from gamma-radiation by induction of granulocyte-colony stimulating factor, *Int J Radiat Biol* 86 (1), 12–21, 2010.

185. Karslioglu, I., Ertekin, M. V., Kocer, I., Taysi, S., Sezen, O., Gepdiremen, A., and Balci, E., Protective role of intramuscularly administered vitamin E on the levels of lipid peroxidation and the activities of antioxidant enzymes in the lens of rats made cataractous with gamma-irradiation, *Eur J Ophthalmol* 14 (6), 478–485, 2004.

186. Kocer, I., Taysi, S., Ertekin, M. V., Karslioglu, I., Gepdiremen, A., Sezen, O., and Serifoglu, K., The effect of L-carnitine in the prevention of ionizing radiation-induced cataracts: A rat model, *Graefes Arch Clin Exp Ophthalmol* 245 (4), 588–594, 2007.

187. Sezen, O., Ertekin, M. V., Demircan, B., Karslioglu, I., Erdogan, F., Kocer, I., Calik, I., and Gepdiremen, A., Vitamin E and L-carnitine, separately or in combination, in the prevention of radiation-induced brain and retinal damages, *Neurosurg Rev* 31 (2), 205–213; discussion 213, 2008.

188. Altas, E., Ertekin, M. V., Gundogdu, C., and Demirci, E., L-carnitine reduces cochlear damage induced by gamma irradiation in guinea pigs, *Ann Clin Lab Sci* 36 (3), 312–318, 2006.

189. Ramos, F. M., Pontual, M. L., de Almeida, S. M., Boscolo, F. N., Tabchoury, C. P., and Novaes, P. D., Evaluation of radioprotective effect of vitamin E in salivary dysfunction in irradiated rats, *Arch Oral Biol* 51 (2), 96–101, 2006.

190. Boerma, M., Roberto, K. A., and Hauer-Jensen, M., Prevention and treatment of functional and structural radiation injury in the rat heart by pentoxifylline and alpha-tocopherol, *Int J Radiat Oncol Biol Phys* 72 (1), 170–177, 2008.

191. Xiao, M., Li, X. H., Fu, D., Latif, N. H., and Ha, C. T., Radioprotective and therapeutic effects of delta-tocotrienol in mouse bone marrow hematopoietic tissue (in vivo study), *Radiat Res* 56th Annual Meeting Radiation Research Society, Maui, Hawaii, September 25–29, 2010.

192. Ghosh, S. P., Kulkarni, S., Hieber, K., Toles, R., Romanyukha, L., Kao, T. C., Hauer-Jensen, M., and Kumar, K. S., Gamma-tocotrienol, a tocol antioxidant as a potent radio-protector, *Int J Radiat Biol* 85 (7), 598–606, 2009.

193. Kulkarni, S. S., Ghosh, S. Hieber, K., Romanyukha, L., Hauer-Jemsen, M., and Kumar, S., Radioprotective efficacy of gamma-tocotrienol and pentoxifylline combination in murine model, *Radiat Res* 56th Annual Meeting Radiation Research Society, Maui, Hawaii, September 25–29, 2010.

194. Jia, D., Koonce, N. A., Griffin, R. J., Jackson, C., and Corry, P. M., Prevention and miti-gation of acute death of mice after abdominal irradiation by the antioxidant N-acetyl-cysteine (NAC), *Radiat Res* 173 (5), 579–589, 2010.

195. Liu, Y., Zhang, H., Zhang, L., Zhou, Q., Wang, X., Long, J., Dong, T., and Zhao, W., Antioxidant N-acetylcysteine attenuates the acute liver injury caused by x-ray in mice, *Eur J Pharmacol* 575 (1–3), 142–148, 2007.

196. Pajonk, F., Riess, K., Sommer, A., and McBride, W. H., N-acetyl-L-cysteine inhibits 26S proteasome function: Implications for effects on NF-kappaB activation, *Free Radic Biol Med* 32 (6), 536–543, 2002.

197. Manda, K., Ueno, M., Moritake, T., and Anzai, K., Radiation-induced cognitive dys-function and cerebellar oxidative stress in mice: Protective effect of alpha-lipoic acid, *Behav Brain Res* 177 (1), 7–14, 2007.

198. Ben-Amotz, A., Rachmilevich, B., Greenberg, S., Sela, M., and Weshler, Z., Natural beta-carotene and whole body irradiation in rats, *Radiat Environ Biophys* 35 (4), 285–288, 1996.

199. Vijayalaxmi, Reiter, R. J., Tan, D. X., Herman, T. S., and Thomas, C. R. Jr., Melatonin as a radioprotective agent: A review, *Int J Radiat Oncol Biol Phys* 59 (3), 639–653, 2004.

200. Erol, F. S., Topsakal, C., Ozveren, M. F., Kaplan, M., Ilhan, N., Ozercan, I. H., and Yildiz, O. G., Protective effects of melatonin and vitamin E in brain damage due to gamma radiation: An experimental study, *Neurosurg Rev* 27 (1), 65–69, 2004.

201. Guan, J., Stewart, J., Ware, J. H., Zhou, Z., Donahue, J. J., and Kennedy, A. R., Effects of dietary supplements on the space radiation-induced reduction in total antioxidant status in CBA mice, *Radiat Res* 165 (4), 373–378, 2006.

202. Manda, K., Ueno, M., and Anzai, K., Memory impairment, oxidative damage and apop-tosis induced by space radiation: Ameliorative potential of alpha-lipoic acid, *Behav Brain Res* 187 (2), 387–395, 2008.

203. Blumenthal, R. D., Lew, W., Reising, A., Soyne, D., Osorio, L., Ying, Z., and Goldenberg, D. M., Anti-oxidant vitamins reduce normal tissue toxicity induced by radio-immuno-therapy, *Int J Cancer* 86 (2), 276–280, 2000.

204. Mills, E. E., The modifying effect of beta-carotene on radiation and chemotherapy induced oral mucositis, *Br J Cancer* 57 (4), 416–417, 1988.

205. Jaakkola, K., Lahteenmaki, P., Laakso, J., Harju, E., Tykka, H., and Mahlberg, K., Treatment with antioxidant and other nutrients in combination with chemotherapy and irradiation in patients with small-cell lung cancer, *Anticancer Res* 12 (3), 599–606, 1992.

206. Lamson, D. W., and Brignall, M. S., Antioxidants in cancer therapy: Their actions and interactions with oncologic therapies, *Altern Med Rev* 4 (5), 304–329, 1999.

207. Sminia, P., van der Kracht, A. H., Frederiks, W. M., and Jansen, W., Hyperthermia, radiation carcinogenesis and the protective potential of vitamin A and N-acetylcysteine, *J Cancer Res Clin Oncol* 122 (6), 343–350, 1996.

208. Korkina, L. G., Afanas'ef, I. B., and Diplock, A. T., Antioxidant therapy in children affected by irradiation from the Chernobyl nuclear accident, *Biochem Soc Trans* 21 (Pt 3) (3), 314S, 1993.

209. Ben-Amotz, A., Yatziv, S., Sela, M., Greenberg, S., Rachmilevich, B., Shwarzman, M., and Weshler, Z., Effect of natural beta-carotene supplementation in children exposed to radiation from the Chernobyl accident, *Radiat Environ Biophys* 37 (3), 187–193, 1998.

210. Misirlioglu, C. H., Demirkasimoglu, T., Kucukplakci, B., Sanri, E., and Altundag, K., Pentoxifylline and alpha-tocopherol in prevention of radiation-induced lung toxicity in patients with lung cancer, *Med Oncol* 24 (3), 308–311, 2007.

211. Ryan, J. L., Heckler, C. E., Williams, J., Morrow, G. R., Pentland, A. P., Curcumin treatment and prediction of radiation dermatitis in breast cancer patients, *Radiat Res* 56th Annual Meeting Radiation Research Society, Maui, Hawaii, September 25–29, 2010.

212. Albanes, D., Malila, N., Taylor, P. R., Huttunen, J. K., Virtamo, J., Edwards, B. K., Rautalahti, M., Hartman, A. M., Barrett, M. J., Pietinen, P., Hartman, T. J., Sipponen, P., Lewin, K., Teerenhovi, L., Hietanen, P., Tangrea, J. A., Virtanen, M., and Heinonen, O. P., Effects of supplemental alpha-tocopherol and beta-carotene on colorectal cancer: Results from a controlled trial (Finland), *Cancer Causes Control* 11 (3), 197–205, 2000.

213. The Alpha-Tocopherol Beta Carotene Cancer Prevention Study Group, The effect of vitamin E and beta-carotene on the incidence of lung cancer and other cancers in male smokers, *N Eng J Med* 330, 1029–1035, 1994.

214. Krinsky, N. I., Antioxidant functions of carotenoids, *Free Radic Biol Med* 7 (6), 617–635, 1989.

215. Hazuka, M. B., Edwards-Prasad, J., Newman, F., Kinzie, J. J., and Prasad, K. N., Beta-carotene induces morphological differentiation and decreases adenylate cyclase activity in melanoma cells in culture, *J Am Coll Nutr* 9 (2), 143–149, 1990.

216. Zhang, L. X., Cooney, R. V., and Bertram, J. S., Carotenoids up-regulate connexin43 gene expression independent of their provitamin A or antioxidant properties, *Cancer Res* 52 (20), 5707–5712, 1992.

217. Carter, C. A., Pogribny, M., Davidson, A., Jackson, C. D., McGarrity, L. J., and Morris, S. M., Effects of retinoic acid on cell differentiation and reversion toward normal in human endometrial adenocarcinoma (RL95-2) cells, *Anticancer Res* 16 (1), 17–24, 1996.

218. Meyskens, F. L. Jr., Role of vitamin A and its derivatives in the treatment of human cancer, in *Nutrients in Cancer Prevention and Treatment*, ed. Prasad, K. N., Santamaria, L., and Williams, R. M., Humana Press, New Jersey, 1995, 349–362.

219. Vile, G. F., and Winterbourn, C. C., Inhibition of adriamycin-promoted microsomal lipid peroxidation by beta-carotene, alpha-tocopherol and retinol at high and low oxygen partial pressures, *FEBS Lett* 238 (2), 353–356, 1988.

220. McCay, P. B., Vitamin E: Interactions with free radicals and ascorbate, *Annu Rev Nutr* 5, 323–340, 1985.

221. Schwartz, J. L., Molecular and biochemical control of tumor growth following treatment with carotenoids or tocopherols, in *Nutrients in Cancer Prevention and Treatment*, ed. Prasad, K. N., Santamaria, L., and Williams, R. M. Humana Press, New Jersey, 1995, 287–316.

222. Prasad, K. N., and Edwards-Prasad, J., Vitamin E and cancer prevention: Recent advances and future potentials, *J Am Coll Nutr* 11 (5), 487–500, 1992.

223. Witschi, A., Reddy, S., Stofer, B., and Lauterburg, B. H., The systemic availability of oral glutathione, *Eur J Clin Pharmacol* 43 (6), 667–669, 1992.

224. Niki, E., Mechanisms and dynamics of antioxidant action of ubiquinol, *Mol Aspects Med* 18 (Suppl.), S63–70, 1997.

225. Hiramatsu, M., Velasco, R. D., Wilson, D. S., and Packer, L., Ubiquinone protects against loss of tocopherol in rat liver microsomes and mitochondrial membranes, *Res Commun Chem Pathol Pharmacol* 72 (2), 231–241, 1991.

226. Devaraj, S., and Jialal, I., Alpha tocopherol supplementation decreases serum C-reactive protein and monocyte interleukin-6 levels in normal volunteers and type 2 diabetic patients, *Free Radic Biol Med* 29 (8), 790–792, 2000.

227. Guenechea, G., Albella, B., Bueren, J. A., Maganto, G., Tuduri, P., Guerrero, A., Pivel, J. P., and Real, A., AM218, a new polyanionic polysaccharide, induces radioprotection in mice when administered shortly before irradiation, *Int J Radiat Biol* 71 (1), 101–108, 1997.

228. Patchen, M. L., MacVittie, T. J., Solberg, B. D., D'Alesandro, M. M., and Brook, I., Radioprotection by polysaccharides alone and in combination with aminothiols, *Adv Space Res* 12 (2–3), 233–248, 1992.

229. Ross, W. M., and Peeke, J., Radioprotection conferred by dextran sulfate given before irradiation in mice, *Exp Hematol* 14 (2), 147–155, 1986.

230. Naidu, N. V., and Reddi, O. S., Effect of post-treatment with erythropoietin(s) on survival and erythropoietic recovery in irradiated mice, *Nature* 214 (5094), 1223–1224, 1967.

231. Vittorio, P. V., Whitfield, J. F., and Rixon, R. H., The radioprotective and therapeutic effects of imidazole and erythropoietin on the erythropoiesis and survival of irradiated mice, *Radiat Res* 47 (1), 191–198, 1971.

232. Ryu, T., Hosaka, N., Miyake, T., Cui, W., Nishida, T., Takaki, T., Li, M., Kawamoto, K., and Ikehara, S., Transplantation of newborn thymus plus hematopoietic stem cells can rescue supralethally irradiated mice, *Bone Marrow Transplant* 41 (7), 659–666, 2008.

233. Perry, A. R., and Iglar, A. F., The accident at Chernobyl: Radiation doses and effects, *Radiol Technol* 61 (4), 290–294, 1990.

234. Carder, T. A., *Handling of Radiation Accident Patients by Paramedical and Hospital Personnel*, 2nd ed., CRC Press, Boca Raton, FL, 1993.

235. Blakely, J., *The Care of Radiation Casualties*, Charlec C. Thomas, Springfield, IL, 1968.

236. Baranov, A., Gale, R. P., Guskova, A., Piatkin, E., Selidovkin, G., Muravyova, L., Champlin, R. E., Danilova, N., Yevseeva, L., and Petrosyan, L., Bone marrow transplantation after the Chernobyl nuclear accident, *N Engl J Med* 321 (4), 205–212, 1989.

237. Kawashima, R., Kawamura, Y. I., Kato, R., Mizutani, N., Toyama-Sorimachi, N., and Dohi, T., IL-13 receptor alpha2 promotes epithelial cell regeneration from radiation-induced small intestinal injury in mice, *Gastroenterology* 131 (1), 130–141, 2006.

238. Talmadge, J. E., Tribble, H., Pennington, R., Bowersox, O., Schneider, M. A., Castelli, P., Black, P. L., and Abe, F., Protective, restorative, and therapeutic properties of recombinant colony-stimulating factors, *Blood* 73 (8), 2093–1103, 1989.

239. Cui, Y. F., Yang, H., Luo, Q. L., Dong, B., Liu, X. L., Xu, H., Mao, B. Z., and Wang, D. W., Radioprotection of recombinant human interleukin-3 and granulocyte-macrophage colony-stimulating factor on peripheral lymphocytes of rhesus monkey irradiated by 3.0 Gy gamma-rays [in Chinese], *Zhongguo Wei Zhong Bing Ji Jiu Yi Xue* 16 (1), 22–25, 2004.

240. Li, N., Zhang, L., Li, H., and Fang, B., Administration of granulocyte colony-stimulating factor ameliorates radiation-induced hepatic fibrosis in mice, *Transplant Proc* 42 (9), 3833–3839, 2010.

241. Okunieff, P., Mester, M., Wang, J., Maddox, T., Gong, X., Tang, D., Coffee, M., and Ding, I., In vivo radioprotective effects of angiogenic growth factors on the small bowel of C3H mice, *Radiat Res* 150 (2), 204–211, 1998.

242. Ebel, J. P., Beck, G., Keith, G., Langendorff, H., and Langendorff, M., Study of the therapeutic effect on irradiated mice of substances contained in RNA preparations, *Int J Radiat Biol Relat Stud Phys Chem Med* 16 (3), 201–209, 1969.
243. Gudkov, S. V., Shtarkman, I. N., Smirnova, V. S., Chernikov, A. V., and Bruskov, V. I., Guanosine and inosine display antioxidant activity, protect DNA in vitro from oxidative damage induced by reactive oxygen species, and serve as radioprotectors in mice, *Radiat Res* 165 (5), 538–545, 2006.
244. Gudkov, S. V., Gudkova, O. Y., Chernikov, A. V., and Bruskov, V. I., Protection of mice against x-ray injuries by the post-irradiation administration of guanosine and inosine, *Int J Radiat Biol* 85 (2), 116–125, 2009.
245. Coeur, A., Rinaldi, R., and Raynfeld, C., Thyroid effects of the absorption of therapeutic doses of iodine compounds in the rat [in French], *Therapie* 17, 621–627, 1962.
246. Brown, S. L., Kolozsvary, A., Liu, J., Jenrow, K. A., Ryu, S., and Kim, J. H., Antioxidant diet supplementation starting 24 hours after exposure reduces radiation lethality, *Radiat Res* 173 (4), 462–468, 2010.

6 Prevention and Mitigation of Late Adverse Effects of High Radiation Doses

INTRODUCTION

Data on the long-term adverse health effects of high doses of ionizing radiation come primarily from the studies on the populations of Hiroshima and Nagasaki who survived atomic bombing, and survivors of cancer treatment involving radiation therapy and/or chemotherapy. The studies on radiation-exposed populations of Hiroshima and Nagasaki revealed increased risk of cancer and noncancerous diseases. The noncancerous diseases included stroke and heart diseases, respiratory diseases, digestive disorders, and myelodysplastic syndromes.[1–8] In addition, the incidence of cancer and mortality was high in these populations.[9–11] The incidence of second primary cancer and noncancerous diseases is also high among survivors of childhood cancer. Noncancerous diseases include endocrine morbidity, growth hormone deficiency, sexual dysfunction, bone growth,[12] increased mortality, kidney complications,[13] cognitive dysfunction,[14] gastrointestinal complications,[15] auditory disorders,[16] early aging,[17] and increased risk of preterm delivery and low-birth-weight baby, ovarian failure, premature menopause and infertility among female survivors of childhood cancer,[18–20] diabetes mellitus,[21] and nonmalignant thyroid diseases.[22] In addition, the risk of second cancer is high in these populations.[23–30] Additional information on the incidence of neoplasms and nonneoplastic diseases among the radiation-exposed populations of Hiroshima and Nagasaki and cancer survivors was described in Chapter 4.

The distinction between prevention and mitigation may not be applicable to cancer prevention among high-risk populations. This is due to the fact that whether the intervening agents are administered before or after exposure to radiation, the primary outcome is the prevention of cancer risk and noncancerous diseases. The concept of bioshield refers to agents or procedures that can provide tissue protection against radiation injuries when administered before and/or after irradiation. Application of the bioshield concept for prevention of the late adverse health effects can be useful for radiation-exposed populations of Hiroshima and Nagasaki and cancer survivors. In addition, the bioshield concept can be applicable also to radiation workers at nuclear power plants who have the potential of receiving high doses of radiation in case of nuclear accidents like the Chernobyl nuclear accident, and more recently the Fukushima nuclear power plant accident.

At present, there are no prevention strategies to reduce the risk of cancer or non-cancerous diseases in radiation-exposed individuals or their offspring, or in cancer survivors who received radiation therapy. There are also no intervention plans to prevent the risk of these diseases in radiation workers at nuclear power plants in the event they are exposed to high doses of radiation following a nuclear accident. As discussed in Chapter 5, among various radioprotective and/or radiomitigating agents, antioxidants satisfy the criteria of an ideal bioshield that can reduce the risk of neoplastic and nonneoplastic diseases among survivors of radiation exposure, and that can provide tissue protection against acute radiation sickness (ARS) induced by high-dose radiation. Some studies on the role of antioxidants in cancer prevention in general are described in the following text.

STUDIES ON THE ROLE OF ANTIOXIDANTS IN CANCER PREVENTION

Radiation-induced cancer cannot be distinguished from that which occurs spontaneously or is induced by chemical carcinogens. Human carcinogenesis is a very complex process with a long latent period (3–30 years) between exposures to radiation and clinically detectable cancer. This implies that preventive strategies can be implemented at any time after irradiation and before cancer becomes detectable. It should be emphasized that irradiated individuals are also exposed to direct and indirect chemical carcinogens, tumor promoters, and cancer-causing viruses that can enhance the carcinogenic effects of radiation. Therefore, cancer formation is the result of interaction between radiation and chemical and biological carcinogens. The identification of biochemical events that can alter activities of genes responsible for cancer formation during the latent period of cancer formation can help to select appropriate intervening agents that can attenuate cancer-causing risks. Increased oxidative stress and chronic inflammation appear to play a central role in the initiation of a series of genetic and epigenetic changes that ultimately lead to cancer formation.[31–35] Among various radioprotective and radiomitigating agents, antioxidants appear to be a more useful bioshield than others, because they inhibit increased oxidative stress and chronic inflammation, and they have been consumed by humans for decades without reported toxicity. Properties of antioxidants that are relevant to cancer prevention in radiation-exposed individual are described next.

PROPERTIES OF ANTIOXIDANTS RELEVANT TO CANCER PREVENTION

Extensive studies on the properties of antioxidants that explain their protective role in reducing the risk of cancer have been published. This issue has been discussed extensively in a review.[31] Antioxidants can neutralize excessive levels of free radicals that increase the risk of cancer. They can prevent the formation of potential carcinogenic substances. For example, vitamin C and vitamin E alone or in combination prevented the formation of nitrosamine in the stomach from nitrites (present in a nitrite-rich diet) and secondary amines.[36] These dietary antioxidants also reduce the levels of fecal mutagens that are formed during digestion of food.[37] The combination of vitamin C and vitamin E was more effective than the individual antioxidant in reducing the levels of fecal mutagens. High levels of antioxidants can prevent conversion of an indirect carcinogen to an active form in the liver, which is needed to increase the risk of cancer. They can prevent the action of tumor promoters, such as estrogen and high fat. Mutations due to gene defect and/or chromosomal damages can increase the risk of cancer. Antioxidants can reduce spontaneous and induced mutations in animal as well as human cells, and thus could play an important role in cancer prevention. For example, vitamins C and E, and beta-carotene, reduced chromosomal damage produced by ionizing radiation and chemical carcinogens.[38–40] In our study, alpha-tocopheryl succinate (vitamin E succinate) reduced radiation-induced chromosomal damage in normal human fibroblasts in culture.[41]

The references for the following statements are provided in a few reviews.[31,42,43] Overexpression of oncogene or mutation in oncogene is considered a risk factor for the development of cancer. Antioxidants can inhibit overexpression of oncogenes, and the expression and levels of mutated oncogenes. A high-fat diet, which is considered a tumor promoter, increases the levels of prostaglandins (PGs) in animal models [44] and may increase the risk of some cancers.[45] Vitamin E succinate inhibited the action of PGE1 on the adenylate cyclase in mammalian cells. Vitamin E and a nonsteroidal anti-inflammatory drug (NSAID), aspirin, inhibited the production of PGs more than the individual agent. Inflammatory reactions release PGs, free radicals, pro-inflammatory cytokines, adhesion molecules, and complement proteins that contribute to the development of cancer and other noncancerous diseases. Antioxidants attenuated the levels of inflammatory products.

Although the host's immune system may not play a direct role in human carcinogenesis, it could play an important role in allowing or rejecting the newly formed cancer cells. Optimally functioning immune cells (natural killer cells) can recognize the newly formed cancer cells and kill them. A weak immune system may allow the newly formed cancer cells to establish themselves in the host; these cells will then grow and metastasize to distant organs. Antioxidants stimulate the humoral and cellular immunity,[46–48] and thus can reduce the risk of developing cancer.

CANCER PREVENTION STUDIES WITH ANTIOXIDANTS

There are substantial amounts of experimental data in cell culture and in animal models that suggest that antioxidants used individually can reduce the risk of radiation-induced cancer. Most human epidemiologic studies reveal that a diet rich in multiple antioxidants is associated with a reduced risk of cancer; however, the use of a single antioxidant in human intervention studies has produced inconsistent results, varying from no effect, beneficial effect, to harmful effect. The reasons for the inconsistent results are discussed in the subsequent section on Human Intervention Studies.

Cell culture studies: Tissue culture systems provide a unique opportunity to evaluate the role of antioxidants in radiation-induced cancer in a cost- and time-effective manner. In addition, detailed studies on the mechanisms of the action of antioxidants at the cellular and genetic levels could not be carried out in animal models owing to the inherent complexity of in vivo systems. The availability of a normal-like murine cell line (CH310T1/2), other mammalian cell cultures of normal cells, as well as immortalized cells provided a new opportunity to investigate the mechanisms of the action of antioxidants or their derivatives in radiation-induced cancer prevention. Not all forms of vitamin E exhibit the same efficacy in cancer prevention studies. In 1982, we identified alpha-tocopheryl succinate (alpha-TS) as the most effective form of vitamin E exhibiting anticancer properties.[49] It has been reported that alpha-tocopheryl succinate, but not alpha-tocopherol or alpha-tocopheryl acetate, reduced the incidence of ionizing radiation–induced transformation of normal-like murine fibroblasts in culture.[50,51] Beta-carotene also reduced the incidence of chemical and ionizing radiation–induced transformation of normal-like murine fibroblasts in culture.[52,53] Natural beta-carotene was more effective than synthetic beta-carotene in reducing the incidence of radiation-induced transformation in vitro. These transformed cells produce tumors when injected into syngeneic mice. Excess amounts of estrogen are considered a tumor promoter. N-acetylcysteine (NAC) markedly reduced estrogen-induced transformation of E6 cells (a normal mouse epithelial cell line) in culture.[54] A number of studies have revealed that breast cancer formation is related to abnormal estrogen oxidation to form an excess of estrogen-3, 4-quinones, which react with DNA to form depurinating adducts and induce mutations. This metabolite of estrogen can induce transformation in normal cells in culture. Thus, NAC can prevent the oxidation of estrogen and thereby reduce the risk of breast cancer. From the cell culture studies, it can be suggested that the addition of natural beta-carotene and alpha-TS to the multiple micronutrient preparation would be necessary for the proposed cancer prevention strategy in radiation-exposed populations.

The exact mechanisms of protection of radiation-induced carcinogenesis in vitro by these antioxidants are unknown; however, I suggest that antioxidants prevent those mutagenic changes that initiate immortalization (preneoplastic state), probably by reducing oxidative damage and chronic inflammation. The results of in vitro studies cannot readily be extrapolated to animals or humans with respect to antioxidant type, dose, or dose schedule. The cell culture models are excellent for demonstrating whether or not antioxidants can prevent radiation-induced cancer in mammalian cells, and for investigating the molecular mechanisms of the action of antioxidants in prevention of radiation-induced cancer. Therefore, the use of a single antioxidant to investigate the incidence of cancer or cancer prevention mechanisms is perfectly valid and essential in cell culture models. The use of a single antioxidant may not be useful for the prevention of human cancer in high-risk populations, such as the radiation-exposed populations of Hiroshima and Nagasaki and cancer survivors, because of the presence of a high internal oxidative environment in these populations in which a single antioxidant may act as a pro-oxidant rather than as an antioxidant.

Animal studies: The role of antioxidants in cancer prevention was demonstrated in animal models long before any human or cell culture study was initiated. However, the studies focused on chemical-induced cancer rather than radiation-induced cancer. Nevertheless, these studies are pertinent to radiation-induced cancer because increased oxidative stress and inflammation are also involved in chemical-induced cancer. A two-stage model of carcinogenesis was primarily developed and utilized to investigate the mechanisms of carcinogenesis and to determine the efficacy of anti-oxidants in cancer prevention. The overwhelming majority of studies performed on this model suggested that supplementation with high doses of individual antioxidants such as vitamin C,[55] vitamin E,[43] retinoids,[56] and carotenoids[57,58] reduced the risk of chemical-induced tumors. Among various forms of vitamin E, alpha-tocopheryl succinate was most effective as an anticancer agent.[43] In the Lady transgenic animal model, the combination of vitamin E, selenium, and lycopene was effective in reducing the risk of prostate cancer, whereas the combination of vitamin E and selenium was ineffective.[59] The data from this animal study also suggested that the use of multiple antioxidants may be necessary to reduce the risk of cancer in general and prostate cancer in particular. In p53 knockout pregnant mice, prenatal supplementation with vitamin E (all-rac-alpha-tocopheryl acetate) reduced postnatal malignancies by reducing the levels of DNA oxidation.[60]

Dihydrolipoic acid, a reduced form of alpha-lipoic aid, significantly reduced tumor incidence and tumor multiplicity in DMBA/TCHQ-induced skin tumor formation.[61] Tetrachlorohydroquinone (TCHQ) is a tumor promoter, and dimethylbenzanthracene (DMBA) is a tumor initiator. Dihydrolipoic acid also markedly inhibited expression of inducible nitric oxide synthase (iNOS) protein and cyclooxygenase-2 (COX-2) activity, and reduced the tumor incidence and tumor multiplicity of DMBA/TPA-induced skin tumors.[62] In mice overexpressing Her2/neu, as an animal model for breast cancer and APCmin mice for intestinal cancer, supplementation with alpha-lipoic acid did not affect the incidence of breast or colon cancer.[63] The reasons for these contradictory results between the transgenic model of cancer and the chemical-induced cancer model in animals remain unknown.

The hereditary human disorder ataxia telangiectasias (AT) is characterized by an extremely high incidence of lymphoid malignancy. Using AT-deficient mice, it was demonstrated that supplementation with NAC increased lifespan and reduced the incidence and tumor multiplicity of lymphoma.[64] However, supplementation with NAC did not change the incidence of liver tumor, but caused a significant decrease in tumor multiplicity in rats treated with DEN/ DEDTC (N-diethyl nitrosamine/diethyldithiocarbamate).[65] On the other hand, using the p53 haploinsufficient Tg.AC (v-H-ras) mouse, which contains activated ras oncogenes and an inactivated p53 tumor suppressor gene, frequently found in human cancer, it was demonstrated that supplementation with NAC did not affect the incidence of benzo(a)pyrene-induced skin tumor; however, it reduced tumor multiplicity.[66] Supplementation with NAC did not affect DMBA-induced mammary tumors in the rodent model.[67] In contrast to DMBA-induced mammary tumors, NAC supplementation reduced the incidence of urethane-induced lung cancer.[68] Vitamin E in combination with NAC was more effective in reducing the incidence of esophageal cancer in esophagogastroduodenal anastomosis (EGDA) rat models than the individual agents.[69] Thus, the effects of NAC in reducing induced cancer in rodent models are variable, depending upon the type of tumor and tumor-inducing agents. It is interesting to note that both NAC and alpha-lipoic acid, when used individually, produced inconsistent results.

Supplementation with coenzyme Q10 reduced azoxymethane-induced aberrant crypt foci and mucin-depleted foci in the colon of male rats.[70] These animal studies suggest that the use of a single antioxidant in cancer prevention may produce inconsistent results.

A few studies have shown that supplementation with individual antioxidants can reduce the risk of radiation-induced cancer. It has been reported that post-irradiation feeding of mice with antioxidants, such as curcumin, ascorbic acid, or eugenol reduced radiation (3 Gy)-induced thymic lymphoma.[71] Dietary supplementation with a mixture of antioxidants or Bowman-Birk protease inhibitor (BBI) concentrate reduced the incidence of cancer induced by space radiation (protons or highly energetic heavy particles [HZE particles]).[72] Administration of TMG [2-(alpha-d-gluco-pyranosyl) methyl-2,5,7,8 tetramethylchroman-6-0l], a water soluble derivative of vitamin E, at a dose of 600 mg/kg of body weight IP after whole-body irradiation of mice with a dose of 7 Gy reduced the incidence of mammary and pituitary cancer among survivors.[73] Dietary supplementation with retinyl acetate reduced radiation-induced skin cancer through altering the expression of proliferating-related genes.[74]

A few studies on animal models found that certain antioxidants at very high doses when used individually may increase the risk of cancer. For example, vitamin E at very high doses (equivalent of 40 g per person per day) increased the risk of chemical-induced cancer in the small intestine of mice.[75] Vitamin C, in the form of sodium ascorbate at a high dose, increased the risk of chemical-induced bladder cancer in rats.[76] It was found that the increased osmolarity of urine caused chronic irritations in the bladder, which may account for the increased risk of chemical-induced cancer following treatments with a high concentration of sodium ascorbate. The use of such high doses of antioxidants in cancer prevention studies is not relevant to humans except that they can produce harmful effects in animals, and possibly in humans.

Most studies published on animal models have utilized a single dietary or endogenous antioxidant and have yielded a sometimes inconsistent protective effect against radiation- or chemical-induced cancer. The efficacy of a mixture of dietary and endogenous antioxidants in reducing the incidence of radiation- or chemical-induced cancer in animal models has seldom been tested. From animal studies, it can be concluded that antioxidants have potential to reduce the risk of cancer in humans.

The variability in the amounts of antioxidants present in the animal's diet may contribute to the inconsistent results with respect to the role of the antioxidant in reducing the risk of cancer following irradiation or exposure to other carcinogens. It should be noted that most animal studies in the past used Purina Chow (Ralston Purina, St. Louis, MO), a standard rodent diet. The antioxidant levels in Purina Chow are known to vary significantly from one batch to another; and this could have an impact in determining the efficacy of supplemented antioxidants in cancer prevention. In my opinion, a well-defined diet that contains multiple dietary and endogenous antioxidants at levels that are higher than the RDA levels for rodents must be used for any cancer-prevention study with antioxidants in animals. The published studies that have failed to use such well-defined diets may not provide accurate results on the efficacy of single or multiple antioxidants in reducing the risk of chemical- or radiation-induced cancer.

Although animal models are useful for determining the efficacy of antioxidants in cancer prevention, the results obtained for these models cannot be extrapolating to humans with respect to the dose, dose schedule, and type of antioxidants because the absorption, tissue distribution, biological turnover, and metabolism of these antioxidants in animals are totally different from those found in humans. Unlike humans, most rodents except guinea pigs make their own vitamin C, which could have an impact on the dose and efficacy of supplemented antioxidants in reducing the risk of radiation-induced cancer.

Human epidemiologic studies: Epidemiologic studies utilize two different experimental designs, retrospective case-control studies and prospective case-control studies. The design of a retrospective case-control study involves analysis of the history of dietary intake through questionnaires and personal interviews of cancer patients and comparison to age- and sex-matched normal subjects. From this comparison, the association between dietary agents and cancer incidence is determined. A prospective case-control study involves analysis of the intake from dietary records of participating normal subjects and then correlates the relationship of dietary agents with cancer incidence in subsequent years. From the dietary data obtained through questionnaires or records, the levels of intake of antioxidants such as vitamins A, C, and E, and beta-carotene, and fat and fiber are estimated using appropriate nutritional computer software. Occasionally, the plasma or serum levels of antioxidants in participating individuals are measured. Using these epidemiologic experimental designs, several studies[13,77,78] concluded that diets rich in antioxidants but low in fat and high in fiber were associated with the reduced risk of cancer.[79] However, no epidemiologic studies have been performed to determine whether a diet rich in antioxidants can reduce the risk of cancer among radiation-exposed individuals.

In the general population, consumption of fruits and vegetables and food items rich in carotene and lycopene may reduce the risk of ovarian cancer. A diet low in fat and high in fiber from fruits and vegetables and regular modest consumption of alcohol are associated with reduced risk of benign prostatic hyperplasia (BPH).[80] It has been estimated that eating fruits and vegetables can reduce the risk of cancer by about 30%.[81] Another epidemiologic study showed that eating one or more apples a day was associated with a reduced risk of colorectal cancer. This effect was not observed with other fruits.[82] When the risk of cancer was correlated with the level of individual antioxidants in the diet or blood, the inverse association between diet and cancer incidence became weak, nonexistent, or reversed. In a prospective study involving 295 cases and 295 control menopausal women, the plasma levels of retinol, retinyl palmitate, alpha-carotene, beta-carotene, beta-cryptoxanthin, lutein, lycopene, total carotenoids, alpha-tocopherol, and gamma-tocopherol were measured. The results showed that beta-carotene, lycopene, and total carotene were lower in cases compared to controls. The risk of developing breast cancer was inversely proportional to the level of beta-carotene in plasma.[83] In another similar prospective study, the level of alpha-carotene, but not other carotenoids, was found to be inversely related to the risk of breast cancer.[84,85] In the Vitamins and Lifestyle (VITAL) cohort study, it was found that long-term intake of beta-carotene, retinol, and lutein was associated with increased risk of lung cancer.[86] In Brazilian women, dietary intake of folate, vitamin B_6, or vitamin B_{12} had no overall association with breast cancer risk; however, dietary intake of high levels of folate was associated with an increased risk of breast cancer in premenopausal women and MTR2756GG genotype.[87]

Increased intake of dietary flavonoids was associated with the reduced risk of lung cancer;[88] however, another study reported no association between individual or multiple flavonoids intake and the risk of breast, ovarian, colorectal, lung, and endometrial cancer.[89] A recent study has reported that intake of lycopene and lycopene products through diet was associated with the decreased risk of prostate cancer.[90] In the Women's Health Initiative (WHI) involving 133,614 postmenopausal women, it was found that dietary intake of antioxidants, carotenoids, and vitamin A were not associated with a reduction in ovarian cancer risk.[91] During 8 years of follow-up involving 56,007 French women, it was found that breast cancer risk was inversely associated with alpha-linolenic acid (ALA) intake from fruits and vegetables, and vegetable oils, but it was positively related to ALA intake from nut mixes and processed foods.[92] This suggests that other protective substances in the food, such as antioxidants, may be necessary to observe the protective effect of ALA on the cancer incidence. In addition, it was observed that the risk of breast cancer was inversely associated with intake of omega-3 in women having the highest levels of omega-6. Thus, epidemiologic studies with dietary intake of antioxidants, fat and fiber, or B vitamins alone have produced conflicting results. This may be due to the fact that each of these nutrients may contribute to reduction in cancer incidence in different amounts; therefore, they cannot be analyzed separately for consistent results. In addition, human diet contains agents that can produce opposite effects on cancer risk. For example, antioxidants and high fiber are considered cancer-protective substances in the diet, whereas diets rich in fat, meat, calories, and nitrites may increase the risk of cancer. Dietary antioxidants can also influence metabolism of ingested or inhaled mutagens and carcinogens.[93,94] These may be the reasons why the results with a diet rich in fruits and vegetables containing high levels of antioxidants, low in fat, and high in fiber consistently produced cancer-protective effects.

Experimental designs of any epidemiologic study have several inherent technical problems that make it difficult to arrive at any definitive conclusions. These limitations include the following: (a) collection of retrospective dietary history data by questionnaire is unreliable, because it is based on the memory of the participants in the study, and because the quantitative and qualitative information on past daily dietary intake are impossible to recall with any degree of accuracy; and (b) dietary records are difficult to express in a quantitative manner because the information on antioxidant intake is based on estimations rather than actual measurements. Thus, the determination of dietary intake of antioxidants or fat and fiber on the basis of a diet history or a dietary record must be considered unreliable until validated by the blood or tissue levels of these nutrients. There are several other confounding factors associated with the lifestyle and environments that could impact cancer incidence in addition to the diet. It is very difficult to account for all of them in the data analysis. It should be emphasized that epidemiologic studies, despite the best experimental design and correct data interpretation, can only infer a direct or inverse relationship between single or multiple nutrients and the risk of cancer. The cause–effect relationship between micronutrient intake and cancer risk can only be established by a well-designed intervention trial with antioxidants in a high-risk population, such as survivors of radiation exposure.

Human intervention studies: High-risk populations, such as heavy tobacco smokers, survivors of radiation exposure, individuals with precancerous lesions, cancer patients who are in remission, and persons with a family history of cancer, are very appropriate models that can be used to evaluate the efficacy of antioxidant supplements on the risk of cancer. The experimental design for a clinical trial must consider not only the statistical, bias, end point, and period of observation issues, but also the selection of an appropriate type and number of micronutrients and their dose and dose schedule, and the internal oxidative environment of the participating subjects.

No studies have been performed on the efficacy of supplemented antioxidants in reducing the cancer risk among radiation-exposed populations. The clinical studies published on other high-risk populations, thus far, were sound with respect to selection of populations, number of patients, and statistical power analysis, but they did not take into consideration the previously discussed scientific rationale with respect to antioxidants. The intervention study rarely utilized endogenous antioxidants. This has resulted in inconsistent results on cancer risk varying from no effect, to beneficial effects, to harmful effects. For example, some studies used only one antioxidant (mostly synthetic form), while others used more than one antioxidant (only the dietary form). The dose range of antioxidants and criteria of end points varied markedly from one study to another. All studies utilized a once-a-day dose schedule. The inconsistent results obtained from the intervention studies have created much misinformation and confusion in the minds of health providers and the public alike; as a consequence, antioxidants are being misused by the public and not recommended by most physicians for cancer prevention.

The extrapolation of existing clinical trial models to determine the efficacy of novel drugs that affect specific molecular targets involved in the etiology of a particular disease should not be applied to the clinical trials in which the efficacy of micronutrients to reduce the risk of cancer is evaluated. Any efforts to do so may produce inconsistent results. Unfortunately, most clinical trials have utilized the experimental design of drugs, which is based on the concept of a single drug–single target effect. The following text discusses intervention studies with antioxidants in high-risk populations.

Tobacco smokers: Heavy tobacco smokers represent an excellent model of a high-risk population in which the efficacy of micronutrients to reduce the risk of cancer can be tested in a well-designed clinical trial. In a large randomized, double-blind, placebo-controlled clinical trial, supplementation with synthetic beta-carotene at a dose of 20 mg/day increased the incidence of lung cancer, prostate cancer, and stomach cancer among male heavy tobacco smokers.[95–97] In my opinion, this trial design was flawed with respect to antioxidants. Heavy tobacco smokers are known to have high levels of internal oxidative environment; therefore, the increase in lung cancer incidence after supplementation with beta-carotene alone could have been predicted, because beta-carotene in the high internal oxidative environment of heavy smokers would be oxidized, and then act as a pro-oxidant rather than as an antioxidant. This possibility was further confirmed by the fact that the same dose of beta-carotene did not affect the incidence of cancer in normal populations who have a lower internal oxidative environment than heavy tobacco smokers.[98] Supplementation with NAC reduced certain biomarkers associated with lung cancer in tobacco smokers,[99] but it remains uncertain whether NAC supplementation alone can affect the incidence of lung cancer. In another study, supplementation with vitamin E (400 IU per day) was associated with a reduced risk of prostate cancer in tobacco smokers, whereas beta-carotene supplementation was associated with a reduced risk of this cancer only in those smokers who have low levels of beta-carotene.[100] Elevated levels of benzo (a) pyrene (B (a) P)-DNA adducts have been associated with a 3-fold increase in the risk of lung cancer in heavy tobacco smokers,[101] but supplementation with dl-alpha tocopherol (400 IU) and vitamin C (500 mg) per day did not reduce the level of B (a) P-DNA adducts in men, but it reduced them in women.[101] Using the population of Alpha-Tocopherol, Beta-Carotene Cancer Prevention (ATBC) study of male Finnish smokers, it was revealed that higher serum alpha-tocopherol concentrations were associated with the reduced risk of pancreatic cancer.[102] In a review of 6 randomized clinical trials and 25 prospective studies, it was concluded that beta-carotene supplementation was not associated with a decreased risk of lung cancer.[103]

Cancer survivors: In a clinical study involving 2592 patients (60% head and neck cancer and 40% lung cancer), a 2-year supplementation with vitamin A (retinyl palmitate, 300,000 IU daily for one year, and 150,000 IU for the second year) and NAC (600 mg daily for 2 years) alone or in combination produced no benefit with respect to secondary primary tumors. [104] This suggested that using one or two antioxidants was not sufficient to reduce the risk of second cancer in cancer survivors. This study utilized unusually high doses of vitamin A that could be toxic after a long-term consumption. Even the dose of NAC in this study is high. These clinical studies have failed to consider the biology of antioxidants and internal oxidative environment of high-risk populations. No studies have been performed to evaluate the role of antioxidants in reducing the risk of cancer among radiation-exposed individuals.

Patients with premalignant lesions: Beta-carotene supplementation in patients with low dietary intake of beta-carotene reduced prostate cancer.[100] High doses of beta-carotene caused regression of leukoplakia (precancerous lesion in mouth).[105,106] Retinoids also caused regression of oral leukoplakia and other cancers.[107]

Supplementation with alpha-tocopherol (50 mg/day) reduced the risk of prostate cancer and colorectal cancer, but increased the incidence of stomach cancer.[95] Supplementation with vitamin E (400 IU) produced no effect on the prostate specific antigen (PSA) levels;[108] however, high serum levels of vitamin E were associated with reduced prostate cancer incidence.[109-111] The result of the combined effect of vitamin E and selenium is consistent with the results of the Selenium and Vitamin E Cancer Prevention Trial (SELECT) study in which the combination of vitamin E and selenium was found to be ineffective in reducing the risk of prostate cancer.[112]

It has been reported that consumption of vitamin E alone increased the incidence of secondary primary cancer or recurrence of the initial tumor after cancer therapy.[113] The causal relationship between vitamin E supplementation and increased mortality as suggested earlier [114] has been contradicted by another statistical analysis.[115] In a recent double-blind, placebo-controlled $2 \times 2 \times 2$ factorial trial of vitamin C (500 mg of ascorbic acid, daily), a natural source of vitamin E (600 IU of alpha-tocopherol, every other day), and beta-carotene (50 mg, every other day), involving 7,627 women free of cancer, it was found that supplementation with individual vitamin C, vitamin E, or beta-carotene had no impact on cancer incidence and cancer mortality during a follow-up period of 9.4 years.[116]

The results obtained from supplementation with single antioxidants should not be extrapolated to the effect of the same antioxidant present in a multiple-antioxidant preparation. Nevertheless, many scientists, researchers, and physicians and some press publications are promoting the idea that supplementation with antioxidants can be deleterious to your health, and should not be taken for cancer prevention. Such misleading promotions have no scientific merit. The use of a single antioxidant to reduce the risk of cancer in high-risk populations, such as heavy tobacco smokers, has no scientific basis for the following reasons: (a) individual antioxidants in a high oxidative environment act as a pro-oxidant, because antioxidants are easily oxidized, and (b) heavy tobacco smokers have a very high internal oxidative environment due to inhalation of smoke and depletion of antioxidant levels. Therefore, the use of a single antioxidant in a high-risk population is expected to increase the risk of cancer. The natural forms of vitamin E and beta-carotene are more effective than their synthetic counterparts. For example, natural beta-carotene reduced radiation-induced transformation, but the synthetic beta-carotene did not.[53] Therefore, the use of synthetic beta-carotene may not produce an optimal effect on cancer prevention among high-risk populations. Cells accumulated more of the natural form of vitamin E than the synthetic form,[117] and alpha-tocopheryl succinate is now considered the most effective form of vitamin E.[118] Therefore, natural forms of vitamin E and beta-carotene should be utilized in any clinical studies with antioxidants. These issues were not considered in the previous clinical studies with antioxidants.

The dose schedule is also very important to enhance the efficacy of multiple micronutrients in reducing the risk of cancer. Most published studies with antioxidants in high-risk populations to develop cancer have utilized a once-a-day dose schedule. Taking antioxidants once per day may create huge fluctuations in the levels of antioxidants in the body. This is due to the fact that biological half-lives (time needed to remove 50% of the antioxidants from the body) of antioxidants markedly vary, depending upon their lipid or water solubility, creating a high degree of fluctuation in tissue antioxidant levels. Indeed, we have shown that a marked alteration in the expression of gene profiles occurs with different levels of vitamin E succinate;[119] therefore, taking antioxidants once a day can create genetic stress in the cells that may compromise the efficacy of antioxidant supplementation after long-term consumption. These factors were not taken into consideration while designing antioxidant trials in high-risk populations in previous studies, and thus they produced inconsistent results.

CANCER RISK AFTER TREATMENT WITH MULTIPLE DIETARY ANTIOXIDANTS

No studies have been performed to evaluate the role of multiple dietary and endogenous antioxidants in reducing the risk of cancer among radiation-exposed individuals. Some studies on other high-risk populations have shown that supplementation with multiple dietary antioxidants without endogenous antioxidants (alpha-lipoic acid, n-acetylcysteine, a glutathione-elevating agent, and co-enzyme Q10, and l-carnitine) may produce inconsistent results. These studies are described in the following text. The administration of multiple dietary antioxidants vitamin A (40,000 IU), vitamin C (2,000 mg), vitamin E (400 IU), zinc (90 mg), and vitamin B$_6$ (100 mg) per day in combination with Bacilli bilie de Calmette-Guerin (BCG) vaccine caused a 50% reduction in the incidence of recurrence of bladder cancer in 5 years, in comparison to control patients who received multiple vitamins containing RDA levels of nutrients and BCG.[120] Supplementation with antioxidants (vitamin A, 30,000 IU; vitamin C, 1,000 mg; vitamin E, 70 mg) per day reduced the incidence of recurrence of colon polyps from 36 to 6 percent;[121] however, consumption of synthetic beta-carotene (25 mg), vitamin C (1,000 mg), and vitamin E (400 mg) per day failed to show any beneficial effects on the recurrence of colon polyps.[122] Daily administration of vitamin C (400 mg) and dl-alpha tocopherol also failed to reduce the incidence of recurrence of colon polyps.[123] In another study, daily supplementation with vitamin C (4,000 mg) and vitamin E (400 mg) also failed to reduce the risk of colon polyps, but when they were combined with a high-fiber diet (more than 12 grams per day) there was a significant reduction in the incidence of recurrence of polyps.[124] This study indicated the importance of a high-fiber diet in combination with antioxidants in cancer prevention.

In the Linxian General Population Nutrition Interventional Trial, a preparation of multiple dietary antioxidants (beta-carotene, vitamin E, and selenium at doses 2–3 times that of the US RDA) reduced mortality by 10%, and cancer incidence by 13%.[125] The beneficial effects of this supplementation on mortality were still evident up to 10 years after the cessation of supplementation, and were consistently greater in younger participants.[126]

The combination of vitamins A, C, and E, omega-3 fatty acids, and folic acid significantly reduced recurrence of adenoma in patients after polypectomy.[127] In a randomized placebo-controlled trial involving 80 untreated patients with prostate cancer, daily supplementation with vitamin E, selenium, vitamin C, and coenzyme Q10 did not affect serum levels of prostate specific antigen (PSA).[128] In a randomized, placebo-controlled trial referred to as the Selenium and Vitamin E Cancer Prevention Trial (SELECT) involving 35,553 healthy men from 427 participating sites in the United States, Canada, and Puerto Rico, with a follow-up period of a minimum of 7 years and a maximum of 12 years, it was observed that selenium (200 mcg/day) or vitamin E (400 IU /day), alone or in combination, did not reduce the risk of prostate cancer.[112]

From the intervention studies discussed previously, it appears that supplementation with multiple dietary antioxidants alone may not be sufficient to produce optimal and consistent reduction in the risk of cancer in high-risk populations. Inclusion of endogenous antioxidants may be necessary in any experimental design to test the efficacy of antioxidants in cancer prevention.

In recent years, there have been trends to perform meta-analysis of published data in which far-reaching conclusions have been made. Since the total number of participating subjects in such analyses becomes huge, numbering in the hundreds of thousands, the conclusions appear impressive and definitive. However, if the meta-analysis is performed on publications that have a flawed experimental design, the conclusions will be the same as those made in the initial publications. These analyses seldom express concerns about the experimental designs that were used in the studies. The publications of such meta-analyses are of no scientific value except that they add to the existing misinformation about the value of micronutrients in cancer prevention.

CANCER RISK AFTER TREATMENT WITH VITAMIN D AND CALCIUM

No studies have been performed to evaluate the role of vitamin D and calcium in reducing the risk of cancer among radiation-exposed individuals. In recent years, the role of vitamin D alone or in combination with calcium has been evaluated, yielding often inconsistent results. The results showed that vitamin D (1000 IU) per day reduced colorectal cancer.[129,130] In another study, vitamin D (400 IU) and calcium (1000 mg) per day produced no effect on colorectal cancer.[131] Administration of vitamin D (400 IU) and calcium (1000 mg) per day reduced colorectal cancer, but in combination with estrogen increased the risk of this cancer.[132] In view of the fact that estrogen is known to have tumor-promoting effects, and that vitamin D has no effect on the tumor-promoting activity of estrogen, the above increase in cancer incidence was expected. This could have been avoided by the addition of antioxidants that are known to reduce the effect of tumor promoters. In a recent review of several clinical studies, it was concluded that supplementation with elemental calcium may have a modest effect in reducing the risk of colorectal cancer; however, this approach was not recommended for reducing the risk of colorectal cancer in the general population.[133] Supplementation with elemental calcium (1000 mg) and vitamin D (400 IU) a day did not reduce the risk of breast cancer.[134] However, in another study, dietary intake of calcium was modestly associated with the reduced risk of breast cancer in postmenopausal women.[79,135] In the Wheat Bran Fiber Trial, higher intake of Ca (1068 mg vs. 690 mg/day) decreased the risk of recurrence of colorectal adenoma by about 45%.[136] The beneficial effect of this treatment was also noted in women with BRCA mutation. It has been reported that Ca and vitamin D supplementation together reduce the recurrence of colorectal adenoma.[137] The reasons for these inconsistencies may be that the contribution of calcium alone or with vitamin D to cancer formation may be small, and therefore, additional micronutrients such as antioxidants and B vitamins may be necessary to produce consistent results. I propose adding vitamin D and calcium to a multiple micronutrient preparation to be used in the proposed cancer prevention strategy in high-risk populations.

CANCER RISK AFTER TREATMENT WITH FOLATE AND B VITAMINS

No studies have been performed to evaluate the role of folate and B vitamins in reducing the risk of cancer among radiation-exposed individuals. A diet rich in folate and vitamins B_6 and B_{12} was associated with a reduced risk of breast cancer and colorectal cancer,[138–142] but had no association with pancreatic cancer. However, supplementation with folic acid alone did not reduce the incidence of colorectal adenoma, and may possibly increase the risk.[143] In other studies,[143,144] consumption of folate did not reduce the incidence of colorectal cancer. In a randomized double-blind, placebo-controlled clinical trial involving 137 patients with polypectomy, supplementation with 5 mg of folate daily reduced the recurrence of colonic adenomas.[145] Dietary intake of high amounts of folate and vitamin B_{12} were independently associated with decreased risk of breast cancer, particularly in postmenopausal women, whereas there was no association between intake of vitamin B_6 and breast cancer risk.[141] In a recent epidemiologic study, it was found that there may be a nonlinear relationship between folate status and the risk of all-cancer mortality, and thus persons with low serum levels of folate may be at risk of cancer.[146] This inconsistency between diet and supplement may be due to the fact that the diet may also be rich in antioxidants in addition to B vitamins and folate; therefore, clinical studies with folate and B vitamins alone are expected to yield different results. I propose combining B vitamins and folate with multiple micronutrients for the proposed cancer prevention strategy.

CANCER RISK AFTER TREATMENT WITH FAT AND FIBER

No studies have been performed to evaluate the role of fat and fiber in reducing the risk of cancer among radiation-exposed individuals. The average American diet contains about 34.1% calories from fat. Animal experiments and human epidemiologic studies have revealed that this level of fat consumption may increase the risk of cancer. The mechanisms of action of the high-fat diet on carcinogenesis are not well understood except that high levels of fat can act as a tumor promoter. In addition, a high-fat diet may increase the blood levels of prostaglandins, which have been implicated in increasing the risk of cancer. In females, a high-fat, low-fiber, Western-style diet appears to be associated with increased levels of plasma and urinary estrogen, which is significant, as this hormone is known to act as a tumor promoter. An intervention study (The Women's Health Initiative Dietary Modification Trial) in which postmenopausal women received a low-fat diet (40% calories from fat) or a high-fat diet (60% calories from fat) showed that a low-fat diet did not reduce the risk of colorectal cancer during the 8.1-year follow-up period.[147] I am not sure if the diet in which fat represents 40% should be considered as a low-fat diet. In addition, the protective effect of low fat alone may be minor; therefore, the cancer protective effect of a low-fat diet alone cannot be assessed in any cancer prevention trial.

The detail references for this section have been provided in a review.[31] The average American diet is low in fiber. Human epidemiologic studies revealed a strong inverse relationship between fiber intake and cancer incidence. High fiber would bind increased amounts of intestinal cholesterol, bile acids, and mutagens and carcinogens that are formed during digestion and eliminate them with the feces. This would then reduce the intestinal absorption, and exposure time of the intestinal cells to these potentially carcinogenic substances. Therefore, it was assumed that only cancer of the intestinal tract would be reduced by this mechanism of action of high fiber. However, the consumption of high fiber reduced the recurrence of breast cancer. Therefore, additional mechanisms of cancer protection by high fiber may exist. Indeed, it has been reported that high fiber intake can generate millimollar levels of butyric acid, a 4-carbon fatty acid, with the help of endogenous bacteria that are present in colon. Butyric acid, a small fatty acid, is absorbed rapidly. Several studies have reported that butyrate and its analog (phenyl butyrate) have exhibited strong anticancer properties against a variety of tumors in vitro and in vivo. Thus, a high-fiber diet may provide protection not only against colon cancer but also against other tumors. Therefore, the inclusion of a low-fat and high-fiber diet is essential for the proposed cancer prevention strategy.

In the Polyp Prevention Trial, dietary intervention, supplementation with reduced fat, and increased consumption of fruits and vegetables and fiber produced no effect on PSA and on the incidence of prostate cancer in normal men.[148] Similarly, adopting a diet low in fat (20% of total calories) and high in fiber (18 g per 1000 kcal) and fruits and vegetables (3.5 serving per 1000 kcal) did not influence the risk of recurrence of colorectal adenomas.[149] In contrast to the previous observations, another study reported that vitamin A from food with or without supplementation and alpha-carotene from food may protect against recurrence of tumors in nonsmokers and nondrinkers.[150] The current dietary guidelines (fruits, vegetables, whole-grain, low-fat dairy, and lean meat) are associated with a decreased risk of mortality from all causes.[151] A dietary supplement (13.5 g/day vs. 2 g/day) with wheat-bran fiber did not protect against recurrence of colorectal adenomas.[152] Testing the effect of a high-fiber diet alone in which the difference between the control and experimental group is small may not yield any significant reduction in cancer incidence because the effect of an additional 10 g of fiber alone on the risk of cancer may be too small to be detected. Such an interventional trial does not appear to have any scientific rationale.

In a multi-institutional, randomized, controlled trial involving 3088 women previously treated for an early stage of breast cancer, supplementation with a diet high in vegetables, fruits, and fiber, and low in fat did not reduce additional breast cancer events or mortality during a 7.3-year follow-up period.[153]The lack of additional micronutrients, such as dietary and endogenous antioxidants, B-vitamins, and calcium with vitamin D, may have contributed to the failure of detecting protective effects of high fiber and low fat in the above study.

ESSENTIAL FACTORS FOR AN EFFECTIVE CANCER PREVENTION STRATEGY

In order to develop an effective cancer prevention strategy among radiation-exposed populations, it is essential know about agents that can increase the risk of radiation-induced cancer and those that can reduce it. Cancer research of the last several decades has identified agents in the environment, diet, and lifestyle that can increase the risk of radiation-induced cancer, and agents in the diet that can reduce it. These issues have been reviewed in several publications and books.[31] Cancer-causing agents (carcinogens) can be divided into two categories: tumor initiators and tumor promoters. Tumor initiators are agents that by themselves are sufficient to cause cancer, whereas tumor promoters by themselves may not cause cancer, but they help tumor initiators to induce cancer at low doses that may not be sufficient to cause cancer. In addition, tumor promoters may enhance the incidence of tumor initiator–induced cancer. Some tumor initiators, such as chemical carcinogens, ozone, tobacco smoking, ultraviolet radiation, cancer-causing viruses, and familial gene defects, and some tumor promoters, such as high levels of estrogen and phorbol esters, can enhance the incidence of radiation-induced cancer. Dietary factors that can reduce the risk of radiation-induced cancer include antioxidants, selenium, and a low-fat and high-fiber diet.

PROPOSED CANCER PREVENTION STRATEGY FOR RADIATION-EXPOSED POPULATIONS

Radiation-exposed populations include survivors of Hiroshima and Nagasaki radiation exposure and their offspring, those exposed to higher doses of radiation after nuclear accidents at the nuclear power plants in Chernobyl, Ukraine, and more recently, in Fukushima, Japan, and cancer survivors who have received radiation therapy. I propose a cancer prevention strategy for the survivors of high-dose radiation exposure who are at high risk for developing cancer and noncancerous diseases. In this strategy, daily supplementation with an appropriately prepared micronutrient preparation and modifications in diet and lifestyle are essential for reducing the cancer risk optimally.

Application of BioShield-R2 for reducing the risk of cancer: As discussed earlier, among various radioprotective and/or radiomitigating agents, BioShield products containing multiple micronutrients, including dietary and endogenous antioxidants, satisfy the criteria of an ideal bioshield because they can provide tissue protection against acute radiation sickness (ARS) induced by high-dose radiation as well as reduce the risk of cancer and nonneoplastic diseases among survivors. The ingredients in BioShield exhibit both radioprotective and/or radiomitigating properties. A few human clinical studies showed that antioxidants protect normal tissues against radiation damage. In addition, they are considered safe in humans when consumed orally on a short- and long-term basis. BioShield-R2 is available commercially and contains dietary antioxidants and their derivatives (vitamin A, natural mixed carotenoids, vitamin C, vitamin E, vitamin E succinate, vitamin D, and mineral selenium), endogenous antioxidants (glutathione elevating agents, such as n-acetylcysteine and alpha-lipoic acid, and L-carnitine, and coenzyme Q10), vitamin D, all B vitamins, mineral selenium, and zinc, but no trace minerals (iron, copper, or manganese) or heavy metals (zirconium and molybdenum). Calcium and vitamin D supplementation at high doses would be included for women above the age of 35 years. Such a preparation of multiple micronutrients is available commercially (www.mypmcinside.com). Radiation-exposed individuals or their offspring can start the supplementation at any time after irradiation.

It has been presumed that the genetic basis of cancer cannot be delayed or prevented; therefore, such individuals wait until the tumor appears. A recent study on the effect of a mixture of dietary and endogenous antioxidants on proton radiation–induced cancer in *Drosophila melanogaster* suggests that antioxidants can reduce the incidence of the genetic basis of cancer. For example, female fruit flies carrying a mutant gene HOP (TUM-1) become very sensitive to developing leukemia-like cancer. Exposure to proton radiation markedly enhanced the incidence of this cancer. Supplementation with the antioxidant mixture though the diet before and after irradiation completely blocked radiation-induced cancer (collaboration with Dr. Sharmila Bhattacharya of NASA at Moffat Field, California). This result obtained from fruit flies cannot be extrapolated to humans, but this study at least suggests that antioxidants have a potential to reduce the risk or delay the appearance of tumors in individuals with a family history of cancer.

Rationale for using multiple micronutrients in BioShield-R2: Multiple micronutrients including dietary and endogenous antioxidants are recommended because many different types of free radicals are produced, and each antioxidant has a different affinity for each of these free radicals, depending upon the cellular environment, and they are distributed differently in organs and within the same cells. The gradient of oxygen pressure varies within the cell and tissues. Vitamin E was more effective as a quencher of free radicals in reduced oxygen pressure, whereas beta-carotene (BC) and vitamin A were more effective in higher oxygen pressures.[154] Vitamin C is necessary to protect cellular components in aqueous environments, whereas carotenoids, vitamins A and E protect cellular components in nonaqueous environments. Vitamin C also plays an important role in maintaining cellular levels of vitamin E by recycling the vitamin E radical (oxidized) to the reduced (antioxidant) form.[155] Also, the DNA damage produced by oxidized vitamin C can be ameliorated by vitamin E. The form and type of vitamin E used in the micronutrient preparation are also important in order to improve its beneficial effects. It has been demonstrated that various organs of rats selectively accumulate the natural form of vitamin E.[117] It has been established that alpha tocopheryl-succinate (alpha-TS) is the most effective form of the vitamin E.[118,156] We have reported that an oral ingestion of alpha-TS (800 I.U./day) for over six months in humans increased plasma levels of not only alpha-tocopherol, but also of alpha-TS, suggesting that alpha-TS can be absorbed from the intestinal tract without hydrolysis to alpha-tocopherol, provided the pool of alpha-tocopherol in the body has become saturated.[118,156] Selenium is a co-factor of glutathione peroxidase and acts as an antioxidant. Therefore, selenium supplementation together with other dietary and endogenous antioxidants is also essential.

Glutathione, one of the endogenously made compounds, represents a potent intracellular protective agent against oxidative damage. It catabolizes H_2O_2 and anions and is very effective (in the presence of glutathione peroxidase) in quenching peroxynitrite.[157] Therefore, increasing the intracellular levels of glutathione is essential for the protection of various organelles within the cells. An oral supplement with glutathione failed to significantly increase plasma levels of glutathione in human subjects,[158] suggesting that this tripeptide is completely hydrolyzed in the gastrointestinal tract. N-acetylcysteine and alpha-lipoic acid increase the intracellular levels of glutathione, and therefore can be used in combination with dietary antioxidants in the proposed cancer prevention strategy.

Coenzyme Q10 is needed by mitochondria to generate energy. In addition, it also scavenges peroxy radicals faster than alpha-tocopherol,[159] and like vitamin C, can regenerate vitamin E in a redox cycle.[160] In addition to the above scientific rationales, clinical trials primarily with one dietary antioxidant, B vitamins, or elemental calcium with or without vitamin D, have produced inconsistent results. It has been shown that a single antioxidant such as vitamin C can stimulate the growth of some cancer cells in culture,[31] and when it is oxidized, it can act as a pro-oxidant.

Therefore, the use of multiple micronutrients containing dietary and endogenous antioxidants is essential for the proposed cancer prevention strategy.

Unique features of Bioshield-R2: The formulation BioShield-R2 does not contain iron, copper, or manganese because they are known to combine with vitamin C to produce free radicals that could reduce the optimal effects of BioShield-R2. In addition, these trace minerals in the presence of antioxidants are absorbed more efficiently. The excretion of these trace minerals or heavy metals from the body is negligible; therefore, they may accumulate in excessive amounts in the body after long-term consumption, which could increase the body stores of free forms of these minerals. The increased body stores of free iron or copper have been associated with the enhanced risk of most chronic human diseases, including cancer. Accumulation of excessive levels of heavy metals could be neurotoxic.

BioShield-R2 contains both vitamin A and natural mixed carotenoids (90% represent beta-carotene). This is due to the fact that beta-carotene, in addition to acting as a precursor of vitamin A, performs unique functions that cannot be produced by vitamin A, and vice versa. For example, beta-carotene increased the expression of the connexin gene, which codes for a gap junction protein that holds two normal cells together,[161] whereas vitamin A did not produce such an effect. Vitamin A produced differentiation in normal and cancer cells, but beta-carotene did not.[107,162] Beta-carotene was more effective in quenching oxygen radicals than most other antioxidants.[57] Thus, the addition of both vitamin A and beta-carotene may enhance the efficacy of micronutrient supplementation in cancer prevention.

BioShield-R2 contains two forms of vitamin E, d-alpha tocopheryl acetate, and d-alpha tocopheryl succinate (alpha-TS). Alpha-TS is now considered the most effective form of vitamin E.[118] Alpha-TS is more soluble than alpha tocopherol and enters the cells easily, where it is converted to alpha-tocopherol, and thus provides intracellular protection against oxidative damage. Alpha-TS can also produce some unique biological effects that cannot be produced by alpha-tocopherol. Therefore, in order to increase the efficacy of vitamin E, the addition of both forms of vitamin E is essential.

BioShield-R2 does not contain herbs or herbal antioxidants. This is due to the fact that certain herbs may interact with prescription and over-the-counter drugs in an adverse manner, and that herbal antioxidants do not produce any unique biological effects that cannot be produced by the antioxidants present in BioShield-R2.

Dose-schedule of BioShield-R2: The recommended BioShield-R2 should be taken orally, divided into two doses, half in the morning and the other half in the evening with a meal. This is because the biological half-lives of micronutrients are highly variable, which can create high levels of fluctuation in the tissue levels of micronutrients. A twofold difference in the levels of certain micronutrients such as alpha-TS can cause a marked difference in the expression of gene profiles (our unpublished data). In order to maintain relatively consistent levels of micronutrients in the body, BioShield-R2 should be taken twice a day.

Suggested modifications in diet and lifestyle: Dietary habits are difficult to alter in humans. Human diets contain both cancer-protective substances and cancer-causing agents (mutagens and carcinogens).[163] Most of the mutagenic and carcinogenic substances that are present in the diet are naturally occurring; however, small amounts of them have been introduced into the diet by the use of pesticides in agriculture production. These issues are usually beyond our control (the exception being exclusive consumption of certified organic produce). Mutagens (compounds that alter genetic activity) are formed during cooking. Browning of vegetables or meats during cooking is an indication of the formation of mutagens. Not all mutations increase the risk of cancer, but all cancers require mutations. Flame-broiled fatty meat, generally preferred by consumers, contains much higher levels of carcinogens like benzo (a) pyrene, than those found in oven-cooked meat. The dietary recommendations of eating fresh fruits and vegetables, and a diet low in fat and high in fiber, are difficult to implement consistently over a long period of time. This issue is difficult to implement because human behaviors are not easily changed. Nevertheless, the dietary recommendations must be a part of the proposed cancer preventive strategy for the survivors of radiation-exposed populations. In addition, recommendations of avoiding exposure to potential carcinogens, tumor promoters, and biological carcinogens should also be adopted.

Dietary recommendations include daily consumption of a low-fat and high-fiber diet with plenty of fresh fruits and vegetables, avoiding excessive amounts of protein, carbohydrate, or calories, restricting intake of nitrite-rich cured meat, charcoal-broiled or smoked meat or fish, caffeine-containing beverages (cold or hot), and pickled fruits and vegetables.

Lifestyle-related recommendations include stopping smoking and chewing tobacco, avoiding second-hand smoke, and overexposure to sun and UV light for tanning, avoiding hyperbaric therapy for energy, restricting intake of alcohol, reducing stress by vacation, yoga, or meditation, and performing moderate exercise 3–5 times a week.

A well-designed clinical trial using the proposed cancer preventive strategy should be initiated in radiation-exposed populations. In the meantime, individuals who have been exposed to any doses of radiation and who are being exposed to low doses of radiation may wish to adopt the proposed cancer prevention strategy in consultation with their physicians.

SAFETY OF MICRONUTRIENTS IN BIOSHIELD-R2

References listed in this section have been described in a review.[164] Antioxidants at doses higher than those present in BioShield-R2 have been consumed by the US population for decades without reported toxicity. However, they could be harmful for some individuals at certain high doses when consumed daily for a long period of time. For example, vitamin A at doses of 10,000 IU or more can cause birth defects in pregnant women, and beta-carotene can produce bronzing of skin at doses of 50 mg or more, which is reversible on discontinuation. Vitamin C as ascorbic acid at high doses (10 grams or more) can cause diarrhea in some individuals; vitamin E at high doses (2,000 IU or more) can induce clotting defects after long-term consumption; vitamin B_6 at high doses may produce peripheral neuropathy; and selenium at doses of 400 mcg or more can cause skin and liver toxicity after long-term consumption. Coenzyme Q10 has no known toxicity, and recommended daily doses are 30–400 mg. N-acetylcysteine doses of 250–1500 mg and alpha-lipoic acid doses of 600 mg are used in humans without toxicity. All ingredients present in BioShield-R2 are safe and come under the category of Food Supplements, and therefore do not require FDA approval for their use in humans.

BIOSHIELD-R2 AND THE RISK OF NONNEOPLASTIC DISEASES

Increased oxidative stress and chronic inflammation may also contribute to the development of nonneoplastic diseases. Therefore, adopting the proposed cancer prevention strategy may also reduce the risk of developing some nonneoplastic diseases.

SUMMARY

The incidence of cancer and noncancerous diseases increases in radiation-exposed populations, which include radiation-exposed individuals and their offspring of Hiroshima and Nagasaki, those exposed to higher doses of radiation after an accident at nuclear power plants in Chernobyl, Ukraine, and most recently, in Fukushima, Japan, and cancer survivors who have received radiation therapy. There is no effective cancer preventive strategy that can easily be implemented at this time for these populations. The common recommendations of modification in diet and lifestyle may not be sufficient to reduce cancer risk optimally. Supplementation with a multiple micronutrient preparation including dietary and endogenous antioxidants together with the diet and lifestyle modifications may be necessary for reducing the risk of cancer in high-risk populations maximally.

The clinical studies published thus far have utilized primarily one dietary antioxidant, sometimes two, and occasionally 3–4 dietary antioxidants alone, folate with or without vitamin B_6, and vitamin B_{12} alone or elemental calcium with or without vitamin D alone, and fat and fiber individually or in combination. The results of these studies have been inconsistent, and varied from beneficial effects, to no effect, to harmful effects. Therefore, there should be a change in the paradigm of the clinical trial while using micronutrients for cancer prevention studies. I have proposed a new paradigm in which appropriately prepared multiple micronutrient preparations, such as BioShield-R2 containing dietary and endogenous antioxidants, B vitamins, elemental calcium with vitamin D, but no iron, copper, manganese, or heavy metals, together with modifications in the diet and lifestyle are utilized. The dose schedule of twice a day is recommended in order to maintain relatively steady tissue micronutrient levels.

The proposed cancer preventive strategy in radiation-exposed individuals has a sound mechanistic basis for reducing the risk of cancer. This strategy can be initiated at any time after irradiation and at any age. The efficacy of a proposed cancer prevention strategy should be tested in radiation-exposed populations by a well-designed clinical trial. Meanwhile, individuals who have been exposed to radiation or who are being exposed to low doses of radiation might like to adopt the proposed cancer prevention strategy in consultation with their physicians. The proposed strategy may also reduce the risk of nonneoplastic diseases.

REFERENCES

1. Iwanaga, M., Hsu, W. L., Soda, M., Takasaki, Y., Tawara, M., Joh, T., Amenomori, T., Yamamura, M., Yoshida, Y., Koba, T., Miyazaki, Y., Matsuo, T., Preston, D. L., Suyama, A., Kodama, K., and Tomonaga, M., Risk of myelodysplastic syndromes in people exposed to ionizing radiation: A retrospective cohort study of Nagasaki atomic bomb survivors, *J Clin Oncol* 29 (4), 428–434, 2011.
2. Shimizu, Y., Kodama, K., Nishi, N., Kasagi, F., Suyama, A., Soda, M., Grant, E. J., Sugiyama, H., Sakata, R., Moriwaki, H., Hayashi, M., Konda, M., and Shore, R. E., Radiation exposure and circulatory disease risk: Hiroshima and Nagasaki atomic bomb survivor data, 1950–2003, *BMJ* 340, b5349, 2010.
3. Baker, J. E., Moulder, J. E., and Hopewell, J. W., Radiation as a risk factor for cardiovascular disease, *Antioxid Redox Signal* 15 (7), 1445–1456, 2010.
4. Little, M. P., Cancer and non-cancer effects in Japanese atomic bomb survivors, *J Radiol Prot* 29 (2A), A43–59, 2009.
5. Termuhlen, A. M., Tersak, J. M., Liu, Q., Yasui, Y., Stovall, M., Weathers, R., Deutsch, M., Sklar, C. A., Oeffinger, K. C., Armstrong, G., Robison, L. L., and Green, D. M., Twenty-five year follow-up of childhood Wilms tumor: A report from the Childhood Cancer Survivor Study, *Pediatr Blood Cancer*, doi: 10.1002/pbc.23090, 2011.
6. Armstrong, G. T., Liu, Q., Yasui, Y., Neglia, J. P., Leisenring, W., Robison, L. L., and Mertens, A. C., Late mortality among 5-year survivors of childhood cancer: A summary from the Childhood Cancer Survivor Study, *J Clin Oncol* 27 (14), 2328–2338, 2009.
7. Mertens, A. C., Yasui, Y., Liu, Y., Stovall, M., Hutchinson, R., Ginsberg, J., Sklar, C., and Robison, L. L., Pulmonary complications in survivors of childhood and adolescent cancer. A report from the Childhood Cancer Survivor Study, *Cancer* 95 (11), 2431–2441, 2002.
8. Chen, M. H., Colan, S. D., and Diller, L., Cardiovascular disease: Cause of morbidity and mortality in adult survivors of childhood cancers, *Circ Res* 108 (5), 619–628, 2011.
9. Ron, E., Preston, D. L., Mabuchi, K., Thompson, D. E., and Soda, M., Cancer incidence in atomic bomb survivors. Part IV: Comparison of cancer incidence and mortality, *Radiat Res* 137 (2 Suppl), S98–112, 1994.
10. Preston, D. L., Ron, E., Tokuoka, S., Funamoto, S., Nishi, N., Soda, M., Mabuchi, K., and Kodama, K., Solid cancer incidence in atomic bomb survivors: 1958–1998, *Radiat Res* 168 (1), 1–64, 2007.
11. Thompson, D. E., Mabuchi, K., Ron, E., Soda, M., Tokunaga, M., Ochikubo, S., Sugimoto, S., Ikeda, T., Terasaki, M., Izumi, S., et al., Cancer incidence in atomic bomb survivors. Part II: Solid tumors, 1958–1987, *Radiat Res* 137 (2 Suppl), S17–67, 1994.
12. Mostoufi-Moab, S., and Grimberg, A., Pediatric brain tumor treatment: Growth consequences and their management, *Pediatr Endocrinol Rev* 8 (1), 6–17, 2010.
13. Lorenzi, M. F., Xie, L., Rogers, P. C., Pritchard, S., Goddard, K., and McBride, M. L., Hospital-related morbidity among childhood cancer survivors in British Columbia, Canada: Report of the Childhood, Adolescent, Young Adult Cancer Survivors (CAYACS) Program, *Int J Cancer* 128 (7), 1624–1631, 2011.
14. Askins, M. A., and Moore, B. D. III, Preventing neurocognitive late effects in childhood cancer survivors, *J Child Neurol* 23 (10), 1160–1171, 2008.
15. Goldsby, R., Chen, Y., Raber, S., Li, L., Diefenbach, K., Shnorhavorian, M., Kadan-Lottick, N., Kastrinos, F., Yasui, Y., Stovall, M., Oeffinger, K., Sklar, C., Armstrong, G. T., Robison, L. L., and Diller, L., Survivors of childhood cancer have increased risk for gastrointestinal complications later in life, *Gastroenterology*, 140 (5), 1464–1471, 2011.
16. Whelan, K., Stratton, K., Kawashima, T., Leisenring, W., Hayashi, S., Waterbor, J., Blatt, J., Sklar, C. A., Packer, R., Mitby, P., Robison, L. L., and Mertens, A. C., Auditory complications in childhood cancer survivors: A report from the childhood cancer survivor study, *Pediatr Blood Cancer* 57 (1), 126–134, 2011.

17. Edelstein, K., Spiegler, B. J., Fung, S., Panzarella, T., Mabbott, D. J., Jewitt, N., D'Agostino, N. M., Mason, W. P., Bouffet, E., Tabori, U., Laperriere, N., and Hodgson, D. C., Early aging in adult survivors of childhood medulloblastoma: Long-term neuro-cognitive, functional, and physical outcomes, *Neuro Oncol* 13 (5), 536–545, 2011.

18. Edgar, A. B., and Wallace, W. H., Pregnancy in women who had cancer in childhood, *Eur J Cancer* 43 (13), 1890–1894, 2007.

19. Green, D. M., Sklar, C. A., Boice, J. D. Jr., Mulvihill, J. J., Whitton, J. A., Stovall, M., and Yasui, Y., Ovarian failure and reproductive outcomes after childhood cancer treatment: Results from the Childhood Cancer Survivor Study, *J Clin Oncol* 27 (14), 2374–2381, 2009.

20. Green, D. M., Nolan, V. G., Kawashima, T., Stovall, M., Donaldson, S. S., Srivastava, D., Leisenring, W., Robison, L. L., and Sklar, C. A., Decreased fertility among female childhood cancer survivors who received 22–27 Gy hypothalamic/pituitary irradiation: A report from the Childhood Cancer Survivor Study, *Fertil Steril* 95 (6), 1922–1927, 2011.

21. Meacham, L. R., Sklar, C. A., Li, S., Liu, Q., Gimpel, N., Yasui, Y., Whitton, J. A., Stovall, M., Robison, L. L., and Oeffinger, K. C., Diabetes mellitus in long-term survivors of childhood cancer. Increased risk associated with radiation therapy: a report for the childhood cancer survivor study, *Arch Intern Med* 169 (15), 1381–1388, 2009.

22. Ron, E., and Brenner, A., Non-malignant thyroid diseases after a wide range of radiation exposures, *Radiat Res* 174 (6), 877–888, 2010.

23. Boivin, J. F., Hutchison, G. B., Zauber, A. G., Bernstein, L., Davis, F. G., Michel, R. P., Zanke, B., Tan, C. T., Fuller, L. M., Mauch, P., et al., Incidence of second cancers in patients treated for Hodgkin's disease, *J Natl Cancer Inst* 87 (10), 732–741, 1995.

24. Taylor, A. J., Little, M. P., Winter, D. L., Sugden, E., Ellison, D. W., Stiller, C. A., Stovall, M., Frobisher, C., Lancashire, E. R., Reulen, R. C., and Hawkins, M. M., Population-based risks of CNS tumors in survivors of childhood cancer: the British Childhood Cancer Survivor Study, *J Clin Oncol* 28 (36), 5287–5293, 2010.

25. Jereczek-Fossa, B. A., Alterio, D., Jassem, J., Gibelli, B., Tradati, N., and Orecchia, R., Radiotherapy-induced thyroid disorders, *Cancer Treat Rev* 30 (4), 369–384, 2004.

26. Rodriguez, M. A., Fuller, L. M., Zimmerman, S. O., Allen, P. K., Brown, B. W., Munsell, M. F., Hagemeister, F. B., McLaughlin, P., Velasquez, W. S., Swan, F., Jr. et al., Hodgkin's disease: Study of treatment intensities and incidences of second malignancies, *Ann Oncol* 4 (2), 125–131, 1993.

27. Cutuli, B., Kanoun, S., Tunon De Lara, C., Baron, M., Livi, L., Levy, C., Cohen-Solal-Lenir, C., Lesur, A., Kerbrat, P., Provencio, M., Gonzague-Casabianca, L., Mege, A., Lemanski, C., Delva, C., Lancrenon, S., and Velten, M., Breast cancer occurred after Hodgkin's disease: Clinico-pathological features, treatments and outcome: Analysis of 214 cases, *Crit Rev Oncol Hematol*, In press.

28. Bhatti, P., Veiga, L. H., Ronckers, C. M., Sigurdson, A. J., Stovall, M., Smith, S. A., Weathers, R., Leisenring, W., Mertens, A. C., Hammond, S., Friedman, D. L., Neglia, J. P., Meadows, A. T., Donaldson, S. S., Sklar, C. A., Robison, L. L., and Inskip, P. D., Risk of second primary thyroid cancer after radiotherapy for a childhood cancer in a large cohort study: An update from the childhood cancer survivor study, *Radiat Res* 174 (6), 741–752, 2010.

29. Tanooka, H., Meta-analysis of non-tumour doses for radiation-induced cancer on the basis of dose-rate, *Int J Radiat Biol* 87 (7), 645–652, 2011.

30. Laverdiere, C., Liu, Q., Yasui, Y., Nathan, P. C., Gurney, J. G., Stovall, M., Diller, L. R., Cheung, N. K., Wolden, S., Robison, L. L., and Sklar, C. A., Long-term outcomes in survivors of neuroblastoma: A report from the Childhood Cancer Survivor Study, *J Natl Cancer Inst* 101 (16), 1131–1140, 2009.

31. Prasad, K. N., Cole, W., and Hovland, P., Cancer prevention studies: Past, present, and future directions, *Nutrition* 14 (2), 197–210; discussion 237–238, 1998.
32. Zhao, W., and Robbins, M. E., Inflammation and chronic oxidative stress in radiation-induced late normal tissue injury: Therapeutic implications, *Curr Med Chem* 16 (2), 130–143, 2009.
33. Kusunoki, Y., Yamaoka, M., Kubo, Y., Hayashi, T., Kasagi, F., Douple, E. B., and Nakachi, K., T-cell immunosenescence and inflammatory response in atomic bomb survivors, *Radiat Res* 174 (6), 870–876, 2010.
34. Popp, W., Plappert, U., Muller, W. U., Rehn, B., Schneider, J., Braun, A., Bauer, P. C., Vahrenholz, C., Presek, P., Brauksiepe, A., Enderle, G., Wust, T., Bruch, J., Fliedner, T. M., Konietzko, N., Streffer, C., Woitowitz, H. J., and Norpoth, K., Biomarkers of genetic damage and inflammation in blood and bronchoalveolar lavage fluid among former German uranium miners: A pilot study, *Radiat Environ Biophys* 39 (4), 275–282, 2000.
35. Muller, K., Kohn, F. M., Port, M., Abend, M., Molls, M., Ring, J., and Meineke, V., Intercellular adhesion molecule-1: A consistent inflammatory marker of the cutaneous radiation reaction both in vitro and in vivo, *Br J Dermatol* 155 (4), 670–679, 2006.
36. Newmark, H., and Mergen, W., Application of ascorbic acid and tocopherols as inhibitors of nitrosamine formation and oxidation in food, in *Criteria of Food Acceptance*, ed. Solms, J. and Hall, R. Forster Publishing, Zurich, 1981, 379.
37. Dion, P. W., Bright-See, E. B., Smith, C. C., and Bruce, W. R., The effect of dietary ascorbic acid and alpha-tocopherol on fecal mutagenicity, *Mutat Res* 102 (1), 27–37, 1982.
38. Duthie, S. J., Ma, A., Ross, M. A., and Collins, A. R., Antioxidant supplementation decreases oxidative DNA damage in human lymphocytes, *Cancer Res* 56 (6), 1291–1295, 1996.
39. Sram, R. J., Dobias, L., Pastorkova, A., Rossner, P., and Janca, L., Effect of ascorbic acid prophylaxis on the frequency of chromosome aberrations in the peripheral lymphocytes of coal-tar workers, *Mutat Res* 120 (2–3), 181–186, 1983.
40. Weitberg, A. B., Weitzman, S. A., Clark, E. P., and Stossel, T. P., Effects of antioxidants on oxidant-induced sister chromatid exchange formation, *J Clin Invest* 75 (6), 1835–1841, 1985.
41. Kumar, B., Jha, M. N., Cole, W. C., Bedford, J. S., and Prasad, K. N., D-alpha tocopheryl succinate (vitamin E) enhances radiation-induced chromosomal damage levels in human cancer cells, but reduced it in normal cells, *J Am Coll Nutr* 21 (4), 339–343, 2002.
42. Prasad, K. N., Vitamins induce cell differentiation, growth inhibition and enhance the effect of tumor therapeutic agents on some cancer cells in vitro, in *Nutrients in Cancer Prevention and Treatment*, ed. Prasad, K. N., Santamaria, L., and Williams, R. M. Humana Press, Totowa, NJ, 1995, 265–288.
43. Prasad, K. N., and Edwards-Prasad, J., Vitamin E and cancer prevention: Recent advances and future potentials, *J Am Coll Nutr* 11 (5), 487–500, 1992.
44. Rao, C. V., and Reddy, B. S., Modulating effect of amount and types of dietary fat on ornithine decarboxylase, tyrosine protein kinase and prostaglandins production during colon carcinogenesis in male F344 rats, *Carcinogenesis* 14 (7), 1327–1333, 1993.
45. Reddy, B. S., Dietary fat, calories, and fiber in colon cancer, *Prev Med* 22 (5), 738–749, 1993.
46. Boutwell, R., Biology and biochemistry of two-steps carcinogenesis, in *Modulation and Mediation of Cancer by Vitamins*, ed. Meyskens, F. J., Prasad, K. N., Karger, Basel, 1983, 2.
47. Delafuente, J. C., Prendergast, J. M., and Modigh, A., Immunologic modulation by vitamin C in the elderly, *Int J Immunopharmacol* 8 (2), 205–211, 1986.

48. Ringer, T. V., DeLoof, M. J., Winterrowd, G. E., Francom, S. F., Gaylor, S. K., Ryan, J. A., Sanders, M. E., and Hughes, G. S., Beta-carotene's effects on serum lipoproteins and immunologic indices in humans, *Am J Clin Nutr* 53 (3), 688–694, 1991.

49. Prasad, K. N., and Edwards-Prasad, J., Effects of tocopherol (vitamin E) acid succinate on morphological alterations and growth inhibition in melanoma cells in culture, *Cancer Res* 42 (2), 550–555, 1982.

50. Radner, B. S., and Kennedy, A. R., Suppression of x-ray induced transformation by vitamin E in mouse C3H/10T1/2 cells, *Cancer Lett* 32 (1), 25–32, 1986.

51. Borek, C., Ong, A., Mason, H., Donahue, L., and Biaglow, J. E., Selenium and vitamin E inhibit radiogenic and chemically induced transformation in vitro via different mechanisms, *Proc Natl Acad Sci USA* 83 (5), 1490–1494, 1986.

52. Pung, A., Rundhaug, J. E., Yoshizawa, C. N., and Bertram, J. S., Beta-carotene and canthaxanthin inhibit chemically- and physically-induced neoplastic transformation in 10T1/2 cells, *Carcinogenesis* 9 (9), 1533–1539, 1988.

53. Kennedy, A. R., and Krinsky, N. I., Effects of retinoids, beta-carotene, and canthaxanthin on UV- and x-ray-induced transformation of C3H10T1/2 cells in vitro, *Nutr Cancer* 22 (3), 219–232, 1994.

54. Venugopal, D., Zahid, M., Mailander, P. C., Meza, J. L., Rogan, E. G., Cavalieri, E. L., and Chakravarti, D., Reduction of estrogen-induced transformation of mouse mammary epithelial cells by N-acetylcysteine, *J Steroid Biochem Mol Biol* 109 (1–2), 22–30, 2008.

55. Cohen, M., and Bhagavan, H. N., Ascorbic acid and gastrointestinal cancer, *J Am Coll Nutr* 14 (6), 565–578, 1995.

56. Hill, D. L., and Grubbs, C. J., Retinoids as chemopreventive and anticancer agents in intact animals (review), *Anticancer Res* 2 (1–2), 111–124, 1982.

57. Krinsky, N. I., Antioxidant functions of carotenoids, *Free Radic Biol Med* 7 (6), 617–635, 1989.

58. Santamaria, L., Bianchi, A., and Mobilio, G., Cancer prevention by carotenoids, in *Nutrition, Growth and Cancer*, ed. Tryfiates, G. P., Alan R. Liss, New York, 1988, 177.

59. Venkateswaran, V., Klotz, L. H., Ramani, M., Sugar, L. M., Jacob, L. E., Nam, R. K., and Fleshner, N. E., A combination of micronutrients is beneficial in reducing the incidence of prostate cancer and increasing survival in the Lady transgenic model, *Cancer Prev Res (Phila Pa)* 2 (5), 473–483, 2009.

60. Chen, C. S., Squire, J. A., and Wells, P. G., Reduced tumorigenesis in p53 knockout mice exposed in utero to low-dose vitamin E, *Cancer* 115 (7), 1563–1575, 2009.

61. Wang, Y. J., Yang, M. C., and Pan, M. H., Dihydrolipoic acid inhibits tetrachlorohydroquinone-induced tumor promotion through prevention of oxidative damage, *Food Chem Toxicol* 46 (12), 3739–3748, 2008.

62. Ho, Y. S., Lai, C. S., Liu, H. I., Ho, S. Y., Tai, C., Pan, M. H., and Wang, Y. J., Dihydrolipoic acid inhibits skin tumor promotion through anti-inflammation and anti-oxidation, *Biochem Pharmacol* 73 (11), 1786–1795, 2007.

63. Rossi, C., Di Lena, A., La Sorda, R., Lattanzio, R., Antolini, L., Patassini, C., Piantelli, M., and Alberti, S., Intestinal tumour chemoprevention with the antioxidant lipoic acid stimulates the growth of breast cancer, *Eur J Cancer* 44 (17), 2696–2704, 2008.

64. Reliene, R., and Schiestl, R. H., Antioxidant N-acetyl cysteine reduces incidence and multiplicity of lymphoma in Atm deficient mice, *DNA Repair (Amst)* 5 (7), 852–859, 2006.

65. Balansky, R. M., Ganchev, G., D'Agostini, F., and De Flora, S., Effects of N-acetylcysteine in an esophageal carcinogenesis model in rats treated with diethylnitrosamine and diethyldithiocarbamate, *Int J Cancer* 98 (4), 493–497, 2002.

66. Martin, K. R., Trempus, C., Saulnier, M., Kari, F. W., Barrett, J. C., and French, J. E., Dietary N-acetyl-L-cysteine modulates benzo[a]pyrene-induced skin tumors in cancer-prone p53 haploinsufficient Tg.AC (v-Ha-ras) mice, *Carcinogenesis* 22 (9), 1373–1378, 2001.

67. Lubet, R. A., Steele, V. E., Eto, I., Juliana, M. M., Kelloff, G. J., and Grubbs, C. J., Chemopreventive efficacy of anethole trithione, N-acetyl-L-cysteine, miconazole and phenethyl isothiocyanate in the DMBA-induced rat mammary cancer model, *Int J Cancer* 72 (1), 95–101, 1997.

68. De Flora, S., Rossi, G. A., and De Flora, A., Metabolic, desmutagenic and anticarcinogenic effects of N-acetylcysteine, *Respiration* 50 Suppl 1, 43–49, 1986.

69. Hao, J., Zhang, B., Liu, B., Lee, M., Hao, X., Reuhl, K. R., Chen, X., and Yang, C. S., Effect of alpha-tocopherol, N-acetylcysteine and omeprazole on esophageal adenocarcinoma formation in a rat surgical model, *Int J Cancer* 124 (6), 1270–1275, 2009.

70. Sakano, K., Takahashi, M., Kitano, M., Sugimura, T., and Wakabayashi, K., Suppression of azoxymethane-induced colonic premalignant lesion formation by coenzyme Q10 in rats, *Asian Pac J Cancer Prev* 7 (4), 599–603, 2006.

71. Dange, P., Sarma, H., Pandey, B. N., and Mishra, K. P., Radiation-induced incidence of thymic lymphoma in mice and its prevention by antioxidants, *J Environ Pathol Toxicol Oncol* 26 (4), 273–279, 2007.

72. Kennedy, A. R., Davis, J. G., Carlton, W., and Ware, J. H., Effects of dietary antioxidant supplementation on the development of malignant lymphoma and other neoplastic lesions in mice exposed to proton or iron-ion radiation, *Radiat Res* 169 (6), 615–625, 2008.

73. Ueno, M., Inano, H., Onoda, M., Murase, H., Ikota, N., Kagiya, T. V., and Anzai, K., Modification of mortality and tumorigenesis by tocopherol-mono-glucoside (TMG) administered after X irradiation in mice and rats, *Radiat Res* 172 (4), 519–524, 2009.

74. Burns, F. J., Chen, S., Xu, G., Wu, F., and Tang, M. S., The action of a dietary retinoid on gene expression and cancer induction in electron-irradiated rat skin, *J Radiat Res (Tokyo)* 43 Suppl, S229–232, 2002.

75. Toth, B., and Patil, K., Enhancing effect of vitamin E on murine intestinal tumorigenesis by 1,2-dimethylhydrazine dihydrochloride, *J Natl Cancer Inst* 70 (6), 1107–1111, 1983.

76. Fukushima, S., Imaida, K., Shibata, M. A., Tamano, S., Kurata, Y., and Shirai, T., L-ascorbic acid amplification of second-stage bladder carcinogenesis promotion by NaHCO3, *Cancer Res* 48 (22), 6317–6320, 1988.

77. Hennekens, C. H., Antioxidant vitamins and cancer, *Am J Med* 97 (3A), 2S–4S; discussion 22S–28S, 1994.

78. Buring, J., and Hennekens, C. H., Antioxidant vitamins in cancer: The Physician's Health Study and Women's Health Study, in *Nutrients in Cancer Prevention and Treatment*, ed. Prasad, K., Santamaria, L., and Williams, R. M., Humana Press, Totowa, NJ, 1995, 223.

79. Koushik, A., Hunter, D. J., Spiegelman, D., Anderson, K. E., Arslan, A. A., Beeson, W. L., van den Brandt, P. A., Buring, J. E., Cerhan, J. R., Colditz, G. A., Fraser, G. E., Freudenheim, J. L., Genkinger, J. M., Goldbohm, R. A., Hankinson, S. E., Koenig, K. L., Larsson, S. C., Leitzmann, M., McCullough, M. L., Miller, A. B., Patel, A., Rohan, T. E., Schatzkin, A., Smit, E., Willett, W. C., Wolk, A., Zhang, S. M., and Smith-Warner, S. A., Fruits and vegetables and ovarian cancer risk in a pooled analysis of 12 cohort studies, *Cancer Epidemiol Biomarkers Prev* 14 (9), 2160–2167, 2005.

80. Kristal, A. R., Arnold, K. B., Schenk, J. M., Neuhouser, M. L., Goodman, P., Penson, D. F., and Thompson, I. M., Dietary patterns, supplement use, and the risk of symptomatic benign prostatic hyperplasia: Results from the prostate cancer prevention trial, *Am J Epidemiol* 167 (8), 925–934, 2008.

81. Rodrigues, M. J., Bouyon, A., and Alexandre, J., Role of antioxidant complements and supplements in oncology in addition to an equilibrate regimen: A systematic review [in French], *Bull Cancer* 96 (6), 677–684, 2009.

82. Jedrychowski, W., and Maugeri, U., An apple a day may hold colorectal cancer at bay: Recent evidence from a case-control study, *Rev Environ Health* 24 (1), 59–74, 2009.

83. Sato, R., Helzlsouer, K. J., Alberg, A. J., Hoffman, S. C., Norkus, E. P., and Comstock, G. W., Prospective study of carotenoids, tocopherols, and retinoid concentrations and the risk of breast cancer, *Cancer Epidemiol Biomarkers Prev* 11 (5), 451–457, 2002.

84. Tamimi, R. M., Hankinson, S. E., Campos, H., Spiegelman, D., Zhang, S., Colditz, G. A., Willett, W. C., and Hunter, D. J., Plasma carotenoids, retinol, and tocopherols and risk of breast cancer, *Am J Epidemiol* 161 (2), 153–160, 2005.

85. Kabat, G. C., Kim, M., Adams-Campbell, L. L., Caan, B. J., Chlebowski, R. T., Neuhouser, M. L., Shikany, J. M., and Rohan, T. E., Longitudinal study of serum carotenoid, retinol, and tocopherol concentrations in relation to breast cancer risk among postmenopausal women, *Am J Clin Nutr* 90 (1), 162–169, 2009.

86. Satia, J. A., Littman, A., Slatore, C. G., Galanko, J. A., and White, E., Long-term use of beta-carotene, retinol, lycopene, and lutein supplements and lung cancer risk: Results from the VITamins And Lifestyle (VITAL) study, *Am J Epidemiol* 169 (7), 815–828, 2009.

87. Ma, E., Iwasaki, M., Junko, I., Hamada, G. S., Nishimoto, I. N., Carvalho, S. M., Motola, J. Jr., Laginha, F. M., and Tsugane, S., Dietary intake of folate, vitamin B_6, and vitamin B_{12}, genetic polymorphism of related enzymes, and risk of breast cancer: A case-control study in Brazilian women, *BMC Cancer* 9, 122, 2009.

88. Tang, N. P., Zhou, B., Wang, B., Yu, R. B., and Ma, J., Flavonoids intake and risk of lung cancer: A meta-analysis, *Jpn J Clin Oncol* 39 (6), 352–359, 2009.

89. Wang, L., Lee, I. M., Zhang, S. M., Blumberg, J. B., Buring, J. E., and Sesso, H. D., Dietary intake of selected flavonols, flavones, and flavonoid-rich foods and risk of cancer in middle-aged and older women, *Am J Clin Nutr* 89 (3), 905–912, 2009.

90. Ellinger, S., Ellinger, J., Muller, S. C., and Stehle, P., Tomatoes and lycopene in prevention and therapy—Is there an evidence for prostate diseases? [in German], *Aktuelle Urol* 40 (1), 37–43, 2009.

91. Thomson, C. A., Neuhouser, M. L., Shikany, J. M., Caan, B. J., Monk, B. J., Mossavar-Rahmani, Y., Sarto, G., Parker, L. M., Modugno, F., and Anderson, G. L., The role of antioxidants and vitamin A in ovarian cancer: Results from the Women's Health Initiative, *Nutr Cancer* 60 (6), 710–719, 2008.

92. Thiebaut, A. C., Chajes, V., Gerber, M., Boutron-Ruault, M. C., Joulin, V., Lenoir, G., Berrino, F., Riboli, E., Benichou, J., and Clavel-Chapelon, F., Dietary intakes of omega-6 and omega-3 polyunsaturated fatty acids and the risk of breast cancer, *Int J Cancer* 124 (4), 924–931, 2009.

93. Anderson, K. E., Pantuck, E. J., Conney, A. H., and Kappas, A., Nutrient regulation of chemical metabolism in humans, *Fed Proc* 44 (1 Pt 1), 130–133, 1985.

94. Conney, A. H., Lysz, T., Ferraro, T., Abidi, T. F., Manchand, P. S., Laskin, J. D., and Huang, M. T., Inhibitory effect of curcumin and some related dietary compounds on tumor promotion and arachidonic acid metabolism in mouse skin, *Adv Enzyme Regul* 31, 385–396, 1991.

95. Albanes, D., Heinonen, O. P., Huttunen, J. K., Taylor, P. R., Virtamo, J., Edwards, B. K., Haapakoski, J., Rautalahti, M., Hartman, A. M., Palmgren, J., et al., Effects of alpha-tocopherol and beta-carotene supplements on cancer incidence in the Alpha-Tocopherol Beta-Carotene Cancer Prevention Study, *Am J Clin Nutr* 62 (6 Suppl), 1427S–1430S, 1995.

96. The Alpha-Tocopherol, Beta Carotene Cancer Prevention Study Group. The effect of vitamin E and beta carotene on the incidence of lung cancer and other cancers in male smokers, *N Engl J Med* 330 (15), 1029–1035, 1994.
97. Omenn, G. S., Goodman, G. E., Thornquist, M. D., Balmes, J., Cullen, M. R., Glass, A., Keogh, J. P., Meyskens, F. L., Valanis, B., Williams, J. H., Barnhart, S., and Hammar, S., Effects of a combination of beta carotene and vitamin A on lung cancer and cardiovascular disease, *N Engl J Med* 334 (18), 1150–1155, 1996.
98. Hennekens, C. H., Buring, J. E., Manson, J. E., Stampfer, M., Rosner, B., Cook, N. R., Belanger, C., LaMotte, F., Gaziano, J. M., Ridker, P. M., Willett, W., and Peto, R., Lack of effect of long-term supplementation with beta carotene on the incidence of malignant neoplasms and cardiovascular disease, *N Engl J Med* 334 (18), 1145–1149, 1996.
99. Van Schooten, F. J., Besaratinia, A., De Flora, S., D'Agostini, F., Izzotti, A., Camoirano, A., Balm, A. J., Dallinga, J. W., Bast, A., Haenen, G. R., Van't Veer, L., Baas, P., Sakai, H., and Van Zandwijk, N., Effects of oral administration of N-acetyl-L-cysteine: A multi-biomarker study in smokers, *Cancer Epidemiol Biomarkers Prev* 11 (2), 167–175, 2002.
100. Kirsh, V. A., Hayes, R. B., Mayne, S. T., Chatterjee, N., Subar, A. F., Dixon, L. B., Albanes, D., Andriole, G. L., Urban, D. A., and Peters, U., Supplemental and dietary vitamin E, beta-carotene, and vitamin C intakes and prostate cancer risk, *J Natl Cancer Inst* 98 (4), 245–254, 2006.
101. Mooney, L. A., Madsen, A. M., Tang, D., Orjuela, M. A., Tsai, W. Y., Garduno, E. R., and Perera, F. P., Antioxidant vitamin supplementation reduces benzo(a)pyrene-DNA adducts and potential cancer risk in female smokers, *Cancer Epidemiol Biomarkers Prev* 14 (1), 237–242, 2005.
102. Stolzenberg-Solomon, R. Z., Sheffler-Collins, S., Weinstein, S., Garabrant, D. H., Mannisto, S., Taylor, P., Virtamo, J., and Albanes, D., Vitamin E intake, alpha-tocopherol status, and pancreatic cancer in a cohort of male smokers, *Am J Clin Nutr* 89 (2), 584–91, 2009.
103. Gallicchio, L., Boyd, K., Matanoski, G., Tao, X. G., Chen, L., Lam, T. K., Shiels, M., Hammond, E., Robinson, K. A., Caulfield, L. E., Herman, J. G., Guallar, E., and Alberg, A. J., Carotenoids and the risk of developing lung cancer: A systematic review, *Am J Clin Nutr* 88 (2), 372–383, 2008.
104. van Zandwijk, N., Dalesio, O., Pastorino, U., de Vries, N., and van Tinteren, H., EUROSCAN, a randomized trial of vitamin A and N-acetylcysteine in patients with head and neck cancer or lung cancer. For the European Organization for Research and Treatment of Cancer Head and Neck and Lung Cancer Cooperative Groups, *J Natl Cancer Inst* 92 (12), 977–986, 2000.
105. Benner, S. E., Winn, R. J., Lippman, S. M., Poland, J., Hansen, K. S., Luna, M. A., and Hong, W. K., Regression of oral leukoplakia with alpha-tocopherol: A community clinical oncology program chemoprevention study, *J Natl Cancer Inst* 85 (1), 44–47, 1993.
106. Garewal, H., Beta-carotene and antioxidant nutrients in oral cancer prevention, in *Nutrients in Cancer Prevention and Treatment*, ed. Prasad, K. S., Santamaria, L., and Williams, R. M., Humana Press, Totowa, NJ, 1995.
107. Meyskens, F. J., Role of vitamin and its derivatives in the treatment of human cancer, in *Nutrients in Cancer Prevention and Treatment*, ed. Prasad, K. S., Santamaria, L., and Williams, R. M., Humana Press, Totowa, NJ, 1995.
108. Hernaandez, J., Syed, S., Weiss, G., Fernandes, G., von Merveldt, D., Troyer, D. A., Basler, J. W., and Thompson, I. M., Jr., The modulation of prostate cancer risk with alpha-tocopherol: A pilot randomized, controlled clinical trial, *J Urol* 174 (2), 519–522, 2005.

109. Alkhenizan, A., and Hafez, K., The role of vitamin E in the prevention of cancer: A meta-analysis of randomized controlled trials, *Ann Saudi Med* 27 (6), 409–414, 2007.

110. Weinstein, S. J., Wright, M. E., Lawson, K. A., Snyder, K., Mannisto, S., Taylor, P. R., Virtamo, J., and Albanes, D., Serum and dietary vitamin E in relation to prostate cancer risk, *Cancer Epidemiol Biomarkers Prev* 16 (6), 1253–1259, 2007.

111. Weinstein, S. J., Wright, M. E., Pietinen, P., King, I., Tan, C., Taylor, P. R., Virtamo, J., and Albanes, D., Serum alpha-tocopherol and gamma-tocopherol in relation to prostate cancer risk in a prospective study, *J Natl Cancer Inst* 97 (5), 396–399, 2005.

112. Lippman, S. M., Klein, E. A., Goodman, P. J., Lucia, M. S., Thompson, I. M., Ford, L. G., Parnes, H. L., Minasian, L. M., Gaziano, J. M., Hartline, J. A., Parsons, J. K., Bearden, J. D. III, Crawford, E. D., Goodman, G. E., Claudio, J., Winquist, E., Cook, E. D., Karp, D. D., Walther, P., Lieber, M. M., Kristal, A. R., Darke, A. K., Arnold, K. B., Ganz, P. A., Santella, R. M., Albanes, D., Taylor, P. R., Probstfield, J. L., Jagpal, T. J., Crowley, J. J., Meyskens, F. L. Jr., Baker, L. H., and Coltman, C. A. Jr., Effect of selenium and vitamin E on risk of prostate cancer and other cancers: The Selenium and Vitamin E Cancer Prevention Trial (SELECT), *JAMA* 301 (1), 39–51, 2009.

113. Bairati, I., Meyer, F., Gelinas, M., Fortin, A., Nabid, A., Brochet, F., Mercier, J. P., Tetu, B., Harel, F., Abdous, B., Vigneault, E., Vass, S., Del Vecchio, P., and Roy, J., Randomized trial of antioxidant vitamins to prevent acute adverse effects of radiation therapy in head and neck cancer patients, *J Clin Oncol* 23 (24), 5805–5813, 2005.

114. Miller, E. R. III, Pastor-Barriuso, R., Dalal, D., Riemersma, R. A., Appel, L. J., and Guallar, E., Meta-analysis: High-dosage vitamin E supplementation may increase all-cause mortality, *Ann Intern Med* 142 (1), 37–46, 2005.

115. Gerss, J., and Kopcke, W., The questionable association of vitamin E supplementation and mortality: Inconsistent results of different meta-analytic approaches, *Cell Mol Biol (Noisy-le-grand)* 55 Suppl, OL1111–1120, 2009.

116. Ahern, T. P., Lash, T. L., Egan, K. M., and Baron, J. A., Lifetime tobacco smoke exposure and breast cancer incidence, *Cancer Causes Control* 20 (10), 1837–1844, 2009.

117. Ingold, K. U., Burton, G. W., Foster, D. O., Hughes, L., Lindsay, D. A., and Webb, A., Biokinetics of and discrimination between dietary RRR- and SRR-alpha-tocopherols in the male rat, *Lipids* 22 (3), 163–172, 1987.

118. Prasad, K. N., Kumar, B., Yan, X. D., Hanson, A. J., and Cole, W. C., Alpha-tocopheryl succinate, the most effective form of vitamin E for adjuvant cancer treatment: A review, *J Am Coll Nutr* 22 (2), 108–117, 2003.

119. Prasad, K. N., Antioxidants in cancer care: When and how to use them as an adjunct to standard and experimental therapies, *Expert Rev Anticancer Ther* 3 (6), 903–915, 2003.

120. Lamm, D. L., Riggs, D. R., Shriver, J. S., vanGilder, P. F., Rach, J. F., and DeHaven, J. I., Megadose vitamins in bladder cancer: A double-blind clinical trial, *J Urol* 151 (1), 21–26, 1994.

121. Roncucci, L., Di Donato, P., Carati, L., Ferrari, A., Perini, M., Bertoni, G., Bedogni, G., Paris, B., Svanoni, F., Girola, M., et al., Antioxidant vitamins or lactulose for the prevention of the recurrence of colorectal adenomas. Colorectal Cancer Study Group of the University of Modena and the Health Care District 16, *Dis Colon Rectum* 36 (3), 227–234, 1993.

122. Greenberg, E. R., Baron, J. A., Tosteson, T. D., Freeman, D. H. Jr., Beck, G. J., Bond, J. H., Colacchio, T. A., Coller, J. A., Frankl, H. D., Haile, R. W., et al., A clinical trial of antioxidant vitamins to prevent colorectal adenoma. Polyp Prevention Study Group, *N Engl J Med* 331 (3), 141–147, 1994.

123. McKeown-Eyssen, G., Holloway, C., Jazmaji, V., Bright-See, E., Dion, P., and Bruce, W. R., A randomized trial of vitamins C and E in the prevention of recurrence of colorectal polyps, *Cancer Res* 48 (16), 4701–4705, 1988.

124. DeCosse, J. J., Miller, H. H., and Lesser, M. L., Effect of wheat fiber and vitamins C and E on rectal polyps in patients with familial adenomatous polyposis, *J Natl Cancer Inst* 81 (17), 1290–1297, 1989.

125. Blot, W. J., Li, J. Y., Taylor, P. R., Guo, W., Dawsey, S., Wang, G. Q., Yang, C. S., Zheng, S. F., Gail, M., Li, G. Y., et al., Nutrition intervention trials in Linxian, China: Supplementation with specific vitamin/mineral combinations, cancer incidence, and disease-specific mortality in the general population, *J Natl Cancer Inst* 85 (18), 1483–1492, 1993.

126. Qiao, Y. L., Dawsey, S. M., Kamangar, F., Fan, J. H., Abnet, C. C., Sun, X. D., Johnson, L. L., Gail, M. H., Dong, Z. W., Yu, B., Mark, S. D., and Taylor, P. R., Total and cancer mortality after supplementation with vitamins and minerals: Follow-up of the Linxian General Population Nutrition Intervention Trial, *J Natl Cancer Inst* 101 (7), 507–518, 2009.

127. Biasco, G., and Paganelli, G. M., European trials on dietary supplementation for cancer prevention, *Ann N Y Acad Sci* 889, 152–156, 1999.

128. Hoenjet, K. M., Dagnelie, P. C., Delaere, K. P., Wijckmans, N. E., Zambon, J. V., and Oosterhof, G. O., Effect of a nutritional supplement containing vitamin E, selenium, vitamin C and coenzyme Q10 on serum PSA in patients with hormonally untreated carcinoma of the prostate: A randomised placebo-controlled study, *Eur Urol* 47 (4), 433–439; discussion 439–440, 2005.

129. Gorham, E. D., Garland, C. F., Garland, F. C., Grant, W. B., Mohr, S. B., Lipkin, M., Newmark, H. L., Giovannucci, E., Wei, M., and Holick, M. F., Vitamin D and prevention of colorectal cancer, *J Steroid Biochem Mol Biol* 97 (1–2), 179–194, 2005.

130. Gorham, E. D., Garland, C. F., Garland, F. C., Grant, W. B., Mohr, S. B., Lipkin, M., Newmark, H. L., Giovannucci, E., Wei, M., and Holick, M. F., Optimal vitamin D status for colorectal cancer prevention: A quantitative meta analysis, *Am J Prev Med* 32 (3), 210–216, 2007.

131. Wactawski-Wende, J., Kotchen, J. M., Anderson, G. L., Assaf, A. R., Brunner, R. L., O'Sullivan, M. J., Margolis, K. L., Ockene, J. K., Phillips, L., Pottern, L., Prentice, R. L., Robbins, J., Rohan, T. E., Sarto, G. E., Sharma, S., Stefanick, M. L., Van Horn, L., Wallace, R. B., Whitlock, E., Bassford, T., Beresford, S. A., Black, H. R., Bonds, D. E., Brzyski, R. G., Caan, B., Chlebowski, R. T., Cochrane, B., Garland, C., Gass, M., Hays, J., Heiss, G., Hendrix, S. L., Howard, B. V., Hsia, J., Hubbell, F. A., Jackson, R. D., Johnson, K. C., Judd, H., Kooperberg, C. L., Kuller, L. H., LaCroix, A. Z., Lane, D. S., Langer, R. D., Lasser, N. L., Lewis, C. E., Limacher, M. C., and Manson, J. E., Calcium plus vitamin D supplementation and the risk of colorectal cancer, *N Engl J Med* 354 (7), 684–696, 2006.

132. Ding, E. L., Mehta, S., Fawzi, W. W., and Giovannucci, E. L., Interaction of estrogen therapy with calcium and vitamin D supplementation on colorectal cancer risk: Reanalysis of Women's Health Initiative randomized trial, *Int J Cancer* 122 (8), 1690–1694, 2008.

133. Weingarten, M. A., Zalmanovici, A., and Yaphe, J., Dietary calcium supplementation for preventing colorectal cancer and adenomatous polyps, *Cochrane Database Syst Rev* (1), CD003548, 2008.

134. Chlebowski, R. T., Johnson, K. C., Kooperberg, C., Pettinger, M., Wactawski-Wende, J., Rohan, T., Rossouw, J., Lane, D., O'Sullivan, M. J., Yasmeen, S., Hiatt, R. A., Shikany, J. M., Vitolins, M., Khandekar, J., and Hubbell, F. A., Calcium plus vitamin D supplementation and the risk of breast cancer, *J Natl Cancer Inst* 100 (22), 1581–1591, 2008.

135. McCullough, M. L., Rodriguez, C., Diver, W. R., Feigelson, H. S., Stevens, V. L., Thun, M. J., and Calle, E. E., Dairy, calcium, and vitamin D intake and postmenopausal breast cancer risk in the Cancer Prevention Study II Nutrition Cohort, *Cancer Epidemiol Biomarkers Prev* 14 (12), 2898–2904, 2005.

136. Martinez, M. E., Marshall, J. R., Sampliner, R., Wilkinson, J., and Alberts, D. S., Calcium, vitamin D, and risk of adenoma recurrence (United States), *Cancer Causes Control* 13 (3), 213–220, 2002.

137. Grau, M. V., Baron, J. A., Sandler, R. S., Haile, R. W., Beach, M. L., Church, T. R., and Heber, D., Vitamin D, calcium supplementation, and colorectal adenomas: Results of a randomized trial, *J Natl Cancer Inst* 95 (23), 1765–1771, 2003.

138. Harnack, L., Jacobs, D. R. Jr., Nicodemus, K., Lazovich, D., Anderson, K., and Folsom, A. R., Relationship of folate, vitamin B-6, vitamin B-12, and methionine intake to incidence of colorectal cancers, *Nutr Cancer* 43 (2), 152–158, 2002.

139. Ishihara, J., Otani, T., Inoue, M., Iwasaki, M., Sasazuki, S., and Tsugane, S., Low intake of vitamin B-6 is associated with increased risk of colorectal cancer in Japanese men, *J Nutr* 137 (7), 1808–1814, 2007.

140. Kune, G., and Watson, L., Colorectal cancer protective effects and the dietary micronutrients folate, methionine, vitamins B_6, B_{12}, C, E, selenium, and lycopene, *Nutr Cancer* 56 (1), 11–21, 2006.

141. Lajous, M., Lazcano-Ponce, E., Hernandez-Avila, M., Willett, W., and Romieu, I., Folate, vitamin B(6), and vitamin B(12) intake and the risk of breast cancer among Mexican women, *Cancer Epidemiol Biomarkers Prev* 15 (3), 443–448, 2006.

142. Zhang, S. M., Role of vitamins in the risk, prevention, and treatment of breast cancer, *Curr Opin Obstet Gynecol* 16 (1), 19–25, 2004.

143. Cole, B. F., Baron, J. A., Sandler, R. S., Haile, R. W., Ahnen, D. J., Bresalier, R. S., McKeown-Eyssen, G., Summers, R. W., Rothstein, R. I., Burke, C. A., Snover, D. C., Church, T. R., Allen, J. I., Robertson, D. J., Beck, G. J., Bond, J. H., Byers, T., Mandel, J. S., Mott, L. A., Pearson, L. H., Barry, E. L., Rees, J. R., Marcon, N., Saibil, F., Ueland, P. M., and Greenberg, E. R., Folic acid for the prevention of colorectal adenomas: A randomized clinical trial, *JAMA* 297 (21), 2351–2359, 2007.

144. Logan, R. F., Grainge, M. J., Shepherd, V. C., Armitage, N. C., and Muir, K. R., Aspirin and folic acid for the prevention of recurrent colorectal adenomas, *Gastroenterology* 134 (1), 29–38, 2008.

145. Jaszewski, R., Misra, S., Tobi, M., Ullah, N., Naumoff, J. A., Kucuk, O., Levi, E., Axelrod, B. N., Patel, B. B., and Majumdar, A. P., Folic acid supplementation inhibits recurrence of colorectal adenomas: A randomized chemoprevention trial, *World J Gastroenterol* 14 (28), 4492–4498, 2008.

146. Yang, Q., Bostick, R. M., Friedman, J. M., and Flanders, W. D., Serum folate and cancer mortality among U.S. adults: Findings from the Third National Health and Nutritional Examination Survey linked mortality file, *Cancer Epidemiol Biomarkers Prev* 18 (5), 1439–1447, 2009.

147. Beresford, S. A., Johnson, K. C., Ritenbaugh, C., Lasser, N. L., Snetselaar, L. G., Black, H. R., Anderson, G. L., Assaf, A. R., Bassford, T., Bowen, D., Brunner, R. L., Brzyski, R. G., Caan, B., Chlebowski, R. T., Gass, M., Harrigan, R. C., Hays, J., Heber, D., Heiss, G., Hendrix, S. L., Howard, B. V., Hsia, J., Hubbell, F. A., Jackson, R. D., Kotchen, J. M., Kuller, L. H., LaCroix, A. Z., Lane, D. S., Langer, R. D., Lewis, C. E., Manson, J. E., Margolis, K. L., Mossavar-Rahmani, Y., Ockene, J. K., Parker, L. M., Perri, M. G., Phillips, L., Prentice, R. L., Robbins, J., Rossouw, J. E., Sarto, G. E., Stefanick, M. L., Van Horn, L., Vitolins, M. Z., Wactawski-Wende, J., Wallace, R. B., and Whitlock, E., Low-fat dietary pattern and risk of colorectal cancer: The Women's Health Initiative Randomized Controlled Dietary Modification Trial, *JAMA* 295 (6), 643–654, 2006.

148. Shike, M., Latkany, L., Riedel, E., Fleisher, M., Schatzkin, A., Lanza, E., Corle, D., and Begg, C. B., Lack of effect of a low-fat, high-fruit, -vegetable, and -fiber diet on serum prostate-specific antigen of men without prostate cancer: Results from a randomized trial, *J Clin Oncol* 20 (17), 3592–3598, 2002.

149. Schatzkin, A., Lanza, E., Corle, D., Lance, P., Iber, F., Caan, B., Shike, M., Weissfeld, J., Burt, R., Cooper, M. R., Kikendall, J. W., and Cahill, J., Lack of effect of a low-fat, high-fiber diet on the recurrence of colorectal adenomas. Polyp Prevention Trial Study Group, *N Engl J Med* 342 (16), 1149–1155, 2000.

150. Steck-Scott, S., Forman, M. R., Sowell, A., Borkowf, C. B., Albert, P. S., Slattery, M., Brewer, B., Caan, B., Paskett, E., Iber, F., Kikendall, W., Marshall, J., Shike, M., Weissfeld, J., Snyder, K., Schatzkin, A., and Lanza, E., Carotenoids, vitamin A and risk of adenomatous polyp recurrence in the polyp prevention trial, *Int J Cancer* 112 (2), 295–305, 2004.

151. Kant, A. K., Schatzkin, A., Graubard, B. I., and Schairer, C., A prospective study of diet quality and mortality in women, *JAMA* 283 (16), 2109–2115, 2000.

152. Alberts, D. S., Martinez, M. E., Roe, D. J., Guillen-Rodriguez, J. M., Marshall, J. R., van Leeuwen, J. B., Reid, M. E., Ritenbaugh, C., Vargas, P. A., Bhattacharyya, A. B., Earnest, D. L., and Sampliner, R. E., Lack of effect of a high-fiber cereal supplement on the recurrence of colorectal adenomas. Phoenix Colon Cancer Prevention Physicians' Network, *N Engl J Med* 342 (16), 1156–1162, 2000.

153. Pierce, J. P., Natarajan, L., Caan, B. J., Parker, B. A., Greenberg, E. R., Flatt, S. W., Rock, C. L., Kealey, S., Al-Delaimy, W. K., Bardwell, W. A., Carlson, R. W., Emond, J. A., Faerber, S., Gold, E. B., Hajek, R. A., Hollenbach, K., Jones, L. A., Karanja, N., Madlensky, L., Marshall, J., Newman, V. A., Ritenbaugh, C., Thomson, C. A., Wasserman, L., and Stefanick, M. L., Influence of a diet very high in vegetables, fruit, and fiber and low in fat on prognosis following treatment for breast cancer: The Women's Healthy Eating and Living (WHEL) randomized trial, *JAMA* 298 (3), 289–298, 2007.

154. Vile, G. F., and Winterbourn, C. C., Inhibition of adriamycin-promoted microsomal lipid peroxidation by beta-carotene, alpha-tocopherol and retinol at high and low oxygen partial pressures, *FEBS Lett* 238 (2), 353–356, 1988.

155. Niki, E., Interaction of ascorbate and alpha-tocopherol, *Ann NY Acad Sci* 498, 186–199, 1987.

156. Carini, R., Poli, G., Dianzani, M. U., Maddix, S. P., Slater, T. F., and Cheeseman, K. H., Comparative evaluation of the antioxidant activity of alpha-tocopherol, alpha-tocopherol polyethylene glycol 1000 succinate and alpha-tocopherol succinate in isolated hepatocytes and liver microsomal suspensions, *Biochem Pharmacol* 39 (10), 1597–1601, 1990.

157. Sies, H., Sharov, V. S., Klotz, L. O., and Briviba, K., Glutathione peroxidase protects against peroxynitrite-mediated oxidations. A new function for selenoproteins as peroxynitrite reductase, *J Biol Chem* 272 (44), 27812–27817, 1997.

158. Witschi, A., Reddy, S., Stofer, B., and Lauterburg, B. H., The systemic availability of oral glutathione, *Eur J Clin Pharmacol* 43 (6), 667–669, 1992.

159. Niki, E., Mechanisms and dynamics of antioxidant action of ubiquinol, *Mol Aspects Med* 18 Suppl, S63–70, 1997.

160. Stoyanovsky, D. A., Osipov, A. N., Quinn, P. J., and Kagan, V. E., Ubiquinone-dependent recycling of vitamin E radicals by superoxide, *Arch Biochem Biophys* 323 (2), 343–351, 1995.

161. Zhang, L. X., Cooney, R. V., and Bertram, J. S., Carotenoids up-regulate connexin43 gene expression independent of their provitamin A or antioxidant properties, *Cancer Res* 52 (20), 5707–5712, 1992.

162. Carter, C. A., Pogribny, M., Davidson, A., Jackson, C. D., McGarrity, L. J., and Morris, S. M., Effects of retinoic acid on cell differentiation and reversion toward normal in human endometrial adenocarcinoma (RL95-2) cells, *Anticancer Res* 16 (1), 17–24, 1996.

163. Ames, B. N., Dietary carcinogens and anticarcinogens. Oxygen radicals and degenerative diseases, *Science* 221 (4617), 1256–1264, 1983.

164. Prasad, K. N., Hovland, A. R., Cole, W. C., Prasad, K. C., Nahreini, P., Edwards-Prasad, J., and Andreatta, C. P., Multiple antioxidants in the prevention and treatment of Alzheimer disease: Analysis of biologic rationale, *Clin Neuropharmacol* 23 (1), 2–13, 2000.

7 Health Risks of Low Doses of Ionizing Radiation

INTRODUCTION

Humans have been exposed to low doses of ionizing radiation from the background since their existence on planet Earth. In addition, humans also receive medical exposures for the diagnosis and treatment of certain human diseases. Radiation workers are exposed to radiation doses higher than nonradiation workers. These workers include those who are responsible for conducting diagnostic radiation procedures and those who are working in the nuclear industries. The crews of commercial flights also receive doses higher than those who do not fly. Frequent flyers are also exposed to radiation doses higher than those who fly infrequently. This chapter discusses sources of radiation exposure to humans, and doses and their potential risk of increasing the incidence of chronic diseases, including cancer.

SOURCES OF BACKGROUND RADIATION

The primary sources of background radiation are cosmic radiation, cosmogenic radionuclides, terrestrial radiation, and man-made radiation (summarized from the fact of US Nuclear Regulatory Commission (NRC) 1994). The level of cosmic radiation varies depending primarily on altitude. For example, the dose rate per year at sea level is 31 mrem (3.1 mSv), at 5000 feet it is 55 mrem (5.5 mSv), at 30,000 feet (normal jetliner) it is 1900 mrem, and at 80,000 feet (spy plane), it is 12,200 mrem.

Cosmic radionuclides arise primarily from the collision of highly energetic cosmic ray particles with atmospheric gases. The most important radionuclide produced is ^{14}C, but others, such as ^{3}H, ^{22}Na, and ^{7}Be are also produced. The ^{14}C in the atmosphere is rapidly oxidized to form $^{14}CO_2$, the majority of which is absorbed by the ocean. These radioactive isotopes are also released during nuclear testing. Nuclear reactors also generate ^{14}C, which can leak out in the case of an accident.

Several radioactive isotopes of varying physical half-lives are found on Earth. They include radioactive potassium (^{40}K with a half-life of 1.28 billion years, emits both beta and gamma radiation) and radioactive rubidium (^{87}Rb with a half-life of 4.7 billion years, emits only beta radiation), and alpha-radiation emitters such as the naturally occurring isotopes ^{238}U with a half-life of 4.7 billion years, ^{235}U with a physical half-life of about 700 million years, and [232]Th with a half-life of 14.05 billion years.

Man-made sources that contribute to background radiation include weapon testing and use, nuclear reactors, and accidents in nuclear power plants. Many short-lived and long-lived radioisotopes are released into the atmosphere when nuclear testing is done or when a nuclear power plant accident occurs. They include primarily ^{131}I with a physical half-life of about 8 days, ^{137}Cs with a physical half-life of 30 years, and ^{90}Sr with a physical half-life of 28.1 years.

The US NRC has estimated the total annual average background dose from natural sources is about 3.1 mSv. Radon and thoron gases contribute about two-thirds of this exposure. Man-made radiation sources, which include medical, commercial, and industrial activities, contribute about 3.1 mSv to the annual exposure of the US population. Thus, the total average annual whole-body dose to the US population may exceed 6.20 mSv. The percent contribution of radiation dose to the background radiation from the various sources is presented in Table 7.1.

The fact sheet of the Office of Radiation Protection of the Washington State Department of Health suggests that tobacco smoking can add an additional 2.8 mSV. The tobacco in cigarettes contains radioactive lead-210 (^{210}Pb), a beta

---

TABLE 7.1

Percentage Contribution of Radiation Dose to the Background Radiation from Various Sources

Radiation Sources	Percentage Contribution
Natural	
Radon and thoron	37
Cosmic (space)	5
Terrestrial (soil)	3
Internal	5
Total	50
Man-made	
Medical procedures	36
Consumer products	2
Nuclear medicine	12
Industrial and occupational	0.1
Total	50.1

Source: Adapted from National Council on Radiation Protection Report No. 160, *Ionizing Radiation Exposure of the Population of the United States*, 2009, http://www.ncrponline.org/Publications/Press_Releases/160press.html.

radiation–producing isotope that precipitates out of the atmosphere and is deposited on tobacco leaves. This radioactive isotope decays to radioactive polonium-210 (^{210}PO), which is an alpha emitter. Tobacco smokers receive radiation doses from ^{210}Pb that deposits on the surface of bones, and ^{210}PO that deposits in the liver, kidney, and spleen. The background radiation levels can vary depending upon altitude and soil rich in naturally occurring radioactive isotopes, such as ^{235}U. About 140,000 people in the Indian states of Kerala and Chenai are exposed to background radiation of about 30 mSv per year, and 15 mSV is contributed by radon gas.

Potential effects of background radiation: Doses from background radiation have been presumed to be safe for human health because humans have adapted to these doses and survived. Antioxidant defense systems that can quench radiation-induced free radicals have helped humans to successfully adapt to background radiation. Any dose of radiation can induce somatic and heritable mutations that can increase the risk of chronic diseases. Mutations that are caused by direct ionization may not be protected by the antioxidant defense system; therefore, they can persist from one generation to another. It should be pointed out that radiation interacts with other mutagens present in the environment, diet, and lifestyle in a synergistic manner to enhance the risk of chronic diseases. Humans suffer from a variety of chronic diseases. It is not possible to separate the contribution of background radiation to human diseases. Based on the fact that radiation is a mutagenic and carcinogenic agent, background radiation must contribute to some extent to the risk of chronic diseases. Continuous exposure to background radiation (3 rem or 0.03 Sv) per generation of 30 years would increase mutations by only about 1–6% of the spontaneous mutation rate in humans. Because of the interaction of radiation with other mutagenic substances present in the environment, diet, and lifestyle, it is difficult to estimate, with any accuracy, the mutation rate induced by background radiation alone. Although all mutations do not lead to cancer, all cancers are preceded by mutations. The rate of spontaneous cancer in humans is probably the result of interaction between background radiation and chemical and biological carcinogens and tumor promoters present in the diet and environment.

DIAGNOSTIC RADIATION PROCEDURES

Growing use of x-ray-based devices for early diagnosis of human diseases has raised concerns about the potential hazards of such procedures in increasing the risk of cancer and somatic and heritable mutations in individuals receiving diagnostic doses of radiation. In 2008, it was estimated that over 60 million computed tomography (CT) scans were performed in the United States.[1] This estimate did not include other diagnostic radiation procedures such as chest x-rays, dental x-rays, fluoroscopic imaging, positron emission tomography (PET), and other nuclear medicine scans. Therefore, it is likely that many more patients were exposed to diagnostic doses of radiation than the current estimates. Diagnostic radiation procedures can deliver up to 70 mSv to patients, depending upon the type of diagnostic procedures.

Recently, dose estimates and cancer risk from cardiovascular imaging have been published.[2] It has been estimated that about 5 billion imaging examinations are

TABLE 7.2

Summary of Estimated Effective Dose from Average Diagnostic Radiation Procedures

Type of Examination	Effective Dose (mSV)
Standard radiography	0.01–0.0
Computed tomography	2.0–20.0
Nuclear medicine	0.3–20.0
Interventional procedure	5.0–70.0

Source: Data were taken from F. A. Mettler Jr., W. Huda, T. T. Yoshizumi, and M. Mahesh, "Effective Doses in Radiology and Diagnostic Nuclear Medicine: A Catalog," *Radiology* 248, no. 1 (2008): 254–263.

performed worldwide each year, and 2 out of 3 involve ionizing radiation.[3] In 2006, the estimated medical radiation exposure dose in the United States reached 3.2 mSv [2], which is more than sixfold higher than that estimated in 2004.[4] Cardiologists prescribe and/or directly perform more than 50% of the radiation imaging examinations, which contribute about two-thirds of the total effective dose to patients.[5] It has been estimated that about 20 million nuclear medicine examinations were performed in 2006, and cardiac examinations accounted for about 57% of all nuclear medicine procedures and 85% of the radiation dose.[6] The effective radiation dose estimates from diagnostic procedures are presented in Table 7.2.[7] Other estimates of doses are presented in Tables 7.3 and 7.4.

RADIATION WORKERS

These risks also exist in radiation workers who are exposed to higher doses of ionizing radiation per year than nonradiation workers. The number of radiation workers has increased proportionally with increased use of diagnostic radiation procedures. The individuals working with radiation sources, referred to as *radiation workers*, are also exposed to radiation doses higher than those received by nonradiation workers. The effective and equivalent doses to radiation workers differ markedly from one procedure to another and from one country to another. In the United States, during a transcatheter aortic valve implantation, radiation workers receive a significantly higher effective dose (about 0.03 mSv) than other individuals who were not involved in this procedure.[8] In another study, the effective doses ranged from 0.02–38.0 uSV for diagnostic catheterizations, 0.17–31.2 uSv for percutaneous coronary interventions, 0.24–9.6 uSv for ablations, and 0.29–17.4 uSv for pacemaker or intracardiac defibrillator implantations.[9] In the Netherlands, the effective dose for neurointerventional procedures was about 6.7 uSv.[10] In Poland, an effective dose for a cardiological angioplastic procedure was 25 mSv, whereas the effective dose to radiation workers from intravascular angioplastic procedures within the abdominal cavity and

TABLE 7.3
Estimated Doses of Ionizing Radiation Delivered during Specific Diagnostic Procedures

Procedure Type	Effective Dose (mSv)
Chest or dental x-ray	0.01
Electron-beam CT (cardiac)	1.0–1.3
Electron-beam CT coronary angiography	1.5–2.0
Catheter coronary angiography	2.1–2.5
Electron-beam CT whole body	5.2
CT (head)	2.0
CT (abdomen)	10.0
Barium enema	7.0
Upper GI exam	3.0
I.V. urogram	2.5
Lumbar spine	1.3
Mammogram	7.0
Passenger from Athens to New York	0.06
Occupational annual dose limit	50.0
General public annual dose limit	1.0
Background annual dose at sea level	1.0

Source: Data were taken from K. N. Prasad, "Rationale for Using Multiple Antioxidants in Protecting Humans against Low Doses of Ionizing Radiation," *Br J Radiol* 78, no. 930 (2005): 485–492.

Note: Occupational and general public dose limit does not include background radiation.

TABLE 7.4
Estimated Doses of Radiation from Selected Radioactive Nuclides Administered Once during Nuclear Medicine Procedures

Procedure Type	Effective Dose (mSv)
18F-Flurodeoxyglucose, 10 mCi	4.80
99mTc-MAA lung scan, 5 mCi	0.60
99mTc-HDP bone scan, 20 mCi	4.00
201Tl Thallium scan, 3 mCi	0.60

Source: From K. N. Prasad, "Rationale for Using Multiple Antioxidants in Protecting Humans against Low Doses of Ionizing Radiation," *Br J Radiol* 78, no. 930 (2005): 485–492.

neuroradiological procedures was 4 mSv.[11] In Nigeria, the average annual dose to radiation workers increased from about 3.6 mSv in 1999 to about 7.7 mSv in 2001.[12]

Crews of commercial flights: The radiation dose received by commercial air crews differs, depending upon the route of travel. The radiation dose rate during a fight from Paris to Buenos Aires is 0.3 mrem/h. Based on 700 hours of flight time per year on a commercial airliner, air crews may receive a total annual effective dose of 210 mrem (2.10 mSv). This is below the 20 mSv recommended by the International Commission on Radiological Protection (ICRP) and the Federal Aviation Administration (FAA). Military and civilian pilots and flight attendants are exposed to increased levels of cosmic ionizing radiation of galactic and solar origin, and secondary radiation produced in the atmosphere, the aircraft structure and its contents, potential chemical carcinogens (fuel, jet engine exhausts, and cabin air pollutants), and electromagnetic fields from cockpit instruments. In May 2000, the European Union (EU) adopted regulations that apply to the air carriers in all 27 nations and require education on the health risks of in-flight radiation. In addition, assessment of radiation doses for all EU flight crewmembers is required. The ICRP has classified aircrews as radiation workers. Depending upon the flight patterns and altitudes, the annual dose range may vary from 0.2 to 5 mSv.[13] The average annual radiation dose equivalent of occupationally exposed adults in the United States is estimated to be about 1.1 mSv. Another study has reported that the average dose rate is 4–5 microSv/h for long-haul pilots and 1–3 microSv/h for short-haul pilots, yielding an annual effective dose of 2–3 mSv for long-haul pilots and 1–2 mSv for short-haul pilots.[14] A different study reported that the average annual dose received varies from 3–6 mSv.[15]

CONSEQUENCES OF EXPOSURE TO LOW DOSES OF RADIATION

It is established that ionizing radiation is a potent mutagen and carcinogen that can induce somatic and heritable mutations, and neoplastic and certain nonneoplastic diseases; however, it is also used in the diagnosis and treatment of certain human diseases. Children are more sensitive to ionizing radiation on all criteria of radiation damage, including cancer, than adults. Also, the time interval between radiation exposure and death in children is longer than in adults, which would increase the risk of expression of deleterious health effects in children more than in adults.[16,17]

Induction of radiation-induced mutations: Low doses of radiation can induce both somatic and heritable mutations. At present, the frequency of mutation in humans is based on mammalian genetics. It has been reported that the frequency of mutations in mice is dependent upon the dose rate and total dose.[17-19] The effect of dose rate on mutations in females is different from that observed in males. The mutation frequency in oocytes at a high dose rate is greater than that in spermatogonia; however, at a lower dose rate, it is much less than that in spermatogonia. Older mice are more sensitive to the dose rate effect than younger mice. The time interval between two fractionated doses is important for the production of mutations in male mice.

Probable steps involved in radiation-induced cancer: Radiation-induced carcinogenesis is no different from that occurring spontaneously or caused by other carcinogens and tumor promoters. It involves a gradual accumulation of multiple mutations over a long period of time. Radiation-induced human cancers have long latent periods; for example, 10 years for leukemia and 30 years or more for solid tumors.[20] This implies that radiation-induced mutations (due to genetic mutations and/or chromosomal damage) that can be detected within 24 hours of radiation exposure are not directly responsible for the development of carcinogenesis in normal human cells because such cells continue to divide and proliferate like nonirradiated normal cells for a long time. However, radiation-induced mutations cause genetic instability in normal cells that can make these cells more sensitive to mutagenic changes caused by increased oxidative stress, which continues to occur as a function of time. Such genetically unstable cells may continue to proliferate and differentiate like nonirradiated cells in spite of carrying genetic abnormalities for a long period of time until the expression of genes that regulate differentiation is mutated. This defect in expression of differentiation genes prevents cells from going through normal patterns of differentiation and death, and thus, they continue to divide. These cells are referred to as *immortal cells*, and they represent the first step in carcinogenesis. Immortalized cells can continue to proliferate and can form adenomas such as polyps in the colon or cysts in the breast. When some key cellular genes, oncogenes or anti-oncogenes, are annotated by continued exposure to increased oxidative stress caused by mutagens, carcinogens, and/or tumor promoters, these cells then become fully transformed and can induce cancer when tested in the appropriate host.[21,22] The existence of a long latent period for radiation-induced cancer provides an opportunity to intervene at any time after radiation exposure with appropriate radioprotective agents in order to reduce the risk of late adverse effects of radiation. Radiation-induced immediate damage to DNA is caused primarily by free radicals, and very little by direct ionization; free radicals are also involved in subsequent mutations in the irradiated cells for a long period of time.

Interactions of radiation with chemical and biological carcinogens and tumor promoters: In addition to low doses of ionizing radiation from background radiation, diagnostic radiation procedures, frequent flying at high altitudes, and working with radiation equipment expose humans to chemical and biological carcinogens and tumor promoters that can enhance the effect of low doses of radiation on neoplastic and nonneoplastic diseases. This has been demonstrated by laboratory experiments primarily on cell culture models. For example, x-radiation-induced transformation was enhanced by about 9-fold [23] by chemical carcinogens in normal mammalian cells. The incidence of transformation by x-irradiation was enhanced by UV light by about 12-fold.[24] X-irradiation also enhanced the level of ozone-[25] and viral-induced[26] transformation in cell cultures. Radiation doses alone may not transform normal fibroblasts, but they do so when combined with a tumor promoter.[27] Ionizing radiation in combination with tobacco smoking increases the risk of lung cancer by about 50%.[20] A low dose of radiation (20 mSV) does not produce detectable levels of mutations as measured by chromosomal damage; however, in the presence of caffeine (which inhibits repair of DNA damage), mutations become detectable.[27] Low doses of radiation (20 and 50 mSV) can act as a mitogen,[28] and lower doses (about 1 mSv) do not activate double-strand DNA break repair mechanisms.[29] This lack of repair after exposure to low doses of radiation can lead to accumulation of mutations that can increase the risk of cancer. A recent study has reported that irradiation of freshly isolated lymphocytes with a dose of 10 mGy of x-rays increased the number of cells with DNA double-strand breaks (DSBs) (Figure 7.1). Treatment of cells with BioShield-R1 before irradiation reduced the number of cells with DSBs. A similar result was obtained when the number of cells with DSBs was determined in the lymphocytes isolated from patients who received BioShield-R1 before a dose of 10 mGy[30] (Figure 7.2). The issue of interaction of radiation with other carcinogens and tumor promoters is often ignored when discussing the risk of radiation-induced cancer in children or adults.

MODELS USED FOR RISK ESTIMATES OF RADIATION-INDUCED CANCER

Two models of risk estimate for radiation-induced cancer have been proposed. The first model proposes that cancer risk in humans following exposure to low doses of radiation (100 mSv or less) may be best estimated by a linear no-threshold relationship, since any dose has the potential to induce cancer. The most recent BEIR report supports this model.[31] This model of risk estimation for radiation-induced cancer can also be applied to children.[32] The second model suggests that there is a threshold dose below which radiation may not induce cancer in humans.[33,34] The second model relies primarily on mathematical modeling. Mathematical modeling may assume certain constant physical factors, such as body weight,[34] which may not reflect the inherent biological variations associated with radiation-induced carcinogenesis. These include differences in radiosensitivity with respect to age, organs, body mass, and differences in the efficacy of repair mechanisms.

CANCER RISKS IN POPULATIONS EXPOSED TO DIAGNOSTIC RADIATION PROCEDURES

Adults and children: The health risk of low doses of radiation include increased incidence of cancer and certain nonneoplastic diseases, such as thyroid diseases, vision disorders, and micronuclei formation in the buccal mucosa following dental-x-rays. Recently, a few excellent reviews have been published in support of the linear no-threshold model for radiation-induced cancer.[31,32] The typical radiation doses for an adult from a chest CT scan can range between 6 and 10 mSV.[35] The average annual dose from background radiation in the United States is approximately 3.1 mSV. Several radiation dose estimates for imaging studies in adults and children have been published.[35,36] A dose of from a CT scan may increase the risk of cancer in children by 1 in 1000 exposed individuals;[37–39] however, another study has reported that the lifetime cancer mortality risks for a one-year-old child exposed to a radiation dose from a CT scan are 0.18% (abdominal CT scan) and 0.07% (head CT scan). This

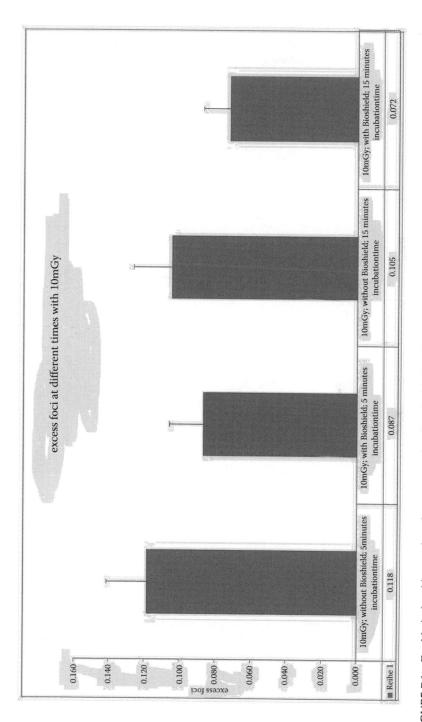

FIGURE 7.1 Freshly isolated human lymphocytes were irradiated with 10 mGy in the presence or absence of Bio-shield-R1 and incubated for 5 min and 15 min. Bio-shield treatment reduced the number of cells with DNA double-stranded breaks (DNA DSBs) 30.

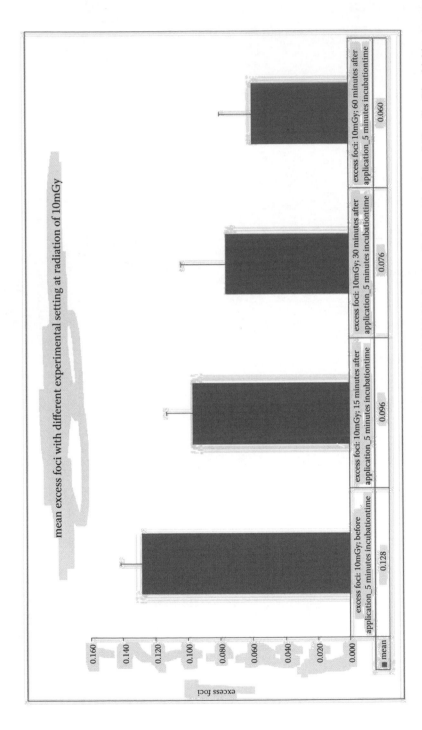

FIGURE 7.2 Freshly isolated lymphocytes from patients who received Bio-shield-R1 before irradiation with a dose of 10 mGy. Bio-shield treatment reduced the number of cells with DNA double-stranded breaks (DNA DSBs) 30.

risk estimate is an order of magnitude higher than for adults. It was further estimated that in 2001, approximately 600,000 abdominal and head CT scans were performed in children under the age of 15, and that 500 of the exposed children might die from cancer attributed to the CT scan.[40] A Canadian study reported that an abdominal CT study in a 5-year-old child may increase the lifetime risk of radiation-induced cancer by approximately 26.1 per 100,000 in female and 20.4 per 100,000 in male patients.[41] Another study performed in Israel estimated an increase of about 0.29% over the total number of patients who are eventually expected to die from cancer.[42] If one considers the fact that patients receiving diagnostic doses of radiation and radiation workers may also be exposed to chemical and biological carcinogens, as well as tumor promoters that enhance radiation-induced cancer risk during their lifetime, the above estimates of cancer mortality risk could be higher.

The analysis of the risk of cancer on the basis of the annual number of diagnostic x-rays taken in the United Kingdom and 14 developing countries revealed that the cumulative risk varied from 0.6% to 1.8%, whereas in Japan, which used the highest number of annual diagnostic x-rays, it was more than 3%.[4] BEIR report VII has estimated that a dose of 15 mSv may increase cancer risk by 1/750 cases. Others have estimated that a coronary multislice CT (MSCT) scan, which delivers patients about 20 mSv, may increase the risk of cancer by 1/500; and for coronary stent by 1/400 for 25 mSv.[18] Estimates of cancer risk are presented in Table 7.5. An epidemiologic study suggested that the increased risk of thyroid cancer was associated with exposure to multiple dental x-rays.[43] This association was independent of age, gender, nationality, and level of education.

If one considers the fact that children may be exposed to chemical and biological carcinogens, as well as tumor promoters that enhance radiation-induced cancer risk, these estimates of cancer mortality risk could be higher for children. It has been

---

TABLE 7.5

Estimated Increases in Cancer Risk from an Exposure to One Computed Tomography (CT) Scan

Source and Dose of Radiation	Increase in Cancer Risk in Exposed Individuals
One CT scan	1 per 1000[37–39]
	1 per 1200[40]
	2.9 per 1000[42]
	0.26 per 1000 female[41]
	0.2 per 1000 male[41]
15 mSv	1 per 750[18]
20 mSV	1 per 500[18]
25 mSv	1 per 400[18]

Note: The superscript number is the reference number.

---

estimated that that the risk of hematopoietic tumors lasts for about 12 years and largely disappears within 30 years after irradiation. On the other hand, solid tumors seldom appear before 10 years after irradiation, and they may continue to appear for 30 years or more.[44]

Cancer risk in children exposed in utero during the atomic bombing of Hiroshima and Nagasaki: This involves the incidence of cancer among children exposed in utero who were born between the days of the atomic bombing (August 6, 1945 in Hiroshima, and August 9, 1945 in Nagasaki) and May 1, 1946. The incidence of cancer between 1950 and 1984 increased by about two to threefold, depending upon the dose.[45] The incidence of breast cancer was higher among girls exposed to atomic bombing under the age of 10 years than those who were older at the time of the bombing.[46] Among those girls who received 0.5 Gy or more, 10 cases of breast cancer were observed, in comparison to the expected 2 cases among nonirradiated girls.[47]

Risk of childhood cancer after irradiation of fetuses: Most of the human data are based on the analysis of radiation-exposed pregnant women of Hiroshima and Nagasaki or of pregnant women who received a radiation dose during a diagnostic radiation procedure or radiation therapy for cancer. The effects of high doses of irradiation on nonneoplastic diseases were discussed in Chapter 4. It has been reported that an increase in crude cancer risk is directly proportional to the number of x-ray films or fetus dose.[48] The cancer risk was greater when fetuses received radiation doses during the first trimester. The authors estimated that 0.01 Gy (1 cGy) delivered to the fetus shortly before birth would cause an increase of 300–800 deaths per million before the age of 10 due to cancer (572 ± 133). This corresponds to an annual absolute risk of 57 cancer deaths per million per rad during the period of observation of 10 years. Further analysis of radiation-induced cancer in children following radiation exposure of the fetuses confirmed the hypothesis of a linear relationship between fetal dose and the relative incidence of cancer;[49–54] however, the magnitude of the effect of fetal irradiation on the incidence of cancer has been questioned.[52] More recently, it was estimated that exposure of fetuses to low levels of radiation increased the risk of leukemia in childhood by an excess relative coefficient of around 50 per Gy (equivalent to an excess absolute risk coefficient of about 3% per Gy).[55] It has been estimated that the proportion of childhood leukemia incidence in Great Britain attributable to natural background is about 15–20%.[56] A significant increase in malignancy has been found even after a dose of 0.25 cGy to the human fetus.[51] Maternal smoking and irradiation with diagnostic doses during pregnancy increases the risk of childhood leukemia by more than 3 times compared to x-irradiation alone.[57] The incidence of leukemia was higher among children irradiated with diagnostic doses of radiation in utero; the incidence was still higher if the mother had previously experienced miscarriage or stillbirth.[58] It has been reported that the incidence of childhood leukemia was 5 times more if the pregnant women received diagnostic doses of radiation during the first trimester; however, if the pregnant women received diagnostic doses of radiation during the second or third trimester, the incidence of cancer was only 1.47 times higher than those who did not receive radiation during pregnancy.[48,49,59] A high incidence of tumors among children after the administration of radioactive iron (^{59}Fe) to pregnant women was reported.[60] The estimated fetus dose was 5–15 cGy. Among 634 children exposed, one leukemia and two sarcoma cases were found. No malignancies were observed in the control group.

Women receiving gonadal doses before conception: The studies on the effects of diagnostic doses of x-rays before conception are not sufficient to make any definitive conclusion. However, a few studies suggest a potential risk of nonneoplastic diseases. For example, it has been reported that young females who received diagnostic doses of radiation (0.5–7 cGy) before conception had 10 times more aneuploidy children, of which 8 were mongoloid, than those who did not receive radiation.[61] A gonadal dose of about 5 cGy to the young women before conception produced an eye defect in subsequent children.[62] A gonadal dose of 5 R (about 5 cGy), which may be received by medical radiation exposure over 30 years, would increase the number of point mutations by about 1%.[63]

The estimated minimal spontaneous frequency of point mutations per generation due to naturally occurring mutagens is about 2%. This mutation rate contributes to less than half of the gross abnormalities, such as mental defects, hematological and endocrine defects, defects in vision and hearing, cutaneous and skeletal defects, and defects in the GI tract that occur after birth. It does not include those mutations whose effects were less drastic.

CANCER RISK AMONG RADIATION WORKERS

Assessment of cancer risk among radiation workers can be estimated only by epidemiologic studies. These studies thus far have produced conflicting results on the risk of neoplastic and nonneoplastic diseases. This may be due to the fact that the latent period after low doses of radiation is very high. In addition, radiation workers who worked between 1940 and the 1980s may have received higher cumulative doses than those working after the 1980s because of advancements in the technology of diagnostic radiation equipment that have reduced exposure to radiation workers. Recently employed radiation workers are difficult to evaluate for cancer risk because of the insufficient time interval between irradiation and the risk of developing cancer. A review of published epidemiologic studies involving 400,000 radiation workers suggested that an accumulated dose of 100 mGy or 100 mSv was significantly associated with an increase in leukemia risk.[64] Another study analyzed the cancer risk among 67,562 Canadian radiation workers (23,580 males and 43,982 females) employed during the time period of 1951 to 1987. The results showed that the thyroid cancer incidence was significantly increased among both males and females, with a combined standardized incidence ratio of 1.74.[65] At present, medical radiation workers in Canada are receiving much lower doses of radiation in comparison to those who were working between 1951 and 1987. The risk of cancer among these radiation workers remains unknown. In an epidemiologic study involving 56,436 US female radiologic technologists who were certified from 1925 to 1980, the incidence of breast cancer was evaluated. The results showed an increased incidence of cancer among those who were working before 1940 but not after that. Improvement in radiation technology and implementation of radiation protection standards have contributed to the reduction of accumulative radiation doses.[66] Another study reported an increased risk of leukemia among radiologists and technologists who were employed before 1950.[67] In a 15-counry collaborative cohort study involving 407,391 nuclear industry workers, the risk of radiation-induced cancer was evaluated.[68] The results showed that among 31 specific types of cancer, a significant association was observed only for lung cancer. However, for multiple myeloma and ill-defined and secondary cancers, a borderline significant association was observed. The same group of investigators found that among 600,000 nuclear industry workers who received an average cumulative dose of about 19.2 mSv, a strong association with improved health was found in most countries.[69] The incidence of thyroid cancer was evaluated among 73,080 radiologic technologists. The increased risk of thyroid cancer was observed among those who worked for more than 5 years prior to 1950. The analysis of cancer risk among the South Korean radiation workers in nuclear power plants revealed no increase in cancer risk.[70] The risk of cataract formation among interventional cardiologists and associated nurses was evaluated in 116 radiation-exposed individuals and 93 unexposed individuals. The results showed that 38% of those who received a cumulative lens dose of about 6.0 mSv developed cataracts, compared to 12% in unexposed individuals. Only 21% of the nurses and technicians who received a cumulative lens dose of 1.5 mSv developed cataracts.[71] The gene expression profiles are markedly altered in the peripheral lymphocytes of radiation workers.[72] Generally, radiation workers may receive whole-body irradiation that is likely to induce heritable mutations that may appear in future generations.

The epidemiologic studies previously outlined provide no conclusive evidence for increased health risks among radiation workers. It should be emphasized that not all radiation workers receive the same cumulative radiation doses. For example, radiation workers involved in interventional cardiology procedures may receive much higher cumulative doses per year than those responsible for chest or dental x-rays. In addition, environmental, dietary, and lifestyle-related factors may influence the radiation-induced carcinogenic processes. It is impossible to account for all external variables that could impact radiation-induced cancer risk in any epidemiologic studies. In my opinion, radiation workers should consider adopting a strategy that can provide biological protection against damage, no matter how small the damage might be.

CANCER RISK IN MILITARY AND CIVILIAN PILOTS AND FLIGHT ATTENDANTS

Several epidemiologic studies have evaluated the risk of cancer in these populations. Most studies suggest an increased risk of prostate, melanoma and other skin cancer, and acute myeloid leukemia in male pilots,[73–76] and breast cancer, melanoma,[75,77,78] and bone cancer[79,80] in female flight attendants. The incidence of acute myeloid leukemia and brain cancer did not change among flight crews from Denmark, Finland, Iceland, Norway, and Sweden.[76] Most epidemiologic studies have revealed that the risk of certain types of cancer increases among pilots and flight attendants. Since they receive whole-body irradiation, the risk of heritable mutations that could appear in future generations may also increase.

CANCER RISK AMONG FREQUENT FLYERS

Frequent flyers may receive extra radiation doses while going through x-ray scanners at the airport and while in the aircraft. They may receive cumulative radiation doses less than flight crews, but the risk of adverse health effects may exist in this population similar to that in flight crews. There are no epidemiologic studies to evaluate the risk of cancer or noncancerous diseases among frequent flyers. Since no radiation doses are considered safe, it is prudent to adopt a strategy to provide biological protection against the adverse health effects of radiation.

RISK OF LOW-DOSE RADIATION-INDUCED NONNEOPLASTIC DISEASES

Developing organisms constitute a highly dynamic system in which rapid cell proliferation, cell migration, and cell differentiation occur. Therefore, it is expected that the radiation response of human embryos as a whole, or specific tissues, would markedly differ depending upon the stage of development. Several animal studies suggest that radiation-induced changes are similar in different species when they are irradiated at an equivalent stage of the development.[44,81] Many studies have shown that low radiation doses are harmful to the human fetus.[48,59,82] The central nervous system and optic tissues in developing fetuses are highly radiosensitive, and small doses (0.05–0.1 Gy) may cause abnormalities in these organs. The extent of damage depends upon total dose, dose rate, linear energy transfer (LET), and mode of radiation delivery (single vs. fractionated dose). The type and number of abnormalities may depend upon the age of the fetus and mode of radiation delivery.

The incidence of nonneoplastic diseases and intermediate health risks measured by certain specific biochemical markers were studied in children living in radiation-contaminated areas near the Chernobyl nuclear accident site. The incidence of thyroid gland enlargement and vision disorders, mostly dry eye syndrome, was closely related to the levels of contamination.[83] Increased levels of oxidized conjugated dienes, products of lipid peroxidation, were found among these children. In another report, increased levels of spontaneous chemiluminescence, an indicator of enhanced oxygen radical activity, in leukocytes of children living in contaminated areas were observed.[84] The accuracy of these intermediate markers for predicting health risks after radiation exposure remains uncertain. Radiation exposure during interventional cardiovascular procedures can induce damage to DNA and can cause chromosomal aberrations.[85] Complex cytogenetic abnormalities among Japanese survivors of the atomic bomb were observed 20 years after radiation exposure. The micronucleus assay in exfoliated buccal cells is considered a useful and minimally invasive assay method for monitoring genetic damage in humans.[86] Dental x-rays can induce formation of micronuclei in buccal cells in both adults and children.[87,88] The significance of this observation in predicting cancer risk following dental x-rays remains to be established.

SUMMARY

Humans are exposed to low doses of ionizing radiation from the background, medical exposures, working in the radiation environment, and flying on commercial jets. The US NRC has estimated the total average annual whole-body dose to the US population is about 6.20 mSv (3.1 mSv from natural sources and 3.1 mSv from manmade sources). The contribution of background radiation to cancer risk is impossible to evaluate because of its interaction with other carcinogens. It appears that the rate of spontaneous cancer in humans is probably the result of interaction between background radiation, and chemical and biological carcinogens and tumor promoters present in the diet, lifestyle, and environment.

Growing use of x-ray-based equipment for early diagnosis of human diseases has raised concerns about potential hazards of such procedures in increasing the risk of cancer and somatic and heritable mutations in patients. These risks also exist in radiation workers who are exposed to higher doses of radiation per year than nonradiation workers. The number of radiation workers has increased proportionally with the increased use of diagnostic radiation procedures.

Developing embryos are extremely sensitive to ionizing radiation; the types and severity of effect depend upon the dose, stage of development, and the mode of delivery of radiation (single dose, fractionation, or protraction). It also appears that the radiosensitivity of developmental stages of animals closely resembles that found in humans. Therefore, the animal data can be used to estimate hazard in humans. The fetus is extremely sensitive to radiation for the criterion of organ defects during the period of organogenesis, which in humans extends for about 2–9 weeks after conception. The radiation exposure after organogenesis can produce other types of damage. Some of the radiation-induced changes in fetuses can be observed at birth, whereas others may appear at a much later date. Some of the functional changes may not be possible to measure at the time of birth. Among organ defects, microcephaly associated with mental retardation is most pronounced after exposure of fetuses during the period of organogenesis. Data on animals and on survivors of the Hiroshima and Nagasaki atom bombs indicate that radiation doses of 0.1 Gy or less during organogenesis may produce brain abnormality. Exposure of fetuses at any stage of development to diagnostic doses of radiation increases the risk of childhood leukemia. The risk appears to be higher during the first trimester than during the second and third trimesters. There appears to be no threshold dose for leukemia among children who received radiation in utero; however, for most organ defects, there appears to be a threshold dose.

REFERENCES

1. Shah, N. B., and Platt, S. L., ALARA: Is there a cause for alarm? Reducing radiation risks from computed tomography scanning in children, *Curr Opin Pediatr* 20 (3), 243–247, 2008.
2. Picano, E., Vano, E., Semelka, R., and Regulla, D., The American College of Radiology white paper on radiation dose in medicine: Deep impact on the practice of cardiovascular imaging, *Cardiovasc Ultrasound* 5, 37, 2007.
3. Picano, E., Sustainability of medical imaging, *BMJ* 328 (7439), 578–580, 2004.

4. Berrington de Gonzalez, A., and Darby, S., Risk of cancer from diagnostic x-rays: Estimates for the UK and 14 other countries, *Lancet* 363 (9406), 345–351, 2004.

5. Bedetti, G., Botto, N., Andreassi, M. G., Traino, C., Vano, E., and Picano, E., Cumulative patient effective dose in cardiology, *Br J Radiol* 81 (969), 699–705, 2008.

6. Amis, E. S. Jr., Butler, P. F., Applegate, K. E., Birnbaum, S. B., Brateman, L. F., Hevezi, J. M., Mettler, F. A., Morin, R. L., Pentecost, M. J., Smith, G. G., Strauss, K. J., and Zeman, R. K., American College of Radiology white paper on radiation dose in medicine, *J Am Coll Radiol* 4 (5), 272–284, 2007.

7. Mettler, F. A. Jr., Huda, W., Yoshizumi, T. T., and Mahesh, M., Effective doses in radiology and diagnostic nuclear medicine: A catalog, *Radiology* 248 (1), 254–263, 2008.

8. Sauren, L. D., van Garsse, L., van Ommen, V., and Kemerink, G. J., Occupational radiation dose during transcatheter aortic valve implantation, *Catheter Cardiovasc Interv*, 2011.

9. Kim, K. P., Miller, D. L., Balter, S., Kleinerman, R. A., Linet, M. S., Kwon, D., and Simon, S. L., Occupational radiation doses to operators performing cardiac catheterization procedures, *Health Phys* 94 (3), 211–227, 2008.

10. Kemerink, G. J., Frantzen, M. J., Oei, K., Sluzewski, M., van Rooij, W. J., Wilmink, J., and van Engelshoven, J. M., Patient and occupational dose in neurointerventional procedures, *Neuroradiology* 44 (6), 522–528, 2002.

11. Staniszewska, M. A., and Jankowski, J., Personnel exposure during interventional radiologic procedures [In Polish], *Med Pr* 51 (6), 563–571, 2000.

12. Ogundare, F. O., and Balogun, F. A., Whole-body doses of occupationally exposed female workers in Nigeria (1999–2001), *J Radiol Prot* 23 (2), 201–208, 2003.

13. Waters, M., Bloom, T. F., and Grajewski, B., The NIOSH/FAA Working Women's Health Study: Evaluation of the cosmic-radiation exposures of flight attendants. Federal Aviation Administration, *Health Phys* 79 (5), 553–559, 2000.

14. Bagshaw, M., Cosmic radiation in commercial aviation, *Travel Med Infect Dis* 6 (3), 125–127, 2008.

15. Blettner, M., Grosche, B., and Zeeb, H., Occupational cancer risk in pilots and flight attendants: Current epidemiological knowledge, *Radiat Environ Biophys* 37 (2), 75–80, 1998.

16. Kleinerman, R. A., Cancer risks following diagnostic and therapeutic radiation exposure in children, *Pediatr Radiol* 36 Suppl 14, 121–125, 2006.

17. Prasad, K. N., *Handbook of Radiobiology*, 2nd Ed., CRC Press, Boca Raton, FL, 1995.

18. Committee in the Biological Effects of Ionizing Radiation, *The Effects of Populations Exposure to Low Levels of Ionizing Radiation: BEIR VII*, National Academy Press, Washington, DC, 2006.

19. Russell, W. L., The effect of radiation dose rate and fractionation on mutation in mice, in *Repair from Genetic Radiation Damage and Differential Radiosensitivity in Germ Cells*, ed. F. H. Sagel, Macmillan Press, New York, 1963, 205–217.

20. Committee on the Biological Effects of Ionizing Radiation, *The Effects of Populations Exposure to Low Levels of Ionizing Radiation: BEIR V*, National Academy Press, Washington, DC, 1990.

21. Prasad, K. N., Cole, W. C., Yan, X. D., Nahreini, P., Kumar, B., Hanson, A., and Prasad, J. E., Defects in cAMP-pathway may initiate carcinogenesis in dividing nerve cells: A review, *Apoptosis* 8 (6), 579–586, 2003.

22. Prasad, K. N., Hovland, A. R., Nahreini, P., Cole, W. C., Hovland, P., Kumar, B., and Prasad, K. C., Differentiation genes: Are they primary targets for human carcinogenesis?, *Exp Biol Med (Maywood)* 226 (9), 805–813, 2001.

23. DiPaolo, J. A., and Donovan, P. J., In vitro morphologic transformation of Syrian hamster cells by UV-irradiation is enhanced by X-irridation and unaffected by chemical carcinogens, *Int J Radiat Biol Relat Stud Phys Chem Med* 30 (1), 41–53, 1976.

24. Borek, C., Zaider, M., Ong, A., Mason, H., and Witz, G., Ozone acts alone and syn-ergistically with ionizing radiation to induce in vitro neoplastic transformation, *Carcinogenesis* 7 (9), 1611–1613, 1986.
25. Pollock, E. J., and Todaro, G. J., Radiation enhancement of SV40 transformation in 3T3 and human cells, *Nature* 219 (5153), 520–521, 1968.
26. Little, J. B., Influence of noncarcinogenic secondary factors on radiation carcinogenesis, *Radiat Res* 87 (2), 240–250, 1981.
27. Puck, T. T., Morse, H., Johnson, R., and Waldren, C. A., Caffeine enhanced measure-ment of mutagenesis by low levels of gamma-irradiation in human lymphocytes, *Somat Cell Mol Genet* 19 (5), 423–429, 1993.
28. Suzuki, K., Kodama, S., and Watanabe, M., Extremely low-dose ionizing radiation causes activation of mitogen-activated protein kinase pathway and enhances prolifera-tion of normal human diploid cells, *Cancer Res* 61 (14), 5396–5401, 2001.
29. Rothkamm, K. and Lobrich, M., Evidence for a lack of DNA double-strand break repair in human cells exposed to very low x-ray doses, *Proc Natl Acad Sci USA* 100 (9), 5057–5062, 2003.
30. Ehrlich, J. S., Brand, R., Uder, M., and Kuefner, M., Effect of proprietary combination of antioxidants/glutathione-elevating agents on x-ray induced DNA double-strand breaks, presented at the Annual Meeting of Society of Cardiovascular Computed Tomography, 2011.
31. Committee on the Biological Effects of Ionizing Radiation, *The Effects of Populations Exposure to Low Levels of Ionizing Radiation: BEIR VII, Phase 2*, The National Academy Press, Washington, DC, 2006.
32. Brenner, D. J., and Sachs, R. K., Estimating radiation-induced cancer risks at very low doses: Rationale for using a linear no-threshold approach, *Radiat Environ Biophys* 44 (4), 253–256, 2006.
33. Cohen, B. L., Cancer risk from low-level radiation, *AJR Am J Roentgenol* 179 (5), 1137–1143, 2002.
34. Bond, V. P., Benary, V., and Sondhaus, C. A., A different perception of the linear, nonthreshold hypothesis for low-dose irradiation, *Proc Natl Acad Sci USA* 88 (19), 8666–8670, 1991.
35. Robbins, E., Radiation risks from imaging studies in children with cancer, *Pediatr Blood Cancer* 51 (4), 453–457, 2008.
36. Hall, E. J., and Garcia, A. J., *Radiobiology for the Radiologist*, Lippincott Williams & Wilkins, Philadelphia, 2006.
37. Rice, H. E., Frush, D. P., Farmer, D., and Waldhausen, J. H., Review of radiation risks from computed tomography: Essentials for the pediatric surgeon, *J Pediatr Surg* 42 (4), 603–607, 2007.
38. Hall, E. J., Lessons we have learned from our children: Cancer risks from diagnostic radiology, *Pediatr Radiol* 32 (10), 700–706, 2002.
39. Hall, E. J., Radiation biology for pediatric radiologists, *Pediatr Radiol*, 2008.
40. Brenner, D., Elliston, C., Hall, E., and Berdon, W., Estimated risks of radiation-induced fatal cancer from pediatric CT, *AJR Am J Roentgenol* 176 (2), 289–296, 2001.
41. Wan, M. J., Krahn, M., Ungar, W. J., Caku, E., Sung, L., Medina, L. S., and Doria, A. S., Acute appendicitis in young children: Cost-effectiveness of UV versus CT in diagnosis—A Markov Decision Analytic Model, *Radiology*, 2008.
42. Chodick, G., Ronckers, C. M., Shalev, V., and Ron, E., Excess lifetime cancer mortal-ity risk attributable to radiation exposure from computed tomography examinations in children, *Isr Med Assoc J* 9 (8), 584–587, 2007.
43. Memon, A., Godward, S., Williams, D., Siddique, I., and Al-Saleh, K., Dental x-rays and the risk of thyroid cancer: A case-control study, *Acta Oncol* 49 (4), 447–453, 2010.

44. Committee on the Biological Effects of Ionizing Radiation, *The Effects of Populations Exposure to Low Levels of Ionizing Radiation: BEIR V*, National Academy Press, Washington, DC, 1980.
45. Yoshimoto, Y., Cancer risk among children of atomic bomb survivors. A review of RERF epidemiologic studies. Radiation Effects Research Foundation, *JAMA* 264 (5), 596–600, 1990.
46. Yoshimoto, Y., Kato, H., and Schull, W. J., Risk of cancer among children exposed in utero to A-bomb radiations, 1950–84, *Lancet* 2 (8612), 665–669, 1988.
47. Miller, R. W., Effects of prenatal exposure to ionizing radiation, *Health Phys* 59 (1), 57–61, 1990.
48. Stewart, A., and Kneale, G. W., Radiation dose effects in relation to obstetric x-rays and childhood cancers, *Lancet* 1 (7658), 1185–1188, 1970.
49. Stewart, A., Webb, J., and Hewitt, D., A survey of childhood malignancies, *Br Med J* 1 (5086), 1495–1508, 1958.
50. Doll, R., Radiation hazards: 25 years of collaborative research. Sylvanus Thompson memorial lecture, April 1980, *Br J Radiol* 54 (639), 179–186, 1981.
51. Newcombe, H. B., and McGregor, J. F., Childhood cancer following obstetric radiography, *Lancet* 2 (7734), 1151–1152, 1971.
52. Shore, F. J., Robertson, J. S., and Bateman, J. L., Childhood cancer following obstetric radiography, *Health Phys* 24 (2), 258–260, 1973.
53. Mole, R. H., Childhood cancer after prenatal exposure to diagnostic x-ray examinations in Britain, *Br J Cancer* 62 (1), 152–168, 1990.
54. Williams, P. M., and Fletcher, S., Health effects of prenatal radiation exposure, *Am Fam Physician* 82 (5), 488–493, 2010.
55. Wakeford, R., Childhood leukaemia following medical diagnostic exposure to ionizing radiation in utero or after birth, *Radiat Prot Dosimetry* 132 (2), 166–174, 2008.
56. Little, M. P., Wakeford, R., and Kendall, G. M., Updated estimates of the proportion of childhood leukaemia incidence in Great Britain that may be caused by natural background ionising radiation, *J Radiol Prot* 29 (4), 467–482, 2009.
57. Stjernfeldt, M., Berglund, K., Lindsten, J., and Ludvigsson, J., Maternal smoking and irradiation during pregnancy as risk factors for child leukemia, *Cancer Detect Prev* 16 (2), 129–135, 1992.
58. Graham, S. L., Levin, L. L., Lilianfield, A. M., Schuman, L. M., Gibson, R., Dowd, J. E., and Hepelmann, L., Preconception, intrauterine and postnatal irradiation as related to leukemia, *Natl Cancer Inst Monogr* 19, 342, 1966.
59. Macmahon, B., Prenatal x-ray exposure and childhood cancer, *J Natl Cancer Inst* 28, 1173–1191, 1962.
60. Hagstrom, R. M., Glasser, S. R., Brill, A. B., and Heyssel, R. M., Long-term effects of radioactive iron administered during human pregnancy, *Am J Epidemiol* 90 (1), 1–10, 1969.
61. Uchida, I. A., Holunga, R., and Lawler, C., Maternal radiation and chromosomal aberrations, *Lancet* 2 (7577), 1045–1049, 1968.
62. Awa, A. A., Bloom, A. D., Yoshida, M. C., Meriishi, S., and Archer, P. G., Cytogenetic study of the offspring of atom bomb survivors, *Nature* 218 (5139), 367–368, 1968.
63. Herskowitz, I. H., Damage to offspring of irradiated women, *Prog Immunol Gynecol* 3, 374, 1957.
64. Daniels, R. D., and Schubauer-Berigan, M. K., A meta-analysis of leukaemia risk from protracted exposure to low-dose gamma radiation, *Occup Environ Med* 68 (6), 457–464, 2010.
65. Zielinski, J. M., Garner, M. J., Band, P. R., Krewski, D., Shilnikova, N. S., Jiang, H., Ashmore, P. J., Sont, W. N., Fair, M. E., Letourneau, E. G., and Semenciw, R., Health outcomes of low-dose ionizing radiation exposure among medical workers: A cohort study of the Canadian national dose registry of radiation workers, *Int J Occup Med Environ Health* 22 (2), 149–156, 2009.

66. Doody, M. M., Freedman, D. M., Alexander, B. H., Hauptmann, M., Miller, J. S., Rao, R. S., Mabuchi, K., Ron, E., Sigurdson, A. J., and Linet, M. S., Breast cancer incidence in U.S. radiologic technologists, *Cancer* 106 (12), 2707–2715, 2006.

67. Yoshinaga, S., Mabuchi, K., Sigurdson, A. J., Doody, M. M., and Ron, E., Cancer risks among radiologists and radiologic technologists: Review of epidemiologic studies, *Radiology* 233 (2), 313–321, 2004.

68. Cardis, E., Vrijheid, M., Blettner, M., Gilbert, E., Hakama, M., Hill, C., Howe, G., Kaldor, J., Muirhead, C. R., Schubauer-Berigan, M., Yoshimura, T., Bermann, F., Cowper, G., Fix, J., Hacker, C., Heinmiller, B., Marshall, M., Thierry-Chef, I., Utterback, D., Ahn, Y. O., Amoros, E., Ashmore, P., Auvinen, A., Bae, J. M., Bernar, J., Biau, A., Combalot, E., Deboodt, P., Diez Sacristan, A., Eklof, M., Engels, H., Engholm, G., Gulis, G., Habib, R. R., Holan, K., Hyvonen, H., Kerekes, A., Kurtinaitis, J., Malker, H., Martuzzi, M., Mastauskas, A., Monnet, A., Moser, M., Pearce, M. S., Richardson, D. B., Rodriguez-Artalejo, F., Rogel, A., Tardy, H., Telle-Lamberton, M., Turai, I., Usel, M., and Veress, K., The 15-Country Collaborative Study of Cancer Risk among Radiation Workers in the Nuclear Industry: Estimates of radiation-related cancer risks, *Radiat Res* 167 (4), 396–416, 2007.

69. Vrijheid, M., Cardis, E., Blettner, M., Gilbert, E., Hakama, M., Hill, C., Howe, G., Kaldor, J., Muirhead, C. R., Schubauer-Berigan, M., Yoshimura, T., Ahn, Y. O., Ashmore, P., Auvinen, A., Bae, J. M., Engels, H., Gulis, G., Habib, R. R., Hosoda, Y., Kurtinaitis, J., Malker, H., Moser, M., Rodriguez-Artalejo, F., Rogel, A., Tardy, H., Telle-Lamberton, M., Turai, I., Usel, M., and Veress, K., The 15-Country Collaborative Study of Cancer Risk Among Radiation Workers in the Nuclear Industry: Design, epidemiological methods and descriptive results, *Radiat Res* 167 (4), 361–379, 2007.

70. Jeong, M., Jin, Y. W., Yang, K. H., Ahn, Y. O., and Cha, C. Y., Radiation exposure and cancer incidence in a cohort of nuclear power industry workers in the Republic of Korea, 1992–2005, *Radiat Environ Biophys* 49 (1), 47–55, 2010.

71. Vano, E., Kleiman, N. J., Duran, A., Rehani, M. M., Echeverri, D., and Cabrera, M., Radiation cataract risk in interventional cardiology personnel, *Radiat Res* 174 (4), 490–495, 2010.

72. Fachin, A. L., Mello, S. S., Sandrin-Garcia, P., Junta, C. M., Ghilardi-Netto, T., Donadi, E. A., Passos, G. A., and Sakamoto-Hojo, E. T., Gene expression profiles in radiation workers occupationally exposed to ionizing radiation, *J Radiat Res (Tokyo)* 50 (1), 61–71, 2009.

73. Buja, A., Lange, J. H., Perissinotto, E., Rausa, G., Grigoletto, F., Canova, C., and Mastrangelo, G., Cancer incidence among male military and civil pilots and flight attendants: An analysis on published data, *Toxicol Ind Health* 21 (10), 273–282, 2005.

74. Band, P. R., Le, N. D., Fang, R., Deschamps, M., Coldman, A. J., Gallagher, R. P., and Moody, J., Cohort study of Air Canada pilots: Mortality, cancer incidence, and leukemia risk, *Am J Epidemiol* 143 (2), 137–143, 1996.

75. Hammer, G. P., Blettner, M., and Zeeb, H., Epidemiological studies of cancer in aircrew, *Radiat Prot Dosimetry* 136 (4), 232–239, 2009.

76. Pukkala, E., Aspholm, R., Auvinen, A., Eliasch, H., Gundestrup, M., Haldorsen, T., Hammar, N., Hrafnkelsson, J., Kyyronen, P., Linnersjo, A., Rafnsson, V., Storm, H., and Tveten, U., Cancer incidence among 10,211 airline pilots: A Nordic study, *Aviat Space Environ Med* 74 (7), 699–706, 2003.

77. Sigurdson, A. J., and Ron, E., Cosmic radiation exposure and cancer risk among flight crew, *Cancer Invest* 22 (5), 743–761, 2004.

78. Rafnsson, V., Tulinius, H., Jonasson, J. G., and Hrafnkelsson, J., Risk of breast cancer in female flight attendants: A population-based study (Iceland), *Cancer Causes Control* 12 (2), 95–101, 2001.

79. Pukkala, E., Auvinen, A., and Wahlberg, G., Incidence of cancer among Finnish airline cabin attendants, 1967–92, *BMJ* 311 (7006), 649–652, 1995.
80. Buja, A. M., Mastrangelo, G., Perissinotto, E., Grigoletto, F., Frigo, A. C., Rausa, G., Marin, V., Canova, C., and Dominici, F., Cancer incidence among female flight attendants: A meta-analysis of published data, *J Womens Health (Larchmt)* 15, 98–105, 2006.
81. Rugh, R., X-irradiation effects on the human fetus, *J Pediatr* 52 (5), 531–538, 1958.
82. Schwarz, G. S., Radiation hazards to the human fetus in present-day society. Should pregnant women be subjected to a diagnostic x-ray procedure?, *Bull NY Acad Med* 44 (4), 388–399, 1968.
83. Ben-Amotz, A., Yatziv, S., Sela, M., Greenberg, S., Rachmilevich, B., Shwarzman, M., and Weshler, Z., Effect of natural beta-carotene supplementation in children exposed to radiation from the Chernobyl accident, *Radiat Environ Biophys* 37 (3), 187–193, 1998.
84. Korkina, L. G., Afanas'ef, I. B., and Diplock, A. T., Antioxidant therapy in children affected by irradiation from the Chernobyl nuclear accident, *Biochem Soc Trans* 21 (Pt 3) (3), 314S, 1993.
85. Andreassi, M. G., Cioppa, A., Manfredi, S., Palmieri, C., Botto, N., and Picano, E., Acute chromosomal DNA damage in human lymphocytes after radiation exposure in invasive cardiovascular procedures, *Eur Heart J* 28 (18), 2195–2199, 2007.
86. Holland, N., Bolognesi, C., Kirsch-Volders, M., Bonassi, S., Zeiger, E., Knasmueller, S., and Fenech, M., The micronucleus assay in human buccal cells as a tool for biomonitoring DNA damage: The HUMN project perspective on current status and knowledge gaps, *Mutat Res* 659 (1–2), 93–108, 2008.
87. Angelieri, F., de Oliveira, G. R., Sannomiya, E. K., and Ribeiro, D. A., DNA damage and cellular death in oral mucosa cells of children who have undergone panoramic dental radiography, *Pediatr Radiol* 37 (6), 561–565, 2007.
88. Ribeiro, D. A, and Angelieri, F., Cytogenetic biomonitoring of oral mucosa cells from adults exposed to dental x-rays, *Radiat Med* 26 (6), 325–330, 2008.
89. Prasad, K. N., Rationale for using multiple antioxidants in protecting humans against low doses of ionizing radiation, *Br J Radiol* 78 (930), 485–492, 2005.

8 Prevention and Mitigation against Radiological Weapons and Nuclear Plant Accidents

INTRODUCTION

There are two major types of radiological weapons: the nuclear bomb and the dirty bomb. The dirty bomb can contain one or more radioactive isotopes that are commercially available. Nuclear bombs include the atom bomb, hydrogen bomb, and neutron bomb. An explosion of a dirty bomb can release radioactive isotopes into the atmosphere emitting gamma radiation and/or beta radiation, depending upon the types of radioactive isotope used in making the bomb. For example, a dirty bomb may contain ^{90}Sr emitting only beta radiation, ^{137}CS emitting only gamma rays, ^{131}I emitting both beta and gamma rays, or a combination of all three radioactive isotopes. On the other hand, an explosion of an atom bomb can release radioactive isotopes into the atmosphere emitting gamma rays, beta radiation, alpha radiation, and neutron radiation that can cause acute radiation sickness (ARS) in humans, if exposed to high doses. The explosion causes radioactive fallout that may contain fission products, such as ^{90}Sr, ^{137}Cs, and ^{131}I, which would contaminate water, soil, and sources of food for a long time, depending upon the half-lives of the radioactive isotopes. Although not important from the point of view of causing ARS, the radioactive carbon (^{14}C with physical half-life of 5730 years) and radioactive hydrogen tritium (^{3}H with a physical half-life of 12.3 years) can also contaminate water, soil, and sources of food. These radioactive isotopes, when ingested, can induce gene mutations.

Some radioactive isotopes, such as polonium-210, emit alpha radiation that is not as penetrating as gamma radiation, but is about 10 times more effective than gamma radiation in causing most types of damage. In order to have an effect, such radioactive isotopes must be ingested through food or drink. This was observed in 2006 when Russian spy Alexander Litvinenko ingested polonium-210 by drinking coffee and died a few weeks later from ARS.

Several countries have nuclear reactors that produce several types of radioactive isotopes, such as ^{14}C, ^{3}H, ^{32}P, and ^{45}Ca, for use in basic and clinical research, and for the diagnosis and treatment of certain human diseases. These radioactive isotopes can be obtained from commercial sources to construct a dirty bomb. A few countries also have stockpiles of nuclear bombs. The threat of unintentional nuclear conflicts, although a remote possibility, exists as long as nations have nuclear weapons in their possession. In addition, the threat of an explosion of a dirty bomb by terrorists has increased in recent years.

The risk of an accident in nuclear power plants can expose large numbers of people to low doses of radiation. In addition, radioactive materials can be released into the atmosphere that would contaminate water, soil, plants, and sources of food. In March 2011, a serious nuclear power plant accident occurred in Fukushima, Japan.

From the previous discussion, it is clear that the risk of radiation exposure to humans from an explosion of a nuclear weapon or a nuclear power plant accident has markedly increased. Therefore, it is imperative that effective preventive and mitigating strategies against radiation injury are developed. At present, approved preventive strategy involves steps that can reduce dose levels. There are no preventive strategies to provide biological protection against low or high doses of radiation. Although some approved mitigating strategies are available to reduce the progression of radiation damage, they are not adequate. This chapter discusses current preventive and mitigating strategies to reduce radiation damage, and includes a proposed micronutrient strategy that can help to improve the efficacy of current recommended steps for prevention and mitigation of radiation injury.

CURRENT PREVENTIVE RECOMMENDATIONS TO REDUCE RADIATION DOSES AFTER EXPLOSION OF A DIRTY BOMB OR A NUCLEAR POWER PLANT ACCIDENT

In the event of detonation of a dirty bomb or a major accident in a nuclear power plant, a few individuals at the site of the explosion or accident may receive high doses of radiation that can cause ARS, leading to death if no treatment intervention is done. Most may be exposed to radioactive isotopes used in the dirty bomb and may receive low doses of radiation that can increase the risk of chronic diseases, including cancer. Following an accident in a nuclear power plant, most may be exposed to radioactive isotopes, primarily iodine (^{131}I with a physical half-life of 8 days, emits both beta radiation and gamma radiation), cesium (^{137}Cs, with a physical half life of 30 years, emits only gamma radiation), and strontium (^{90}Sr, with a physical half-life of 29.1 years, emits primarily beta radiation). In addition, water, soil, plants, and sources of food may become contaminated with these radioactive isotopes and will expose individuals living in the vicinity of the accident site, which would increase the risk of chronic diseases, including cancer. Generally speaking, the explosion of a dirty bomb does not cause mass casualties, whereas an atom bomb does.

General recommendations: Federal, state, and local agencies recommend several simple steps for reducing radiation doses delivered to you soon after the explosion of a dirty bomb.

If you are outside and close to the incident, take the following steps:

1. Cover your nose and mouth with a cloth as soon as possible and go to the nearest building that has not been damaged by the explosion in order to reduce the risk of breathing in radioactive dust particles or smoke. The

building's walls will shield you from the gamma radiation and radioactive particles or dust that might be coming from outside.

2. Soon after entering the building, remove all your outer clothing as soon as possible because it may be contaminated with radioactive dust. Place all discarded clothing in a plastic bag; seal and store it in a separate room where no one is allowed to enter. If a bag is not available, store all contaminated clothing in a separate room away from human contact. Removing outer clothing may eliminate up to 90 percent of radioactive dust and thereby reduce the level of radiation dose. Unfortunately, this procedure has no impact on radiation doses that have already been delivered to various internal organs before removing the clothing.

3. Radiation experts from federal or local agencies will examine the contaminated clothing to determine the type of radioactive materials, the type of radiation (gamma, beta, or alpha radiation), and the dose rate per minute or per hour. They will be responsible for disposing of all contaminated materials safely. Do not dispose of any contaminated clothing yourself.

4. Soon after removing your contaminated clothing, take a shower, shampoo your hair, and then wash your body thoroughly with soap and water.

5. Do not touch objects thrown off by an explosion—they might be radioactive.

6. Do not panic. Tune to your local or national radio or television news for more instructions and follow them carefully.

In addition to the steps previously outlined, the following suggestions should be followed:

1. If you are outside and close to the incident, you may have suffered physical injuries from the explosion or received significant amounts of radiation. If injured or nauseous, you should go to the hospital as soon as possible. While the symptoms of nausea and vomiting may indicate that you have been exposed to high doses of radiation, these symptoms can also occur for various other reasons. Therefore, the best course of action is to see your doctor as soon as possible.

2. If you are inside and close to the incident, and if the walls and windows are not broken, stay in the building and wait for further advice from federal or local agencies.

3. To keep radioactive dust or powder from getting inside, close all windows, outside doors, and fireplace dampers. Turn off fans as well as heating and air conditioning systems, which bring in air from the outside.

4. If the walls or windows are broken, go to an interior room where there is no possibility of air entering from outside. If the building has been damaged, quickly go to an undamaged building. If you have to go outside, make sure to cover your mouth and nose with a cloth. Once inside the building, follow the steps described previously.

If you are traveling in a car and happen to be in the area of the explosion and close to the site of the incident, take these steps:

1. Stop the car as quickly and as safely as possible. Close the windows and turn off the engine, air conditioner, heater, and vents to prevent radioactive contaminated air from entering the vehicle.
2. Cover your nose and mouth with a cloth to avoid breathing in radioactive dust or smoke. If you are close to your home, office, or a public building, go there immediately and follow the procedures as described earlier. If this is not possible, stay in the car until you are told it is safe to get back on the road.
3. Listen to local or national radio for further instructions and follow them carefully.

If you are with your children or pets near the incident site, follow these following steps:

1. Stay together and take the steps described earlier.
2. If your children are with another family, they should stay there until they are told that it is safe to go home.
3. If they are in school, they should stay there until advised by the school official that it is safe to go home. Schools have emergency plans and shelters, and they should also protect your children.
4. If your pets are outside at the time of incident, bring them inside as soon as possible and take the same precautions that you would for yourself.
5. Wash your pets with soap and water thoroughly to remove any radioactive dust, and keep them with you at all times.

The explosion of a dirty bomb or a major accident in a nuclear power plant will definitely contaminate the soil, water, plant, and sources of food and water. Do not consume any food or water that was near the incident site until it has been examined by radiation experts and deemed safe for consumption. In the meantime, eat only food that has been sealed and drink only bottled water. Soil contamination can cause long-term hazards to food and the water supply, and this could affect both humans and animals.

When to take potassium iodide: Radioactive iodine (^{131}I) is one of the radioisotopes released into the atmosphere in the event of a major accident in a nuclear power plant. It can also be released into the atmosphere following an explosion of a dirty bomb that has ^{131}I as one of the radioactive isotopes. Iodine selectively accumulates in the thyroid gland. Therefore, take potassium iodide (a source of nonradioactive iodine) as soon as possible in order to block the accumulation of ^{131}I in the thyroid gland, and thus, protect the gland against radiation-induced cancer as well as noncancerous diseases. It should be emphasized that the doses of potassium iodide should not exceed that recommended by your doctor. In reality, it is difficult to predict whether a dirty bomb will contain radioactive iodine or not. It should be pointed out that ^{131}I is one of the radioisotopes released into the atmosphere after explosion of an atom bomb.

1. Taking potassium iodide before ingestion of radioactive iodine will saturate the thyroid gland with nonradioactive iodine, thus preventing the

accumulation of radioactive iodine in the thyroid gland and thereby protecting the thyroid gland from radiation damage.
2. Potassium iodide does not protect the thyroid gland from other radioactive materials.
3. Potassium iodide also does not protect other organs in the body from gamma radiation and beta radiation coming from the radioactive iodine.

Contamination of soil, food, food sources, and water: The following steps are suggested:

1. Water and exposed food near the site of the incident definitely will be contaminated with radioactive isotopes; therefore, do not consume such items.
2. The radiation experts from a federal or local agency will monitor food and water quality for safety and keep the public informed.
3. The consumption of sealed water or food is safe.

Detection of radiation dose levels: In order to determine the dose level in the environment, the following steps are recommended: You cannot see, smell, feel, or taste radiation; therefore, you will not be able to know whether or not you have been exposed. However, the first responders (police and firefighters) will quickly check the radiation levels in your local environment by using a special radiation detector and then advise you if the levels pose any danger to you and your community.

Potential risk of low levels of radiation: Low levels of radiation exposure may increase the risk of cancer, birth defects (in pregnant women), somatic mutations, and heritable mutations. Somatic mutations can increase the risk of diseases during your lifetime, whereas heritable mutations can increase the risk of diseases in future generations.

Public safety readiness exercises: Several cities across the United States occasionally hold public safety readiness dirty bomb event exercises. These exercises simulate a dirty bomb explosion, detection, response, and evacuation scenario. Such exercises are very useful in preparing the public, government officials, and first responders (police, firemen, doctors, and nurses) to meet the challenges that may occur following the explosion of a dirty bomb. However, they do not address the issue of protecting individuals before the explosion of a dirty bomb, nor do they assist in preparing the authorities to treat those individuals who have been exposed to low doses of radiation or who will likely be exposed to low doses of radiation by ingesting radioactive materials through contaminated air, water, or food.

Various types of protective gear have been developed for first responders involved in cleaning up radioactive materials at the explosion site and surrounding areas. These protective suits are useful in protecting against radioactive dust present in the air and can reduce the levels of exposure to gamma radiation being emitted from the radioactive sites. Unfortunately, these suits cannot offer full protection from radiation; certain amounts of gamma radiation will still be able to penetrate the material. That's why first responders involved in cleanup are allowed to work only for certain amounts of time; they're asked to leave the site after a short time in order to minimize their level of radiation exposure. Again, this strategy does not address the issue of protecting first responders prior to radiation exposure or treating those who have been exposed to low doses of radiation during the cleanup process.

Radiation detectors are very useful in detecting radiation levels and identifying the types of radioactive materials present. Similarly, the radioactive badges worn by first responders are very useful in estimating the level of doses received while cleaning contaminated sites. Unfortunately, none of these measures are useful in reducing radiation damage.

Application of current recommendations after a nuclear power plant accident: In the event of a major nuclear plant accident, radioactive isotopes including ^{131}I, ^{90}Sr, and ^{137}Cs are released into the atmosphere. Individuals can receive high to low doses of radiation, depending upon their location at the time of the accident. In the event of a nuclear power plant accident, current steps that are recommended to reduce radiation dose levels should be followed.

Limitations of current preventive recommendations: The current recommendations for reducing the radiation dose levels from radioactive materials following a dirty bomb explosion or a nuclear power plant accident have the following limitations:

1. They do not address the issue of how to improve the body's defense system before irradiation with high doses of radiation in order to reduce radiation damage, such as the symptoms of ARS after irradiation.
2. They do not address the issue of how to reduce the long-term adverse health effects of radiation, such as cancer, birth defects, cataracts, infertility, and somatic and heritable mutations among survivors of radiation exposures.

More information regarding current preventive recommendations can be obtained from the following sources:

- The Federal Emergency Management Agency (FEMA): http://www.fema.gov; (202) 646-4600
- The Conference of Radiation Control Program Directors Inc. (CRCPD): http://www.crcpd.org; (502) 227-4543
- The Nuclear Regulatory Commission (NRC): http://www.nrc.gov; (301) 415-8200
- The Radiation Emergency Assistance Center/Training Site (REAC/TS): http://orise.orau.gov/reacts; (865) 576-3131
- The US Department of Energy (DOE): http://www.energy.gov; (800) 342-5363

PROPOSED PREVENTIVE BIOLOGICAL RADIATION PROTECTION STRATEGY AFTER EXPLOSION OF A DIRTY BOMB OR A NUCLEAR POWER PLANT ACCIDENT

Preventive biological radiation protection refers to agents or procedures that reduce ARS when administered before irradiation to reduce radiation damage. Although current recommendations are expected to reduce the dose levels following an explosion of a dirty bomb or a nuclear power plant accident, no recommendations have been made to reduce the symptoms of ARS, and the late adverse health effects, such as cancer and nonneoplastic diseases, among survivors of high-dose-induced radiation injury, or among those who are likely to receive radiation doses higher than the background radiation. In order to complement the recommended steps to reduce the dose level, I propose to utilize a preparation of micronutrients containing high but nontoxic doses of multiple dietary and endogenous antioxidants orally before irradiation in order to provide biological protection against radiation damage. This micronutrient preparation, referred to as BioShield-R2, is commercially available. Taking this micronutrient preparation before irradiation may reduce the symptoms of ARS and may even protect individuals from dying. This issue was discussed in Chapter 6. Taking the same preparation of micronutrients after irradiation may reduce the adverse health effects among survivors of high doses of radiation as well as those who are exposed to low doses of radiation. This issue was also discussed in detail in Chapter 6.

CURRENT AND PROPOSED PREVENTIVE RECOMMENDATIONS TO REDUCE RADIATION DOSES AFTER EXPLOSION OF AN ATOM BOMB

If there is a warning of an imminent atomic bomb explosion, shelter in the basement of a house would effectively act as a shield against radiation exposure. The physical events that occur following the explosion of an atom bomb are in part different from those caused by the detonation of a dirty bomb. The explosion of an atom bomb releases a wave of intense heat near the hypocenter that causes severe burns and lethality. No preventive countermeasures can be developed against such an event, because everything is instantly vaporized due to the intense heat. The shockwave created by the blast can also cause instant death. Those who escape the blast may receive lethal doses of radiation that can cause death within 24 hours from central nervous system (CNS) syndrome, 14 days from gastrointestinal (GI) syndrome, or 60 days from bone marrow syndrome. Some individuals receiving doses that can produce bone marrow syndrome may survive, while others who receive lower doses will definitely survive. The survivors of radiation exposures may have increased risk of lung, neurological, and ear disorders as well as cancer and some noncancerous diseases. In addition, pregnant women exposed to low doses of radiation may have an increased risk of birth defects in their children, and there may be increased levels of somatic and heritable mutations in both men and women.

Since radioactive fallout occurs after the explosion of an atom bomb, the same steps that were recommended by the federal and local agencies to reduce radiation dose levels in the event of an explosion of a dirty bomb should be followed. In addition, the proposed biological radiation protection strategies discussed in Chapter 5 should be adopted in order to reduce acute and long-term adverse health effects of high or low doses of radiation. Generally, lead shielding, putting a greater distance between people and radiation sources, and minimizing the time spent at the site with high radiation levels are the best recommendations for reducing radiation dose levels. However, these physical radiation protection strategies of reducing radiation dose levels are not always practical or relevant, because they cannot protect tissue during or after irradiation.

CURRENT AND PROPOSED MITIGATION STRATEGY TO REDUCE DAMAGE AFTER IRRADIATION WITH LOW OR HIGH DOSES OF RADIATION

Mitigation measures refer to procedures or agents that can reduce the progression of ARS and chronic diseases including cancer when administered after irradiation. Physicians, physicists, radiation biologists, nurses, and other health care professionals, and their private or public hospitals will assume the responsibility of treating individuals exhibiting the symptoms of ARS. In the event of an explosion of a nuclear weapon or a major accident in a nuclear power plant, radiation exposures may result in a few or mass casualties. In order to provide guidelines for management of a large number of radiation-exposed individuals, the Strategic National Stockpile Radiation Working Group developed a consensus document.[1,2] In addition, the Centers for Disease Control (CDC) has developed a self-study training program for clinicians called "Radiological and Nuclear Terrorism: Medical Response to Mass Casualties," which is available at the website http://www.bt.cdc.gov/radiation/masscasualties/ training.asp. This document can also be obtained via e-mail at cdcinfo@cdc.gov. Based on these documents, the following recommendations can be made in the event of a nuclear explosion:

1. Stabilization of patients should be considered a first priority.
2. Decontaminate patients as soon as possible.
3. Estimate radiation doses that individuals may have received before arriving at the medical facility. If this is not available, individual radiation dose can be estimated by determining the time of onset and severity of nausea, vomiting, decline in peripheral lymphocyte counts over several hours or days after irradiation, and appearance of chromosomal aberrations (including dicentric and ring forms) in peripheral blood lymphocytes.
4. Documenting clinical signs and symptom characteristics of bone marrow syndrome, GI syndrome, and CNS syndrome, or skin responses over time is essential in order to provide a rational basis for the selection of therapeutic agents and expected prognosis of the irradiated individuals.
5. The potential therapeutic agents may include replacement therapy (antibiotics and transfusion of blood and electrolytes when indicated), which is most suitable for treating individuals exposed to doses that produce bone marrow syndrome. However, for those individuals who are exposed to doses that produce GI syndrome, additional agents such as certain cytokines, bone marrow or stem cell transplant, and multiple antioxidants may be needed.
6. Patients exhibiting CNS syndrome are likely to die within 2 to 3 days, depending upon the total dose, and more frequently within 24 hours. Such patients should be isolated without any treatment in order to save resources

to treat those who may have been exposed to bone marrow syndrome or GI syndrome.

7. Consider all open wounds to be contaminated.
8. If surgery is indicated, perform it within 36 to 48 hours, prior to the onset of thrombocytopenia, leukopenia, and immunosuppression.

The mitigation of radiation damage in radiation-exposed individuals is based on three principles: (1) removal of ingested radioactive materials from the patient's body as soon as possible, (2) reducing the progression of radiation-induced acute damage, and (3) reducing the risk of radiation-induced chronic diseases.

Removal of ingested radioactive materials: Following the ingestion of radioactive isotopes through the air, water, or food, the extent of damage depends upon the type of radioactive isotopes, their concentrations, and their levels of accumulation in the organs. The best strategy is to remove the radioactive isotopes from the body as soon as possible. The success of this strategy, however, depends upon the type and form of radioactive isotope and time after ingestion. For example, if radioactive hydrogen ^{3}H (tritium) is ingested in the form of ^{3}H-thymidine, it will be quickly become part of the DNA molecule, which is difficult to remove from the body. This could increase the risk of cancer of organs containing dividing cells, such as bone marrow, intestines, testes, and ovaries. On the other hand, if radioactive hydrogen is ingested in the form of radioactive water, it is quickly eliminated by drinking a lot of fluids. Except for ^{131}I, other radioactive isotopes once ingested are difficult to remove from the body by consuming a nonradioactive counterpart isotope because of their toxicity. Even for ^{131}I, taking nonradioactive potassium iodide is effective in preventing the accumulation of ^{131}I in the thyroid gland, if consumed orally within two hours of ingestion of radioactive iodine. Ingestion of ^{90}Sr will cause an accumulation of this radioactive metal in the bone, resulting in an increased risk of bone cancer. The concentration of ^{90}Sr can be diluted by consuming nonradioactive strontium soon after ingestion; however, strontium could be toxic. The CDC has recommended 10% magnesium sulfate for removing ingested ^{60}Co, ^{32}P, and ^{226}Ra from the GI tract. In addition, anti-emetic agents can also be used, if needed. Replacement of fluids and electrolytes may be needed in case of excess loss of fluid. In addition, chelating agents, such as diethylenetriaminepentaacetate (DTPA) may be used to remove certain radioactive elements such as Ca and Zn. However, DTPA is considered nephrotoxic and therefore can only be used under strict supervision of doctors.

Reducing the progression of radiation-induced ARS: If individuals are present in the vicinity of a dirty bomb explosion or away from the center of an atomic explosion, they may receive high doses of radiation, causing symptoms of ARS. These doses can produce bone marrow syndrome (50% to 100% lethality within 60 days with doses of 3–5 Gy), GI syndrome (100% lethality within 14 days with doses of 6–40 Gy), and CNS syndrome (100% lethality within 24 hours with doses of 50 Gy or above). Some Food and Drug Administration (FDA)-approved treatment methods and agents, such as replacement therapy for bone marrow syndrome, and replacement therapy plus certain growth factors, such as granulocyte-colony stimulating factor (G-CSF), and bone marrow transplants for GI syndrome are available to treat these irradiated individuals. I have proposed that a micronutrient preparation containing multiple dietary and endogenous antioxidants referred to as BioShield-R3 be consumed orally 24 hours after irradiation in order to improve the efficacy of standard therapy in the management of irradiated individuals exposed to lethal doses of radiation. This issue was discussed in detail in Chapter 5.

Reducing the risk of radiation-induced chronic diseases: At this time, there are no approved biologics or chemicals for mitigating the incidence of radiation-induced chronic diseases, including cancer. I have proposed the use of a micronutrient preparation containing multiple dietary and endogenous antioxidants referred to as BioShield-R2, which can be administered at any time after irradiation in order to reduce the risk of neoplastic and nonneoplastic diseases. This issue was discussed in detail in Chapter 6.

USE OF BIOSHIELD-R2 AND BIOSHIELD-R3

It is impossible to predict when an explosion of a nuclear weapon may occur. Despite the best education and training with respect to the safety of nuclear power plants, human error or natural disaster can cause major nuclear accidents such as seen in Chernobyl in Ukraine, and more recently, in Fukushima, Japan. Therefore, in addition, to public safety readiness exercises that address the issue of radiation dose reduction, I suggest that BioShield-R2 (for prevention and mitigation of late adverse effects of radiation) and BioShield R3 (for mitigation of ARS), which would provide biological protection against radiation injury, should be given serious consideration for adoption. This proposed micronutrient strategy for biological radiation protection should complement the current recommendations for prevention and mitigation of radiation damage.

SUMMARY

Federal and local agencies have recommended simple steps for reducing radiation dose levels following an explosion of a dirty bomb or an atom bomb. These steps include reducing inhalation, ingestion, or absorption through the skin of radioactive isotopes, and shielding from gamma radiation. They have also provided contact information for additional advice related to the levels of radiation exposure and their effects on human health. The same steps should be followed in the event of a major accident in a nuclear power plant. At present, there are no prevention strategies to provide biological protection against damage caused by low or high doses of radiation. I have proposed a preparation of micronutrients containing multiple dietary and endogenous antioxidants referred to as BioShield-R2, which may reduce symptoms of ARS, and prevent some irradiated individuals from dying when administered before irradiation.

Although some approved agents and procedures to mitigate the symptoms of ARS are available, they are not adequate. I have proposed a preparation of micronutrients containing multiple dietary and endogenous antioxidants referred to as BioShield-R3, which may reduce symptoms of ARS and prevent some irradiated individuals from dying when administered orally after irradiation. The addition of proposed micronutrient preparations to the standard therapy may improve the management of irradiated individuals exhibiting symptoms of ARS more than that produced by the individual agents. Standard therapy is not suitable for reducing the risk of radiation-induced chronic diseases. The proposed use of BioShield-R2 may also reduce the risk of radiation-induced neoplastic and nonneoplastic diseases. In addition to implementing current recommendations for reducing the radiation dose levels, I suggest that use of BioShield-R2 (for prevention and mitigation of late adverse effects of radiation) and BioShield-R3 (for mitigation of acute radiation sickness) should be given serious consideration.

REFERENCES

1. Waselenko, J. K., MacVittie, T. J., Blakely, W. F., Pesik, N., Wiley, A. L., Dickerson, W. E., Tsu, H., Confer, D. L., Coleman, C. N., Seed, T., Lowry, P., Armitage, J. O., and Dainiak, N., Medical management of the acute radiation syndrome: Recommendations of the Strategic National Stockpile Radiation Working Group, *Ann Intern Med* 140 (12), 1037–1051, 2004.
2. Coleman, C. N., Hrdina, C., Bader, J. L., Norwood, A., Hayhurst, R., Forsha, J., Yeskey, K., and Knebel, A., Medical response to a radiologic/nuclear event: Integrated plan from the Office of the Assistant Secretary for Preparedness and Response, Department of Health and Human Services, *Ann Emerg Med* 53 (2), 213–222, 2009.

9 Prevention and Mitigation of Damage after Low Radiation Doses

INTRODUCTION

Recent analysis of human data suggests that radiation doses of a few mSv and up delivered in a single dose to patients receiving diagnostic radiation procedures or in cumulative doses to radiation workers may induce increased risk of cancer. Crews of commercial or military aircraft are also exposed to low doses of cosmic radiation during their flight time. The main purpose of the current recommendations for radiation protection in humans is to reduce dose levels as much as possible. These recommendations include reducing distance from the radiation source and time of radiation exposure whenever possible, shielding with high atomic number elements, such as lead, and adopting the principles of ALARA (as low as reasonably achievable). These steps can be defined as physical strategies for radiation protection. There is no doubt that adopting the current physical protection strategy can reduce the level of radiation doses to patients and radiation workers. However, these recommendations do not address the issue of biological protection against radiation injury before or after irradiation. At present, there is no radiation protection strategy to reduce the risk of late adverse health effects of low doses of radiation, such as neoplastic and nonneoplastic diseases, and somatic and heritable mutations. The development of a biological radiation protection strategy would complement the effectiveness of current physical radiation protection strategies. This chapter discusses the potential value of a micronutrient preparation containing multiple dietary and endogenous antioxidants in prevention and mitigation of late adverse health effects of low doses of radiation when administered orally before and/or after irradiation.

APPLICATION OF THE CONCEPT OF A BIOSHIELD FOR PREVENTION OF RADIATION DAMAGE

The concept of a bioshield as described in Chapter 5 includes chemicals or biologics that can prevent radiation damage when administered before irradiation, and/or can mitigate radiation injury when administered after irradiation. An ideal bioshield must satisfy three criteria: (1) provide radiation protection in laboratory experiments (cell culture and/or animals) when administered before or after irradiation, (2) protect normal tissues in humans when administered orally before or after irradiation, and (3) is safe when administered orally at radioprotective and radiation mitigating doses in humans on a short- and long-term basis. Among the various radioprotective and radiation mitigating agents that are available, only a preparation with multiple micronutrients containing dietary and endogenous antioxidants appears to satisfy all three criteria of an ideal bioshield. This issue was discussed in detail in Chapter 5.

Individuals are frequently exposed to low doses of radiation while undergoing diagnostic radiation procedures or while traveling in a commercial or military aircraft. Military aircraft can fly at altitudes as high as 80,000 feet, which is about twice the height of commercial aircraft; therefore, crews of military aircraft receive radiation doses higher than those of commercial aircraft. On the other hand, radiation workers are exposed to low doses of radiation daily during their working hours, more than nonradiation workers. Epidemiologic studies suggest that both groups of people (radiation workers and crews of military or commercial aircraft) are at increased risk of cancer and noncancerous diseases, and somatic and heritable mutations, although some continue to question the validity of this conclusion. Epidemiologic studies are never considered conclusive until supported by laboratory studies and/or actual intervention trial in humans. Since low-dose radiation studies to determine the risk of cancer and noncancerous diseases cannot be directly performed in normal human subjects, we have to rely on the laboratory studies together with epidemiologic data for our assessment of risks in humans.

The effects of low doses of radiation on cancer risk is difficult to evaluate in any intervention studies because of the very long latent period between exposure and development of cancer; however, it can be evaluated on certain biochemical and genetic markers of cancer and noncancerous disease risk factors, such as oxidative damage and pro-inflammatory cytokines, expression of changes in gene profiles related to cell proliferation, differentiation, apoptosis, DNA double-strand breaks (DSBs), and chromosomal damage. Indeed, increased oxidative damage was found in the peripheral blood of the children who lived in the radioactive contaminated area of the Chernobyl nuclear plant accident, where the background radiation was higher than for those children living away from the accident site.[1,2] The alterations in expression of gene profiles have been observed among radiation workers of the nuclear industry.[3] Human peripheral lymphocytes, when irradiated with a dose of 10 mGy (10 mSv) (equivalent to a dose delivered to patients in a single CT scan) showed an increased number of cells with DNA DSBs compared to unirradiated control cells.[4] A similar result was obtained when patients received a dose of 10 mGy and then assayed DNA DSBs in the peripheral lymphocytes after irradiation[4] (Figure 8.2 in Chapter 8). The DNA DSBs, if not prevented or repaired, can lead to neoplastic diseases. Therefore, the implementation of the concept of a bioshield for the prevention and mitigation of radiation damage caused by low doses of radiation should be given serious consideration. I present specific recommendations for using a bioshield depending upon the type of populations exposed to low doses of radiation.

BIOSHIELD-R1 FOR INDIVIDUALS RECEIVING DIAGNOSTIC RADIATION PROCEDURES AND FREQUENT FLYERS

There is growing evidence from laboratory and human studies that supports the view that low doses of radiation cause increased risk of cancer and noncancerous diseases as well as heritable mutations; however, many physicians, such as radiologists, nuclear medicine specialists, and cardiologists involved in delivering diagnostic radiation procedures question the validity of these studies. It is possible that these physicians lack adequate knowledge of radiation biology and genetics; therefore, they do not appreciate the potential somatic and genetic hazards of low doses of radiation. They may be so busy that they do not have adequate time to keep themselves informed on recent research on low-dose effects. They may also be concerned that telling patients about any potential risk of radiation may discourage them from undergoing needed diagnostic procedures. Some who accept the potential risks of diagnostic procedures ignore such risk by arguing that more people die in car accidents and from misuse and overdose of prescription and over-the-counter drugs each year than expected cancer deaths from diagnostic radiation procedures. These arguments cannot be considered seriously because unlike other causes of accidental death, exposure to low doses of radiation increases the risk of cancer and heritable mutations, which can increase the risk of genetic diseases in future generations. In addition, cancer disease causes severe financial burden and long-term emotional effects that cannot be measured. Most radiation biologists and geneticists and other radiation scientists agree that no radiation dose can be considered safe. The current efforts for radiation protection are focused on reducing the radiation dose levels as much as possible. I propose a micronutrient strategy for prevention and mitigation of radiation injury that would complement the current strategies of dose reduction in reducing the potential risk of radiation injury, no matter how small that damage might be.

A commercial patented preparation of multiple dietary and endogenous antioxidants, referred to as BioShield-R1 is available. Taking one dose of this dietary supplement 15–30 minutes before any diagnostic radiation procedure and another dose a few hours after the procedure may reduce late adverse health effects, including cancer. In a recent study [4] it was demonstrated that a dose of 10 mGy or 10 mSv of x-ray induced an increased number of cells with DNA DSBs in the human peripheral lymphocytes irradiated in culture or in vivo (patients receiving 10 mSv of x-ray during a CT scan). Pretreatment of human peripheral lymphocytes in culture before irradiation or patients before a CT scan with BioShield-R1 markedly reduced the number of cells with DSBs. BioShield-R1 was ineffective when administered after irradiation in vitro as well as in vivo. It is also possible that BioShield-R1, by attenuating free radicals during irradiation, may have prevented other types of radiation-induced DNA and cellular damage. It has been estimated that 2/3 of x-ray damage is caused by free radicals.[5] The multiple antioxidants present in BioShield-R1 can attenuate at least 2/3 of the radiation damage. One dose of BioShield-R1 after a diagnostic radiation procedure is suggested because long-lived free radicals exist after irradiation, which also must be neutralized. The doses of antioxidants in BioShield-R1 are safe and cost-effective. Patients who are undergoing diagnostic radiation procedures should consider taking BioShield-R1 after consultation with their doctors.

No epidemiologic studies to evaluate the risk of radiation-induced ill health among frequent flyers have been performed; however, the health risks that exist among those receiving diagnostic radiation procedures may also be applicable to frequent flyers. Therefore, it is suggested that frequent flyers may wish to consider taking one dose of BioShield-R1 before going through security, and another dose after arrival at their destination, after consultation with their physicians.

BIOSHIELD-R2 FOR RADIATION WORKERS AND CREWS OF COMMERCIAL AND MILITARY AIRCRAFT

Some epidemiologic studies among radiation workers at the medical facilities and the nuclear industry revealed an increased risk of cancer and cataract, while others have disputed them. Nobody can dispute the fact that even a small dose of radiation to the gonads can induce heritable mutations that may appear in future generations. Radiation is known to interact with other carcinogens in the environment, diet, and lifestyle in a synergistic manner to enhance the risk of cancer. Therefore, it is very difficult to estimate cancer risk induced by radiation alone. Radiation workers also follow the current recommendations of reducing the dose levels. Additionally, they are provided with the film badges that monitor their daily radiation dose in order to assure that workers do not receive annual accumulative doses of more than the maximum permissible dose (MPD) of 50 mSv. The concept of MPD is interpreted inaccurately by many radiation workers. They believe that as long as you are within the MPD, you do not have to be concerned about the risk of radiation damage. This false sense of security is widely believed by most radiation workers. This is unfortunate because no radiation dose can be considered safe. The concept of MPD states that if you receive cumulative doses of 50 mSv within one year, you are unlikely to suffer from serious adverse health effects during your lifetime. This does not say it is totally safe. This also does not address the issue of heritable mutations. If we have taken a job as a radiation worker, we have to accept the potential health risks of low doses of radiation. At present, there are no strategies for biological protection against radiation-induced health risks among radiation workers. Supplementation with a preparation of micronutrients containing dietary and endogenous antioxidants may reduce the health risks of low doses of radiation.

A commercial patented preparation of micronutrients referred to as BioShield-R2 is available. This preparation contains multiple dietary and endogenous antioxidants, all B vitamins, mineral selenium, zinc, calcium, and magnesium. The rationale for using multiple antioxidants was described in Chapter 5. BioShield-R2 contains no herbs, herbal antioxidants, or heavy metals, such as zirconium, molybdenum, and vanadium. Herbs were not included because some of them can interact with prescription and nonprescription drugs in an adverse manner. Heavy metals were not added because they may accumulate large amounts in the body after long-term consumption. Heavy metals are known to be neurotoxic. Other antioxidants, such as resveratrol, curcumin, omega-3, and genistein, were not added because they do not produce any unique effects that cannot be produced by antioxidants present in BioShield-R2. In addition, trace minerals, such as iron, copper, and manganese were also not added to BioShield-R2, because these minerals are known to interact with vitamin C and produce excessive amounts of free radicals. They are absorbed from the intestinal tract better in the presence of antioxidants than in their absence. Increasing the body's store of free iron can increase the risk of chronic diseases including cancer. Excess accumulation of copper is associated with Alzheimer's disease. Manganese miners who accumulate excessive amounts of this mineral develop Parkinson's disease.

The ideal situation would be for a worker to take BioShield-R2 before starting to work with radiation sources and continue throughout the worker's lifetime. However, because of a long latent period for cancer development after irradiation, it would be beneficial to start taking BioShield-R2 at any time after becoming a radiation worker, after consultation with their physician. BioShield-R2 should be taken orally twice a day (half of the dose in the morning with a meal and another half in the evening with a meal).

Crews of commercial as well as military aircraft should consider taking BioShield-R2 on the same dose schedule as radiation workers in consultation with their physicians. It is expected that consuming BioShield-R2 as suggested may reduce the risk of radiation-induced cancer and noncancerous diseases in both radiation workers and crews of commercial and military aircraft. Additional benefits of BioShield-R2 include improved immune function and overall well-being.

The latent period between irradiation and the appearance of cancer is very long, often more than 20–30 years. Radiation induces genetic instability in the cells that become more sensitive to additional mutations caused by other carcinogens during the latent period. One of the actions of all carcinogens is mediated via free radicals and chronic inflammation. In order to reduce the risk of cancer as well as heritable mutations, these biochemical events should be quenched. Antioxidants are the only group of chemicals that at nontoxic doses can neutralize free radicals and reduce inflammation.[6,7] Therefore, supplementation with multiple antioxidants may be one of the rational choices for reducing the health risks of low doses of radiation among radiation workers.

SUMMARY

Recent epidemiologic studies suggest that diagnostic radiation procedures can increase the risk of cancer. The risk of cataract increases among interventional cardiologists and associated technologists. Some radiologists have questioned these results. There are no epidemiologic studies to evaluate the risk of cancer or other diseases among frequent flyers who receive doses of radiation while going through security x-ray scanners and while on the flight. However, radiation-induced adverse health effects may exist among frequent flyers similar to those who receive diagnostic radiation procedures. Epidemiologic studies to evaluate the risk of cancer and noncancerous diseases among radiation workers have produced inconsistent conclusions. Some have suggested there is a direct association between radiation doses received and the risk of cancer, while others have reported no such effects. Radiation can also induce heritable mutations that can increase the risk of diseases in future generations. This issue is seldom raised while discussing the effects of chronic exposure to low doses of radiation among radiation workers. Like radiation workers, crews of commercial as well as military aircraft receive low doses of cosmic radiation while flying at high altitudes (about 37,000 feet for commercial aircraft and about 80,000 feet for military aircraft). Current recommendations for radiation protection include steps that can reduce dose levels; however, there are no strategies to provide biological protection for any group of individuals who receive low doses of radiation. I have proposed an antioxidant strategy that may prevent and mitigate radiation injury after exposure to low doses of radiation. I have suggested that individuals who are scheduled to receive diagnostic radiation procedures may want to consider taking one dose of BioShield-R1 15 to 30 minutes before the procedure and another dose a few hours after the procedure, after consultation with their doctors. Frequent flyers may wish to consider taking one dose of BioShield-R1 orally before going through security, and another dose after arrival at their destination, after consultation with their doctor. On the other hand, radiation workers might consider taking BioShield-R2 orally twice a day (half in the morning and another half in the evening) for their entire lifespan, after consultation with their physician. Crews of commercial and military aircraft may also consider taking BioShield-R2 in a manner similar to that described for radiation workers. It is expected that adoption of the proposed micronutrient strategy for prevention and mitigation of the risk of cancer and noncancerous diseases produced by low doses of radiation could be worthwhile.

REFERENCES

1. Ben-Amotz, A., Yatziv, S., Sela, M., Greenberg, S., Rachmilevich, B., Shwarzman, M., and Weshler, Z., Effect of natural beta-carotene supplementation in children exposed to radiation from the Chernobyl accident, *Radiat Environ Biophys* 37 (3), 187–193, 1998.
2. Korkina, L. G., Afanas'ef, I. B., and Diplock, A. T., Antioxidant therapy in children affected by irradiation from the Chernobyl nuclear accident, *Biochem Soc Trans* 21 (Pt 3) (3), 314S, 1993.
3. Fachin, A. L., Mello, S. S., Sandrin-Garcia, P., Junta, C. M., Ghilardi-Netto, T., Donadi, E. A., Passos, G. A., and Sakamoto-Hojo, E. T., Gene expression profiles in radiation workers occupationally exposed to ionizing radiation, *J Radiat Res (Tokyo)* 50 (1), 61–71, 2009.
4. Ehrlich, J. S., Brand, R., Uder, M., and Kuefner, M., Effect of proprietary combination of antioxidants/glutathione-elevating agents on x-ray induced DNA double-strand breaks, presented at the Annual Meeting of Society of Cardiovascular Computed Tomography, 2011.
5. Hall, E. J., *Radiobiology for the Radiologists*, J. B. Lippincott, Philadelphia, 1994.
6. Abate, A., Yang, G., Dennery, P. A., Oberle, S., and Schroder, H., Synergistic inhibition of cyclooxygenase-2 expression by vitamin E and aspirin, *Free Radic Biol Med* 29 (11), 1135–1142, 2000.
7. Devaraj, S., and Jialal, I., Alpha tocopherol supplementation decreases serum C-reactive protein and monocyte interleukin-6 levels in normal volunteers and type 2 diabetic patients, *Free Radic Biol Med* 29 (8), 790–792, 2000.

10 Implementation Plans for Prevention and Mitigation of Radiation Injury

INTRODUCTION

The recommendations for radiation damage prevention and mitigation, no matter how effective and useful they may be, must have detailed implementation plans. Such recommendations would remain of theoretical interest without an effective implementation plan. A recent accident in the Fukushima nuclear power plant in Japan has provided an urgent need to develop a detailed implementation plan for radiation protection by all nuclear industries throughout the world, in order to protect power plant workers and the civilian population residing near the plants in the event of an accident. In addition, the threat of an explosion of a dirty bomb by terrorists and unintentional nuclear conflicts exist; therefore, federal and state agencies responsible for radiation protection should also develop detailed implementation plans for providing adequate protection to their citizens in the event such nuclear events occur.

Although x-ray-based diagnostic procedures have been useful in early diagnosis of human diseases and improved treatment outcomes, their excessive use has alarmed many radiation scientists because of increased risk of cancer among patients receiving such procedures. Workers exposed to radiation at medical facilities, nuclear power plants, and in nuclear submarines, as well as crews of commercial and military aircraft flying at high altitudes who receive cosmic radiation doses higher than those who are not exposed to ionizing radiation, also have increased risks of developing cancer.

All medical facilities involved in diagnostic radiation procedures, all nuclear powered facilities, and all national and international airlines have varying levels of implementation plans for providing patients, radiation workers, and flight crews and attendants with adequate protection against the potential hazards of low and high doses of radiation. This chapter describes existing implementation plans that include education, exercises and drills, and stockpiling of radioprotective and radiation mitigating agents.

RADIATION PROTECTION STRATEGIES

Radiation protection strategies can be divided into two categories: (1) a physical strategy that involves increasing the distance between the radiation source and workers, reducing the time of exposure, and shielding and adopting of the ALARA (as low as reasonably achievable) concept for reducing the radiation dose levels; and (2) a biological strategy that involves the use of standard therapy when indicated and oral administration of a micronutrient preparation containing multiple dietary and endogenous antioxidants before and/or after irradiation. The implementation of a physical strategy of radiation protection would reduce only the radiation dose levels. For biological radiation protection, agents used in standard therapy and micronutrient preparations (BioShield-R1, -R2 and -R3) that would prevent and mitigate acute and late radiation injury after irradiation with low or high doses of radiation are commercially available. Therefore, it is essential that implementation of both physical and biological radiation protection strategies should receive serious consideration from federal, state, and local agencies. The implementation of such programs would provide optimal efficacy in the management of radiation injury after exposure to low or high doses of radiation. Based on the analysis of several published studies on prevention and mitigation of radiation damage, scientific rationale, and mechanisms of radiation injury, an oral supplementation with micronutrients containing antioxidants in combination with standard therapy appears to be one of the rational choices for providing biological protection.

IMPLEMENTATION PLANS FOR RADIATION PROTECTION STRATEGIES

Implementation plans for radiation protection include: (1) educational programs on the effects of ionizing radiation, (2) exercises and drills on current and proposed recommendations, and (3) stockpiling of radiation preventive and mitigating agents for biological protection.

EDUCATIONAL PROGRAMS ON THE EFFECTS OF IONIZING RADIATION

The levels of education on the effects of ionizing radiation depend upon the scenario of radiation exposure. In case of a nuclear accident or explosion, the first responders, who include police, firefighters, and members of hazardous materials (HAZMAT) response teams, will be asked to decontaminate radioactivity from the affected areas and advise people to take steps to reduce radiation dose levels as much as possible. The doctors and nurses at medical facilities will be responsible for managing radiation injury. Therefore, these groups of people must receive sufficient education on the effects of low and high doses of radiation as well as potential therapeutic agents that can be used to treat individuals exposed to lethal doses of radiation and those who survive radiation exposure.

Radiological or nuclear terrorism: The risk of nuclear accidents at power plants as well as nuclear terrorism exists in the United States and abroad. The Centers for Disease Control and Prevention (CDC) published a program on Public Health Planning for Radiological and Nuclear Terrorism on their website (http://www.bt.gov/radiation/publichealth.asp). This program offers the most up-to-date information on state and local public health emergency response planning for radiological and nuclear terrorism incidents for public health officials at the state and local levels. The program describes the public health role in conducting monitoring programs for those who may have been exposed to radiation, the role of communications in public health response to a radiological terrorism incident, and discusses key considerations for distribution of pharmaceutical countermeasures against radiation damage by the strategic national stockpiles following a radiological terrorism incident. The target audience for this program may include state, local, and territorial health agency senior officials, emergency planners, epidemiologists, environmental health specialists, public health physicians and nurses, public health students, sanitarians, HAZMAT members, and first responders. No fee is charged for participation in this program.

The Radiation Emergency Medical Management Program of the US Department of Health and Human Services also conducts training and educational courses on radiation biology and radiological terrorism. Classroom radiation training courses include the health effects of ionizing radiation and radiation emergency assistance. These courses are also provided at the websites of the CDC and the Federal Emergency Management Agency (FEMA). The CDC covers the following topics:

1. Public health planning for radiological and nuclear terrorism
2. Public health response to radiological and nuclear threat
3. Radiological or nuclear terrorism: Medical response to mass casualties
4. Radiological terrorism: Just-in-time training for hospital clinicians
5. Preparing for radiological population monitoring and decontamination
6. Medical response to nuclear and radiological terrorism
7. The role of public health in a nuclear or radiological terrorist incident
8. Thyroid disease management and radiation exposure
9. Hanford; The psychological dimensions of radiation exposure

In addition to these topics, courses also cover the following aspects of radiation biology:

1. Part 1: Physical principles of ionizing radiation
2. Part 2: Biologic effects of radiation
3. Part 3: Cardinal principles of radiation protection
4. Part 4: Principles of patient radiation protection and ALARA
5. Part 5: National Council on Radiation Protection (NCRP) and US Food and Drug Administration (FDA) regulations and occupational exposure management

Nuclear power plant emergency: FEMA has developed emergency programs to respond before and after nuclear power plant accidents (http://www.fema.gov/hazard/nuclear/nu_before.shtm). The public information materials regarding a nuclear power plant emergency before it happens can be obtained from the power company that operates your local nuclear power plant or your local emergency service office. If you live within ten miles of the nuclear power plant, you should receive these materials annually from the power plant company or your state or local government.

Before a nuclear power plant emergency occurs, it is important to become familiar with procedures, such as concepts of distance, shielding, and time, which may reduce the radiation dose level. The more distance you keep between you and the source of radiation, the lesser amount of radiation dose you will receive. Shielding with dense materials, such as lead or a building with windows and doors closed, will reduce your radiation dose level. The less time you spend near the radiation source, the smaller the radiation dose you will receive.

It is important to become familiar with terms that are used after a nuclear power plant accident. If you get a notification of an *unusual event* in a nuclear power plant, this would mean that a small problem has occurred at the plant, but no radiation leak is expected, and no action on your part will be necessary. If you get a notification of an *alert*, this would mean that a small problem has occurred at the plant, and small amounts of radiation could leak inside the plant. This will not affect you and no action is required. *Area sirens* may be sounded if a plant site area emergency has occurred. Listen to the radio or television for safety information. If a *general emergency* is declared, this would indicate that radiation could leak outside the plant and may spread into the atmosphere. The siren will sound. Listen to local radio or television and be prepared to follow instructions promptly. Evacuation from your residence may be necessary.

FEMA has provided the following guidelines for individuals when a nuclear power plant emergency occurs:

1. Keep a battery-powered radio with you at all times and listen to the radio for specific instructions.
2. Close and lock doors and windows.

If you are told to evacuate:

1. Keep car windows and vents closed
2. Use recirculated air

If you are advised to remain indoors:

1. Turn off the air conditioner, ventilation fans, furnace, and other air intakes.
2. Go to a basement or other underground area, if possible.
3. Do not use the telephone unless absolutely necessary.

If you expect you have been exposed to nuclear radiation:

1. Change clothes and shoes.
2. Put exposed clothing in a plastic bag.
3. Seal the bag and place it out of the way.
4. Take a thorough shower.

Keep food in covered containers or in the refrigerator. Food not previously covered should be washed before being put into containers.

Chapter 8 discussed, in detail, the procedures recommended after an explosion of a dirty bomb or an atom bomb in order to reduce the radiation exposure.

Aircrews: The Federal Aviation Administration (FAA) recommended limits for aircrews is a 5–year average effective dose of 20 mSv per year. The FAA also provides software for estimating the amount of galactic cosmic radiation received on a flight. Two versions of the CARI program, CARI-6 and CARI-6M, can be downloaded from this website: http://www.cami.jccbi.gov/radiation.html. The downloadable version of CARI-6 is more sophisticated than the interactive version. In addition to calculating a flight radiation dose, this also allows the user to store and process multiple flight profiles and to calculate dose rates at user-specified locations in the atmosphere. CARI-6M does not require a great-circle route between origin and destination airports; it allows the user to specify the flight path by entering the altitude and geographic coordinates of waypoints. The FAA has published a document entitled, What Aircrews Should Know About Their Occupational Exposure to Ionizing Radiation.

Aircrews should receive adequate education on the basic radiation effects, annual radiation dose limits to crews, actual annual radiation dose level received by aircrews, and the potential health risks of such doses. This document is available to the public through the National Technical Information Service, Springfield, Virginia 22161.

Radiation workers at medical facilities and nuclear power plants: The radiation workers at medical facilities include radiologists, physicians working with nuclear medicine, and radiation oncologists and their respective nursing and technologist staffs. The levels of education about radiation biology vary markedly among these groups. For example, residents in radiation oncology are taught radiation biology in more detail with an emphasis on high-dose radiation effects than residents in diagnostic radiology or in nuclear medicine. None of them receive sufficient amounts of knowledge on the late adverse effects of low levels of radiation on neoplastic and nonneoplastic diseases, and somatic and heritable mutations. The cardinal principle of radiobiology, which states clearly that no dose of radiation should be considered totally safe, is seldom emphasized. The decision on the use of radiation in humans is always based on risk versus benefit. Some but not all radiation workers are aware of this fact; however, patients who receive diagnostic doses of radiation in procedures are seldom told about the potential risk of radiation. For example, several articles have been published showing increased risk of cancer following a radiation dose from a single CT scan; however, most radiologists are questioning the validity of this conclusion. These radiologists believe that there is no such risk; therefore, there is no need to alarm patients who then may decide not to undergo needed diagnostic procedures. Since they did not have adequate education in radiation biology, this may in part account for their resistance to accept the idea that diagnostic doses of radiation can increase the risk of cancer, and if gonads are irradiated during the procedure, heritable mutations are induced that can increase the risk of diseases in future generations. The radiation workers at nuclear power plants are also exposed daily to low doses of radiation that could increase the risk of cancer.

The concept of maximum permissible dose (MPD) is poorly understood and interpreted by radiation workers at the medical facilities or at the nuclear power plant. I have taught radiation biology to residents in radiology and dentists, and have interacted with many colleagues in the radiation safety office who have sufficient understanding of the physics of radiation biology, but have very little understanding of the hazards of low doses of radiation. Many interpret the MPD as a safe radiation dose. When individuals accidently receive doses of radiation that are a little higher than expected, they are told that they are within the MPD; therefore, do not worry about it. This type of statement may create an environment in which there is very little concern about the health risk of low doses of radiation. Those radiation scientists or physicians who believe that low doses of radiation pose no health risk often argue that more people die from a car accidents or toxicity of prescription and drug overdoses each year, than from cancer induced by low doses of radiation. This may be true with one major difference. Low doses of radiation induce heritable mutations that can increase the risk of diseases in future generations, whereas others do not produce such an effect.

At present, all countries except the United States follow the recommendations made by the International Commission on Radiological Protection in 1990, Publication 60, regarding the maximum permissible dose for radiation workers. The annual dose limit is 20 mSv, averaged over 5 years. However, in order to provide greater flexibility, a dose of 50 mSv in a single year would be tolerable as long as the average dose over 5 years was 20 mSv or less. The United States National Council on Radiation Protection and Measurements (NCRP) follows the earlier recommendation of 50 mSv annually for radiation workers.

In summary, radiation workers should be taught radiation biology in more detail with an emphasis on the effects of low doses of radiation. This is essential in order to develop respect for, rather than fear of, radiation. It should be pointed out that humans accept many risks in order to improve the quality of life, but they should also take all possible training and precautions to avoid them.

Current recommendations for radiation protection involve only the steps to reduce the dose levels. At present, there is no strategy to provide biological protection to radiation workers in order to reduce the risk of neoplastic and nonneoplastic diseases. The proposed micronutrient strategy for biological radiation protection should be given serious consideration for implementation while developing an implementation plan for prevention and mitigation of radiation damage.

EXERCISES AND DRILLS

Exercises and drills are performed annually for emergency preparedness and response to a nuclear weapon explosion or a major nuclear power plant accident. These are essential in order to identify appropriate responsible agencies at the federal, state, and local levels, and to simplify the decision-making process. The United States Nuclear Regulatory Commission (USNRC), FEMA, nuclear power plant operators, state and local officials, as well as volunteers and first responders (police, firefighters, and medical response personnel) have produced comprehensive emergency preparedness programs that would protect the public in the event of a radiological emergency (explosion of a dirty bomb or a major nuclear power plant accident) (http://www.nrc.gov/about-nrc/emerg-preparedness/faq.html). The mock scenario may involve explosion, fire, and radioactive leak at a nuclear plant facility. Individuals may be injured or may receive varying levels of radiation dose. The surrounding areas may become contaminated with radioactive isotopes. Preparedness drills may involve first responders who decontaminate the area wherever possible, and medical personnel to take care of injured individuals and ready them for transport to the emergency room (ER) of the nearest hospital and treat them as needed. In addition, an evacuation plan, if needed, is also implemented during a practice drill. Persons located within a 10-mile radius of a nuclear power plant are notified by means of sirens, tone-alert radios, and similar alert mechanisms. Individuals living near a nuclear power plant are advised annually regarding the NRC- and FEMA-approved procedures they should follow in the event of a major nuclear power plant accident. State and local offices and agencies and FEMA should be consulted for detailed plans for preparedness. The Department of Energy (DOE) has established the Federal Radiological Monitoring and Assessment Center (FRMAC) to coordinate all radiological measurements from various field teams. The Federal Bureau of Investigation (FBI) becomes involved only if criminal activities have occurred or are threatened on or near the nuclear power plant facility.

STOCKPILING OF RADIATION PREVENTIVE AND MITIGATING AGENTS FOR BIOLOGICAL PROTECTION

At present, there are no approved nontoxic drugs that can be taken orally before irradiation in order to reduce acute or late adverse health effects in case of accidental exposure to low or high doses of radiation. Radiation workers at nuclear power plants are at risk for exposure to high doses of radiation in the event a major nuclear accident occurs. They are routinely exposed to very low doses of radiation daily. In the event of an explosion of a dirty bomb, a few individuals near the site of the explosion may receive lethal doses of radiation; most individuals residing near the site of an atom bomb explosion are likely to receive high doses of radiation that can cause mass casualties. Therefore, it becomes essential to stockpile some promising radiation preventive and mitigating agents.

Stockpiling of potassium iodide (KI): Stockpiling of KI is recommended because consuming KI orally will prevent the uptake of radioactive iodine (^{131}I) by the thyroid gland, and thus block radiation damage to the thyroid gland. In the event of a major nuclear power plant accident or an explosion of an atom bomb, ^{131}I represents one of the radioactive isotopes released into the atmosphere. Potassium iodide does not protect other organs against gamma-ray emitting radionuclides. Therefore, stockpiling of KI alone is not sufficient to prevent or mitigate radiation injury following an explosion of a dirty bomb, atom bomb, or a major nuclear power plant accident. Stockpiling of other useful radioprotective and radiation mitigating agents should be considered.

Stockpiling of a chelating agent and 10% magnesium sulfate: Stockpiling of diethylenetriaminepentaacetate (DTPA) is recommended to remove ^{65}zinc and ^{45}Ca from the body. However, potential nephrotoxicity of DTPA may limit its usefulness in removing certain radioactive isotopes. An oral intake of 10% magnesium is recommended in the event it becomes necessary to remove radioactive cobalt-60, radium-226, and phosphorus-32 from the intestinal tract.

Stockpiling of agents used in replacement therapy: Agents used in replacement therapy include antibiotics, blood transfusion, and electrolytes. This therapy is effective only against radiation doses ($^{30}LD_{50}$) that produce bone marrow syndrome, and can save all irradiated individuals who will die from this dose without replacement therapy. Bone marrow syndrome can also cause 100% lethality, if radiation doses are high. Replacement therapy may not be sufficient to save all irradiated individuals from dying after irradiation with high doses of radiation. Agents that are used in replacement therapy should be stockpiled. Replacement therapy does not reduce the risk of long-term adverse health effects of low or high doses of radiation. In addition, replacement therapy cannot be used for prevention of radiation injury, because it is administered only after irradiation when indicated.

Stockpiling of growth factors: Stockpiling of certain growth factors, such as granulocyte colony-stimulating factor (G-CSF), is recommended because it stimulates the proliferation of hematopoietic cells. G-CSF is also used in the management of individuals exposed to radiation doses that produce gastrointestinal (GI) syndrome. This agent, in combination with replacement therapy, would be useful in mitigation radiation damage when administered after irradiation. Therefore, certain growth factors should be stockpiled. It should be pointed out that replacement therapy in combination with a growth factor would be more effective in reducing adverse health effects of radiation, such as increased risk of cancer and noncancerous diseases, than the individual agents. The growth factor cannot be used for prevention of radiation damage, because it is effective only after irradiation when indicated.

Proposed stockpiling of micronutrient preparations: It is well established that low or high doses of ionizing radiation cause damage by generating excessive amounts of free radicals during and after irradiation and acute and chronic inflammation after irradiation. Dietary and endogenous antioxidants are known to neutralize free radicals and reduce inflammation. Therefore, taking multiple antioxidant supplements before irradiation and/or after irradiation appears to be one of the rational choices for prevention and mitigating damage from low or high doses of radiation. The rationale for using multiple dietary and endogenous antioxidants for prevention and mitigation of radiation damage was discussed in detail in Chapters 5–7 and 9. I have proposed the use of a commercial preparation, BioShield-R2, which can be taken orally daily before irradiation. This would maintain high levels of antioxidants in the body. If a nuclear accident or dirty bomb explosion occurs, most individuals may be exposed primarily to low doses of radiation; however, a few may receive lethal doses of radiation. Those who are taking daily BioShield-R2 and are exposed to low or high doses of radiation would suffer less initial damage than those who were not taking any antioxidant preparation. In order to reduce the risk of late adverse health effects of radiation, they may want to continue BioShield-R2 throughout their lifespan. Those who are not taking BioShield-R2 before receiving low or high doses of radiation should consider taking it 24 hours after irradiation, and continue throughout their lifespan. On the other hand, those who were not taking BioShield-R2 before irradiation and are exposed to lethal doses of radiation should consider taking BioShield-R3 at about 24 hours after irradiation for a period of 60 days, and then switch to BioShield-R2 for the rest of their life. In addition, the use of bioshields before and/or after irradiation may improve the efficacy of replacement therapy and growth factors during management of the acute phase of injury. The unique features of BioShield-R2 and BioShield-R3 were discussed in Chapters 5–7 and 9. I propose that bioshields be stockpiled in addition to those items listed in the previous paragraphs.

RESPONSIBLE AGENCIES FOR STOCKPILING RADIATION PREVENTIVE AND MITIGATING AGENTS

Potassium iodide (KI): The Bioterrorism Preparedness and Response Act of 2002 requires that KI tablets be made available through the Strategic National Stockpile (SNS) within the Health and Human Services (HHS) agency to state and local governments within 20 miles of nuclear power plants for stockpiling and distribution in sufficient quantities. Within HHS, the Office of Public Health Emergency Preparedness (OPHEP) is responsible for implementing the policy. FEMA acts as the single point of contact for receipt of KI applications from state, local, and tribal governments. FEMA's regional offices, state, local, or tribal government's plans are reviewed and the decision is forwarded to HHS. The Nuclear Regulatory Commission (NRC) maintains its current program for distribution and approval of the plan. FEMA also submits the plan to the NRC for their approval. Further information can be obtained from this website: http://books.nap.edu/catalog/10868.html. Requests for potassium iodide should be submitted to the Chief, Nuclear and Chemical Hazards Branch, Federal Emergency Management Agency, Room 202, 500 C Street, SW, Washington, DC 20472.

Micronutrient preparations (BioShield R2 and BioShield-R3): At present, multiple micronutrient preparations are not being considered for stockpiling. In view of recent laboratory and limited human data showing the benefits of micronutrient preparations containing dietary and endogenous antioxidants in biological protection against low and high doses of radiation FEMA may wish to review data on this issue, and give serious consideration to stockpiling micronutrient products, such as BioShield-R2 and BioShied-R3, for prevention and mitigation of radiation damage.

SUMMARY

The implementation plans of current and proposed prevention and mitigation of radiation injury are presented in this chapter. The implementation plans include educational programs on the effects of ionizing radiation, exercises and drills, and stockpiling of radioprotective and radiation mitigating agents. The names of the federal, state and local agencies responsible for implementing plans are listed in this chapter. Current recommendations for prevention of radiation injury in the event of a major nuclear power plant accident or an explosion of a nuclear weapon are based on a physical radiation protection strategy that reduces radiation dose levels. At present, there are no biological radiation protection strategies for prevention of radiation damage against exposure to low or high doses of radiation.

Potassium iodide (KI) is the only approved agent for prevention of thyroid disorders against radioactive iodine (^{131}I), which is one of the major components released into the atmosphere following the explosion of an atom bomb or a major nuclear power plant accident. By an act of Congress, KI is allowed to be stockpiled. I have proposed that a micronutrient preparation, BioShield-R2, containing dietary and endogenous antioxidants, may prevent radiation damage when administered orally before irradiation. There are a few agents available for reducing acute radiation sickness (ARS) in humans. These include agents used in replacement therapy, such as antibiotics, electrolytes, blood, and certain growth factors, such as G-CSF. However, the replacement therapy is effective only against doses that produce bone marrow syndrome killing 50% of the irradiated individuals within 60 days (^{60}LD$_{50}$). The addition of growth factors to the replacement therapy may further improve the survival rate of individuals who have received higher doses that produce bone marrow syndrome. I have proposed that addition of a micronutrient preparation BioShield-R3, which contains dietary and endogenous antioxidants, when administered orally 24 hours after irradiation in combination with standard therapy may improve the efficacy of standard therapy in improving the survival rate of individuals exhibiting bone marrow syndrome as well as GI syndrome. At present, there are no effective treatments of individuals exhibiting GI syndrome that can protect irradiated individuals from dying. Also, there are no effective strategies to reduce the risk of neoplastic and nonneoplastic diseases among survivors of radiation exposures. I have proposed that BioShield-R2 may be effective in reducing long-term adverse health effects of radiation. I have suggested that FEMA may want to review data on micronutrient preparations, such as BioShield-R2 and BioShield-R3, and give serious consideration to stockpiling them for prevention and mitigation of radiation damage in humans.

11 Health Risks of Nonionizing Radiation and Their Prevention and Mitigation

INTRODUCTION

All radiation is a form of energy with a wide electromagnetic spectrum. An electromagnetic field is associated with two components, the electric field and the magnetic field, that often work together to create an electromagnetic field. Electric fields arise from differences in voltage: the higher the voltage, the greater the strength of the field. The strength of an electric field is measured in volts per meter (V/m). Magnetic fields arise when the electric current flows: the greater the current strength, the stronger the magnetic field. The strength of the magnetic field is measured in amperes per meter (A/m), but more commonly it is measured in microtesla (μT). The electromagnetic field is characterized by its frequency or its corresponding wavelength. Electromagnetic radiation can be grouped in two major classes: ionizing radiation and nonionizing radiation. Ionizing radiation has a higher frequency and shorter wavelength and enough energy to knock out one electron from one of the orbits of the atom. Nonionizing radiation has a lower frequency and higher wavelength and has enough energy to move atoms in a molecule around or cause them to vibrate, but not enough to remove an electron from one of the orbits of the atom. The electromagnetic field is present in the atmosphere and is generated by humans. Human-made sources of electromagnetic radiation of the nonionizing radiation type include cell phones, microwave, and television and radio-transmitters. Radio frequencies are measured in watts per square meter (W/m2). The previous 10 chapters have discussed in detail various aspects of ionizing radiation.

This chapter discusses the types of nonionizing radiation and their impact on human health. Nonionizing radiation includes ultraviolet (UV) radiation, infrared, visible light, sound waves, and radio waves, each of which is associated with a defined wavelength. For example, radio waves have a wavelength of 10^2–10^4 cm, microwaves 1 cm, infrared 10^{-2} cm, visible light 10^{-5} cm, and ultraviolet radiation 10^{-6} cm. Nonionizing radiation with a very long wavelength has a very low frequency. Radio wave frequencies have wavelengths between 1 and 100 meters and frequencies in the range of 1 million to 100 million hertz or cycles per second, whereas microwaves have wavelengths of about 1 cm and frequency of about 2.5 billion hertz. Among the types of nonionizing radiation, UV radiation has been most extensively studied with respect to its impact on human health.

ULTRAVIOLET RADIATION

In addition to emitting galactic radiation in the form of ionizing radiation, the sun also emits nonionizing radiation in the form of UV radiation, infrared, visible light, and heat. UV radiation has been classified into three groups, depending upon their wavelength. They are UVA, UVB, and UVC. UVA radiation has a wavelength in the range of 320–400 nm, and is not absorbed by the ozone layer. UVB has a wavelength in the range of 290–320 nm and is mostly absorbed by the ozone layer. UVC has a wavelength in the range of 100–290 nm and is completely absorbed by the ozone layer and atmosphere (http://www.epa.gov/sunwise/doc/uvradiation.html). The levels of UV radiation reaching the Earth's surface depend upon the stratospheric ozone layer, time of day, time of year, latitude, altitude, weather conditions, and reflection. Thinning of the ozone layer would allow more UV radiation to reach the Earth's surface. At noon, when the sunlight has the shortest distance to travel through the atmosphere, the levels of UV radiation on the Earth are at their peak. In the early morning and late afternoon, the sunlight passes through the atmosphere at an angle, and therefore the intensity of UV radiation reaching Earth is reduced. The angle of sun with respect to the Earth varies with the seasons, causing variations in the levels of UV radiation reaching the Earth's surface. The intensity of UV radiation is highest in the summer. The sunlight is strongest at the equator, where the distance between the sun and Earth is shortest. The intensity of UV radiation increases with an increase in the altitude. Cloud cover may decrease the intensity of UV radiation on Earth, depending upon the thickness of the clouds. Surfaces like snow, sand, pavement, and water reflect much of the UV radiation that reaches them, and thus, the intensity of UV radiation even in the shade could be deceptively high.

Other sources of UV radiation: These sources include tanning booths, mercury vapor lighting (often found in sports stadiums), some halogen, fluorescent, and incandescent lights, and some types of lasers.

UV index: The UV index, developed by the National Weather Service and the Environmental Protection Agency (EPA) indicates the strength of solar UV radiation on a scale of 1 (low) to 11+ (extremely high). This information provides a guideline for predicting the intensity of UV radiation in order for individuals to take appropriate action to prevent UV radiation–induced injury.

BIOLOGICAL EFFECTS OF UV RADIATION

UV radiation has both beneficial and harmful effects, depending upon the level of UV radiation and period of exposure to UV radiation. Humans are primarily exposed to the UVA type of UV radiation, but they are also exposed to small amounts of UVB. UVA radiation has a longer wavelength that can enter the middle layer of the skin (the dermis), whereas UVB radiation, which has a short wavelength, can penetrate only the outer layer of the skin (the epidermis). Certain oral and topical medicines, such as antibiotics, birth control pills, and benzoyl peroxide products used in the treatment of acne, as well as some cosmetics such as alpha-hydroxy acids (AHAs) increase the sensitivity of the skin to UV radiation. The Federal Drug Administration (FDA) suggests that cosmetics containing AHAs should state on the label that it can increase the skin response to UV light. UV radiation produces both harmful effects, such as photoaging, immune suppression, and melanoma and non-melanoma skin cancers, and beneficial effects, such as synthesis of vitamin D. Using hairless mice, it was demonstrated that chronic exposure to smoke in combination with UV radiation enhanced the skin damage, including the incidence of squamous cell carcinoma.[1] UV radiation–induced acute erythema and sunburn reactions are the most important factors, in combination with cumulative UV radiation, that cause photoaging, precancerous lesions, and skin cancers.

UV radiation-induced photoaging: Numerous studies have been published on the harmful effects of UV radiation on human health, especially the increased rate of skin aging (photoaging or premature aging), and the enhanced risk of melanoma and nonmelanoma skin cancers. The exact pathogenesis of UV radiation–induced photoaging of the human skin remains unknown. However, UV radiation exposure of the skin increases oxidative stress, leading to inflammatory reactions and premature aging. Other cellular and molecular pathways, including excessive production of free radicals, inflammation, degradation of extracellular matrix (ECM) proteins, and increased activity of matrix metalloproteinase (MMP), neutrophil-derived proteases, and deletion of mitochondrial DNA (mtDNA), have been proposed. Synthesis of ECM proteins and their degradation by MMP are part of the dermal remodeling that results from chronic exposure of skin to UV radiation. The early stages of UV radiation–induced skin aging are characterized by the increased degradation of elastic fiber proteins and activity of extracellular matrix proteases.[2] It has been reported that neither UVB nor UVA altered synthesis of ECM proteins, but UVA radiation increased the production of MMP-1 and MMP-3 in dermal fibroblasts in culture.[3] Cytokines interleukin-1beta (IL-1beta) and tumor necrosis factor-alpha (TNF-alpha) markedly decreased fibrillin mRNA levels but increased the synthesis of MMP-1, -3, and -9. However, the synthesis of collagen was not affected by UVB, UVA, or IL-1beta treatment. The hallmark of photoaged skin is solar elastosis, which is probably an end product of elastic fiber degradation. It is known that exposure of human skin to UV radiation, infrared radiation, or heat leads to an influx of neutrophils. These neutrophils have high levels of proteolytic enzymes capable of degrading collagen, ECM proteins, and elastic fiber. Indeed, it has been shown that UV radiation–exposed skin exhibited neutrophil-derived proteolytic activity of elastase and MMPs, suggesting that neutrophils are major contributors to the UV radiation–induced photoaging process of the skin.[4] It has been suggested that UV radiation–induced mtDNA deletion may lead to inadequate energy production in dermal fibroblasts that can cause functional and structural alterations in the skin leading to photoaging.[5] Solar urticaria is a rare photosensitivity skin disorder that is induced by UV radiation. Both UVA radiation and UVB radiation are effective in inducing solar urticaria.[6]

UV radiation–induced cancer: UV radiation is known to increase the risk of skin cancers (basal cell carcinoma, squamous cell carcinoma, and cutaneous melanoma). In addition to UVB radiation, UVA radiation also acts as a human skin carcinogen.[7] However, it was shown that UVB and not UVA is responsible for inducing melanoma.[8] In the United States, more than 2 million new cases of skin cancers are detected each year,[9] and about 68,130 cases (38,870 men and 29,260 women) of cutaneous melanoma were diagnosed in 2010 (http://emedicine.medscape.com/article/1100753-overview). Melanoma represents only about 4% of all skin cancers, but it is responsible for more than 73% of skin cancer deaths. Exposure to UV radiation from sunlight or artificial sources at a younger age increases the risk of developing skin cancer. Using hepatocyte growth factor/scatter factor (HGF/SF) transgenic mice, it was demonstrated that irradiation of neonatal mice with UV radiation significantly increased the incidence of melanoma.[10] Additional UV radiation exposure to adult mice significantly increased melanoma multiplicity. It has been estimated that about 25% of sun exposure occurs before 18 years of age. Excessive acute exposures to sun (painful sunburn) have been implicated in the pathogenesis of squamous cell carcinoma, basal cell carcinoma, and cutaneous malignant melanoma. Chronic exposure to UV radiation is the most important risk factor for the development of actinic keratoses and squamous cell carcinoma. An epidemiologic study has reported that acute UV radiation exposure or painful sunburn before the age of 20 years was associated with an increased risk of squamous cell carcinoma, nodular basal cell carcinoma, multifocal superficial basal cell carcinoma, melanoma and its precursors, melanocytic nevi and atypical nevi, as well as actinic keratoses. On the other hand, lifetime chronic sun exposure appears to be associated with a lower risk of malignant melanoma, but it increased the risk of melanocytic nevi and atypical nevi that are precursor of cutaneous melanoma.[11] It has been reported that among 11,478 white German children of preschool age who were exposed to painful sunburns and who took vacation in foreign countries with a sunny climate, the incidence of nevi, a precursor of malignant melanoma, was higher than those who did not have above characteristics.[12] A meta-analysis of published epidemiologic studies revealed that occupational UV radiation exposure significantly increased the risk of developing basal cell carcinoma[13] as well as squamous cell carcinoma.[14] Intentional UV radiation exposure has been recommended by different institutions to enhance the vitamin D level among those elderly individuals who exhibit vitamin D deficiency; however, excessive exposure to sunlight or tanning must be avoided in order to reduce the risk of skin cancers.[15] It has been reported that lower levels of UVB radiation and vitamin D were associated with increased rates of pancreatic cancer,[16] bladder cancer,[17] and brain cancer.[18]

An analysis of several epidemiologic studies revealed that tanning (sun bed) also increased the incidence of malignant melanoma and squamous cell carcinoma, but not basal cell carcinoma.[19] In 2008, the World Health Organization (WHO) listed UV radiation as a group 1 carcinogen. In spite of this warning, millions of people, especially in Western countries, use indoor tanning beds every year. A review of several published studies suggests that use of indoor tanning beds represent a significant and avoidable risk factor for the development of melanoma and nonmelanoma skin cancers.[20] In 2009, about 18.1% of women and 6.3% of men used indoor tanning in the United States. Among them only about 13.3% of women and 4.2% of men suggested that avoiding the use of tanning bed could reduce their risk of skin cancer.[21] It is interesting to note that some individuals continue to use tanning beds in spite of awareness of risks associated with the use of such facilities.

UV radiation–induced multiple sclerosis: An epidemiologic study conducted in France revealed a strong association between multiple sclerosis (MS) prevalence and annual mean UVB radiation and average winter UVB radiation.[22] Another epidemiologic study in Australia showed that low maternal exposure to UV radiation in the first trimester is independently associated with subsequent risk of MS in offspring.[23] It was suggested that vitamin D deficiency induced by exposure to low levels of UV radiation may account at least in part for the increased incidence of MS.[24]

UV radiation–induced suppression of immune function: Animal studies have revealed that UVB radiation produces selective suppression of normal immune responses.[25] The published studies further showed that production of UV radiation–induced cytokines by epidermal cells may be an important factor in immune suppression.[26] It has been reported that UV radiation reduced the competency of bone marrow–derived CD11c+ cells in mice, which was associated with increased interleukin-10 (IL-10) and prostaglandin E2 (PGE2) secretion.[27] UV radiation–induced immune suppression could impact the risk of other chronic diseases.

UV radiation–induced mutations: Both UVA radiation and UVB radiation can induce gene mutations in human skin.[28] Beta-transducin repeat-containing proteins (beta-TrCP), an E3 ubiquitin ligase receptor, appears to play an important role in carcinogenesis.[29] UV radiation caused an increase in beta-TrCP mRNA and protein levels in human cells. It has been reported that inhibition of beta-TrCP function caused a decrease in UVB radiation–induced edema, hyperplasia, and inflammatory response, and an increase in UVB radiation–induced apoptosis in mouse epidermal cells.[30]

MECHANISMS OF UV RADIATION–INDUCED DAMAGE

UV radiation–induced free radicals are strongly involved in photoaging and skin cancer, and their increased levels in UV-irradiated skin has been demonstrated by electron spin resonance (ESR).[31] UV radiation causes DNA damage that can lead to photoaging and cancer in the skin. UVB radiation is absorbed mostly by the DNA of the epidermis keratinocytes; therefore, this form of UV radiation is more relevant for the induction of skin cancer, whereas UVA radiation generates excessive amounts of free radicals in the skin that can damage DNA as well as cause photoaging and immune suppression, increasing the risk of cancer.[32] UVB radiation interacts directly with DNA, producing two major products of DNA damage, cyclobutane-pyrimidine dimers (CPDs) and (6-4) pyrimidine-pyrimidone. A recent study has shown that interferon-gamma (IFN-gamma) may be involved in the activity of UV radiation–induced melanoma.[33] UVB-induced melanocyte activation is characterized by aberrant growth and migration, and a high level of INF-gamma in mice. These effects were abolished by antibody-mediated inhibition of INF-gamma. INF-gamma was produced by macrophages recruited to neonatal skin after exposure to UVB radiation. The frequent mutation or deletion in INK4A/ARF locus located at chromosome 9p21 is observed in human and animal tumors.[34,35] It has been suggested that INK4A/ARF also plays a significant role in UV radiation–induced melanoma. This was confirmed by the fact that deletion of INK4A/ARF in HGF/SF transgenic mice markedly slowed the development of melanoma. Individuals carrying mutation in INK4A/ARF will be more sensitive to UV radiation–induced melanoma than those with a corresponding wild-type gene.[36] It has been further demonstrated that nuclear excision repair loss, together with inactivation of INK4A/ARF, may be responsible for the induction of UV radiation–induced melanoma through mutation in the Kras gene.[37] The transcriptional factor Nrf2 and its cytoplasmic anchor protein, Kelch-like-ECH-associated protein 1 (Keap1), play a protective role against UV radiation–induced damage by regulating the cellular antioxidant response in the skin. Therefore, deficiency of Nrf2 would increase the rate of UV radiation–induced photoaging. Indeed, it has been demonstrated that UVB-irradiated Nrf2-deficient mice showed an increased rate of photoaging, such as coarse wrinkle formation, loss of skin flexibility, epidermal thickening, and deposition of extracellular matrix in the upper dermis. In addition, the level of lipid peroxidation product 4-hydroxy-2-nonenal increased and the level of glutathione decreased in these UV-irradiated mice.[38] However, there was no significant difference in the incidence of skin cancers between UVB-irradiated Nrf2-deficient and Nrf2-sufficient mice, suggesting that the mechanisms of UVB radiation–induced photoaging and skin cancers may in part be different.[39] It has been reported that UVA radiation, but not UVB radiation, causes nuclear translocation and accumulation of Nrf2 by a factor of 6.5 in comparison to unirradiated controls.[40] This effect of UVA radiation was enhanced by the photosensitizer hematoporphyrin. The disruption of Nrf2 in dermal fibroblasts derived from Nrf2 or Keap1 knockout mice increased the number of apoptotic cells following UVA irradiation, whereas the disruption of Keap1 decreased the number of apoptotic cells in comparison to unirradiated wild-type Nrf2 dermal fibroblasts. These results suggest that Nrf2-Keap1 pathway plays an important role in the protection of the skin against UVA radiation.[40] It has been shown that the p53 gene could protect skin cells from UVB radiation–induced DNA damage. UV-induced DNA damage activates the mechanisms of removal of DNA damage, delay in cell cycle progression, and DNA repair or apoptosis by transcriptional activation of p53-related genes, such as p21 and bax. Several studies have reported that UVB irradiation induces mutation in the p53 gene.[41] Indeed, a high-frequency p53 mutation was reported in premalignant actinic keratosis lesions that are considered precancerous for developing squamous cell carcinoma. Thus, it was suggested that p53 mutation may be involved in converting premalignant lesions to squamous cell carcinoma.

PROTECTION AGAINST UV RADIATION–INDUCED DAMAGE TO THE SKIN

Sunscreen: Based on the data available, it was concluded that sunscreen ingredients or products do not pose a human health concern, and that the regular use of appropriate broad-spectrum sunscreen products before prolonged exposure to sunlight may reduce the risk of skin cancers.[42,43]

Antioxidants: In addition to sunscreen, antioxidants also provide protection against UV radiation–induced skin damage. Topical application of commonly used antioxidants, such as vitamin C, vitamin E, carotenoids, coenzyme Q10, or n-acetylcysteine, protects the skin against UV radiation–induced photoaging as well as carcinogenesis. It has been reported that treatment of individuals with the combination of antioxidant and sunscreen was most effective against UV radiation–induced damage to skin.[44]

Coenzyme Q10: It has been shown that coenzyme Q10 protected against UVA radiation–induced damage to human keratinocytes in culture by reducing oxidative damage and to human dermal fibroblasts by decreasing the levels of MMPs. In a clinical study, it was observed that 1% of coenzyme Q10 cream for five months reduced the wrinkle score grade determined by a dermatologist.[45] This effect of coenzyme Q10 is mediated through the inhibition of Interleukin-6 (IL-6) production. It has been shown that a novel coenzyme Q10 nanostructured lipid carrier (CoQ10-NLC) preparation was more effective than coenzyme Q10 emulsion in reducing photoaging, because it displayed a stronger capability to penetrate the skin after topical application, and exhibited greater antioxidant properties.[46] Treatment of human dermal fibroblast cells in culture with coenzyme Q10 or colorless carotenoids, phytoene, and phytofluene, suppressed UV radiation–induced elevation of inflammatory markers, such as prostaglandin E2, and IL-6, and matrix metalloproteinase-1 (MMP-1). The combination of carotenoids and coenzyme Q10 was more effective in reducing the levels of these inflammatory mediators than the individual agents.[47]

N-acetylcysteine (NAC): A single neonatal dose of UV radiation to transgenic mice for hepatocyte growth factor induced melanoma in adult, but administration of n-acetylcysteine (NAC) (7 mg/ml in mothers' drinking water) through nursing until 2 weeks after birth significantly delayed the development of melanoma by decreasing oxidative stress.[48] UV irradiation increases oxidative stress in nevi in humans. A single oral dose of NAC (1200 mg) prevented UV radiation–induced oxidative stress in nevi, which could reduce the risk of developing melanoma.[49] Incubation of human fibroblasts in the presence of 6 mM NAC protected UV radiation–induced DNA damage by scavenging free radicals and by donating cysteine for the synthesis of glutathione.[50]

Vitamin C and vitamin E: Topical application of high doses of a mixture of vitamin C and E inhibited UV radiation–induced erythema, sunburn, and tanning as well as photoaging and skin cancer.[51] Topical application of alpha-tocopherol alone decreased UVB-induced skin cancer in mice, whereas photoproducts of alpha-tocopherol, tocopherol dimers and trimers may participate in prevention of UV radiation–induced photoaging.[52] It has been demonstrated that treatment of mouse keratinocytes in culture with alpha-tocopherol or alpha-tocopheryl acetate reduced UVB radiation–induced damage by inhibiting the activation of transcriptional factor NF-kappaB.[53] A randomized, double-blind human study revealed that topical application of a mixture of ascorbic acid, alpha-tocopherol, and melatonin was more effective than the individual agents in providing photoprotection against UV radiation–induced erythema in humans.[54] The use of a colloidal thickened microemulsion of vitamin E and vitamin C allowed better absorption of both vitamin E and vitamin C by the reconstructed human epidermis.[55] It has been found that gamma-tocopherol inhibited cyclooxygenase-2 (COX-2) independent of its antioxidant activity, suggesting that this form of vitamin E may also have a role in protecting against UV radiation-induced photoaging and skin cancers.[56] Alpha-tocopherol-protected human keratinocytes in culture against UVB radiation–induced cyclobutane pyrimidine dimer, which is involved in induction of skin cancer.[57] In a randomized, double-blind, placebo-controlled clinical study involving young females, it was observed that oral consumption of a mixture of vitamin E, vitamin C, carotenoids, selenium, and proanthocyanidins (Seresis) slowed down the development of UVB radiation–induced erythema. The levels of MMP-1 and MMP-9 also decreased in antioxidant-treated groups.[58]

Selenium: Oral administration of sodium selenite through drinking water before UV irradiation markedly reduced the incidence of skin cancers in hairless mice.[59] Topical application of selenomethionine at a concentration of 0.002%, 0.02%, and 0.05% markedly reduced UV radiation–induced acute damage.[60] Pretreatment of human keratinocytes and melanocytes in culture with sodium selenite or seleno-methionine reduced UVB radiation–induced cell death; however, sodium selenite was more effective than selenomethionine in reducing UVB radiation–induced skin damage.[61] The combination of topical application of alpha-tocopherol in combination with selenomethionine plus oral supplementation with alpha-tocopheryl acetate was most effective in reducing UVB radiation–induced skin cancer in hairless mice.[62]

Carotenoids: Oral administration of dietary carotenoids such as beta-carotene and lycopene has been shown to reduce UV radiation–induced photoaging in humans.[63–65] Dietary supplementation with beta-carotene reduced UVA-induced production of free radicals in mouse skin.[66]

Herbs and vegetables antioxidants: *Cordyceps sinensis* (Berk) is a fungus exhibiting anticancer and cytoprotective properties. It has been reported that pre-treatment of normal human fibroblasts in culture with polysaccharide-rich *Cordyceps* mycelial components lowered the levels of cyclobutane-pyrimidine dimers (CPDs), a product of DNA damage.[67] The investigators suggested that *Cordyceps* may be a use-ful agent to provide photoprotection and lower the risk of basal cell carcinoma, the main skin cancer caused by CPDs. It has been shown that topical application of green tea polyphenols (GTPs) or oral administration of green tea through drinking water prevented UV radiation–induced nonmelanoma skin cancers. This protective effect of green tea is mediated, at least in part, through rapid repair of DNA damage. The repair of DNA damage by GTPs is mediated through the induction of interleukin-12 (IL-12), which possesses the ability to repair DNA damage.[68] Other polyphenols, such as grape seed proanthocyanidins, resveratrol, silymarin, and genistein pro-tected skin against UV radiation in mice by reducing inflammation, oxidative stress, and enhancing repair of DNA.[69] A cosmetic cream containing anthocyanins, isolated from TNG73 purple sweet potato (0.61 mg per 100 g cream), was effective in reduc-ing UV radiation–induced skin damage. Although anthocyanins absorbed both UVA radiation and UVB radiation, they were particularly effective against UVB radia-tion.[70] Plant flavonoids also produced similar effects against UV radiation damage to the skin.[65,71] Topical application of 0.5% solutions of isoflavon compounds, such as genistein, daidzein, or biochanin A, protected pig skin from solar-stimulated UV radiation–induced photoaging; however, it was less effective than that produced by a topical application of a solution containing 15% ascorbic acid, 1% alpha-tocopherol, and 0.5% ferulic acid.[72] Isoflavone extract from soy or isoflavone isolated from red clover (*Trifolium pretense*) protected against UV radiation–induced photoaging, inflammation, and immune suppression.[73,74] Pretreatment of human reconstituted skin with pomegranate-derived products protected against UVB radiation–mediated skin damage. This pretreatment reduced UVB-induced cyclobutane pyrimidine dimers, 8-hydroxy-2-deoxyguanosine, protein oxidation, and proliferating cell nuclear antigen (PCNA) protein expression as well as collagenase (MMP-1), gelatinase (MMP-2, MMP-9), stromelysin (MMP-3), marilysin (MMP-7), elastase (MMP-12), and tropoelastin.[75] Topical application of silymarin, isolated from fruits and seeds of milk thistle (*Silybum marianum* L. Gaertn), inhibited UV radiation–induced skin cancers through inhibiting oxidative stress and inflammation.[76] The goji berry juice (*Lycium barbarum*) has been recognized in traditional Chinese medicine for various therapeutic purposes because of its antioxidant and immune-modulating effects. It has been demonstrated that drinking goji berry juice (1–10%) markedly reduced the effect of acute solar-stimulated UV radiation on the mice skin.[77] Resveratrol treatment increased the viability of human keratinocyte cells irradiated with UBA radiation and protected them from UVA radiation–induced oxidative stress. It also increased the level of Nfr2 protein and its accumulation in the nucleus by degrading Keap1 protein, which then upregulated antioxidant enzymes.[78] Treatment of human keratinocytes in culture with *Prunella vulgaris* or its main phenolic acid component, rosmarinic acid, before or after UVB radiation, prevented damage by reducing oxidative stress and DNA damage. In addition, the above treatment also decreased release of IL-6.[79] UVA and UVB radiation caused decomposition of quercetin, and decomposition products may enhance the skin response to UV radiation; however, addition of quercetin enhanced the photostability of butyl methoxydibenzoylmethane (BMDBM) and octyl methoxycinnamate (OMC).[80,81] The protective effect of quercetin was due to inhibition of UVA- and UVB-induced generation of excessive amounts of free radicals.[82] It has been suggested that treatment of human keratinocytes in culture with quercetin increased protein levels of the transcriptional factor Nrf2, increased expression of antioxidant genes, and reduced the production of reactive oxygen species following UVA irradiation.[83] This suggests that Nfr2 plays an important role in protection of the skin from UVA radiation damage by quercetin.

Pycnogenol is an extract of the bark of the French maritime pine tree (*Pinus pinaster*). It contains a mixture of water-soluble antioxidants, bioflavonoids, particularly proanthocyanidins, and organic acids. Topical application of Pycnogenol provided dose-dependent protection from solar-stimulated UV radiation–induced acute inflammation, immunosuppression, and carcinogenesis in hairless mice skin.[84] Pycnogenol protected against squamous cell carcinoma in the skin of hairless mice produced by exposure to tobacco smoke and UV radiation.[1]

Combined effect of sunscreens and antioxidants: Topical application of a mixture of antioxidants containing vitamin C, ferulic acid, and phloretin protected skin against UV radiation–induced photoaging in humans, and it was suggested that an antioxidant mixture may increase the effectiveness of sunscreens in providing photoprotection for human skin in a synergistic manner.[85,86] It was demonstrated that a mixture of vitamin C and vitamin E or vitamin C alone, when added to a commercial UVA sunscreen (oxybenzone), caused a greater than additive protective effect against photoaging to be observed.[87] Topical application of natural or synthetic antioxidants before UV radiation provides significant photoprotection to the skin. The addition of antioxidants to commercial sunscreens may improve their efficacy in photoprotection against UVA radiation.[88]

Melanocyte-stimulating hormone: Five patients with solar urticaria received a single subcutaneous dose of 16 mg afamelanotide, an analog of melanocyte-stimulating hormone (MSH), implanted in wintertime. This treatment increased the level of melanin in the skin, which reduced the solar urticaria response to UV radiation.[6] Melanization of the skin by MSH treatment may be responsible for protection against UV radiation.

OTHER FORMS OF ELECTROMAGNETIC RADIATION

Electromagnetic radiation varies in its frequency. The major categories of electromagnetic frequency are described next.

Electromagnetic radiation of extremely low frequency (ELF) fields: The electric power supply and all other appliances using electricity are the main sources of ELF fields and have frequencies up to 300 Hz.

Electromagnetic radiation of intermediate frequency (IF) fields: Computer screens, antitheft devices, and security systems are the main sources of IF fields, which have frequencies of 300 to 10 MHz.

Electromagnetic radiation of radiofrequency (RF) fields: Cell phones, radio, television, radar, and the microwave oven are the main sources of RF fields, which have frequencies of 10 MHz to GHz. At radio frequencies, electric and magnetic fields are closely interrelated and their levels of power densities are measured in watts per square meter (W/m2). Among sources of electromagnetic radiation of radio-frequency (RF) fields, cell phones have drawn significant attention because of the effects on human health of electromagnetic radiation emitted during use.

CELL PHONES

The cell or mobile phone technology and its use have exploded during the last decade throughout the world. It is estimated that about 4–5 billion people use cell phones at this time. The cell phone converts voices to impulses, which are transmitted over radio waves at frequencies ranging from about 800 to over 1,900 MHz. The biological effects of the radio frequencies emitted by cell phones have been the subject of intensive debate, and no firm conclusions on the adverse health effects of this form of radiation can be drawn at this time.

Cell phones and brain cancer risk: The fact that electromagnetic radiation at radio frequencies from cell phones can be absorbed into the brain has prompted concerns that regular use of a cell phone for a long period of time may increase the risk of acoustic neuroma and other brain tumors. The effects of cell phone use on cancer risk have been investigated using primarily epidemiologic methodologies. The results of these studies on cancer incidence have been inconsistent, varying from no effect to adverse effect.[89–91] Reviews of several studies on the effect of cell phone use on the risk of brain tumors revealed that regular use of cell phones for a period of 10 years or more was associated with an increased risk of acoustic neuroma and glioma.[92,93] Other epidemiologic studies reported no such association between cell phone use and risk of brain tumor.[94,95] In a recent case-control study performed in Japan, it was found that use of cell phones for 20 minutes or more for five years was associated with increased risk of acoustic neuroma.[96] In another epidemiologic study, regular use of cell phones was associated with an increased risk of benign parotid gland tumors.[97] A recent epidemiologic study evaluated the incidence of brain cancer among users of cell phones in England during the period of 1998 to 2007. The results showed that the rate of increase in brain cancer may be less than 1 additional case per 100,000 people during that period.[98] This study did not mention whether the cell phone users were using them excessively. Excessive use of cell phones among adolescents has become apparent only recently. Therefore, the conclusion of this study remains uncertain until confirmed by studies in which excessive use of cell phones is correlated with the risk of brain cancers. A review of epidemiologic studies concluded that there was no causal relationship between cell phone use and increased risk of brain tumor;[99,100] however, an increased risk of glioma was observed at the highest exposure levels.[100] A recent analysis of the relationship between cell phone subscriptions and the incidence of brain tumors in the United States has revealed that there may be a linear relationship between cell phone use and brain tumor incidence.[101]

In an epidemiologic study performed at Oxford, England, it was found that maternal occupational (such as sewing machinist, textile industry worker) exposure to extremely low-frequency electromagnetic radiation during pregnancy had no impact on cancer incidence in children.[102] A similar observation was made in another study.[103] Exposure to extremely low-frequency electromagnetic fields (50 Hz) was not associated with an increased risk of cancer, but it was associated with an increased risk of amyotrophic lateral sclerosis (ALS).[104]

It is known that epidemiologic studies reveal an association rather than a causal relationship between cell phone use and cancer risk. A causal relationship can only be established by intervention studies in which humans are exposed to high-frequency electromagnetic radiation emitting cell phones (about 1900 MHz) daily for a specified period of time (20–30 min/day) for at least 10 years, and then the incidence of brain tumors is determined during a follow-up period of an additional 10 years. Unfortunately, the results of epidemiologic studies are often propagated by the media and some professionals as a causal relationship between cell phone use and adverse health effects. At present, most epidemiologic studies support the view that the excessive use of cell phones may increase the risk of brain cancer; however, some have questioned this conclusion. Epidemiologic studies can be more conclusive if they are supported at least by the results of laboratory investigation. Very few laboratory studies have been published on the effect of electromagnetic radiation emitting cell phones on cellular carcinogenesis. At this time, there are not sufficient data to make any definitive conclusion about the health risk of cell phone use; therefore, additional intervention studies with appropriate daily exposure time and radio frequency should be performed. Because of long latent time periods between exposures to radio-frequency electromagnetic radiation–emitting cell phones and the development of adverse health effects, especially cancer, and because of its potential interaction with other carcinogenic and mutagenic agents during this period, conclusive data from the epidemiologic studies alone are difficult to obtain.

The current controversies regarding the effects of cell phone use on cancer risk are analogous to those encountered in early 1960s on the carcinogenic potential of low-dose ionizing radiation. The denial that low-dose ionizing radiation is not carcinogenic persisted for decades until recently when it was accepted as a human carcinogen by federal agencies. The author hopes that the controversy regarding the effects of cell phone use on human health will be settled sooner because of its potential health implications around the world involving billions of people. Based on the analysis of several epidemiologic studies, the World Health Organization (WHO) in May, 2011, stated that cell phone usage may be a "possible human carcinogen." Younger individuals may be more sensitive to the overuse of cell phones than adults for the development of brain tumors. At this time, it is not necessary or prudent to recommend any limitation on the use of cell phones, but caution should be maintained regarding overuse of this communication technology, especially by younger individuals.

Cell phones and noncancerous disease risk: An epidemiologic study on an Egyptian population living near a cell phone base station revealed that they were at increased risk of developing neuropsychiatric problems such as headache, memory changes, dizziness, tremors, depressive symptoms, and sleep disturbance compared to a control population.[105] In a study involving 4156 young adults (20–24 years of age) the relationship between the use of cell phones and stress, sleep disturbances, and symptoms of depression was evaluated. The results revealed that overuse of cell phones was associated with sleep disturbances and symptoms of depression among men and symptoms of depression among women.[106] Another study reported no effects of cell phone usage on sleeping patterns; however, people residing near the mobile phone base station may suffer from sleep disturbances.[107] The analysis of 196 young adult users of cell phones revealed that chronic stress, low emotional stability, depression, and extraversion were associated with the use of this technology, especially among females at younger ages.[108] Another study reported that exposure to continuous low-frequency electromagnetic fields emitted from the mobile base station did not produce any adverse health effects among people residing near the station.[109] An analysis of the relationship between cell phone use and the incidence of tinnitus suggested that the use of cell phones over a long period of time (4 years or more) is significantly associated with an increased risk of tinnitus.[110] Long-term (more than 3 years) and excessive use of cell phones may cause damage to the cochlea as well as the auditory cortex.[111] Others have observed no such effects from the radio-frequency radiation-emitting cell phones.[112,113]

Cell phones and genetic risk: The laboratory studies with animal and cell culture models are very few. The radio-frequency radiation emitted from a cell phone produced no effect on cancer incidence in mice.[114] Exposure of mammalian cells in culture to 835-MHz radio-frequency electromagnetic radiation field slightly enhanced the levels of chromosomal aberrations induced by a chemical (ethylmethanesulfonate).[115] A review of cytogenetic studies in vitro and in vivo suggests that exposure to radio-frequency emitting cell phones does not cause any genetic or cytogenetic damage in vitro or in vivo.[116] It should be mentioned that radio frequencies and exposure times per day and total exposure periods used in these studies have varied widely.

MECHANISMS OF CELL PHONE–INDUCED DAMAGE AND ITS PROTECTION BY ANTIOXIDANTS

Several animal studies suggest that exposure of guinea pigs to an 1800-MHz electro-magnetic radiation–emitting cell phone for 10 or 20 minutes per day for seven days increased the levels of malondialdehyde (MDA) and nitric oxide, and decreased the activities of superoxide dismutase (SOD), myeloperoxidase, and glutathione peroxi-dase in the liver.[117] In another study, using rats, it was demonstrated that exposure to a 900-MHz electromagnetic radiation–emitting cell phone increased oxidative stress in the kidney; however, pretreatment with melatonin or caffeic acid phenethyl ester protected the kidney from electromagnetic radiation–induced damage.[118] Elecromagnetic radiation–emitting cell phones also induced increased oxidative damage in the brain of rats, and pretreatment with melatonin significantly reduced oxidative damage.[119] It has been demonstrated that exposure to a 900-MHz radio frequency–emitting cell phone for 30 min/day for a period of 30 days increased oxidative stress in the endometrium of rats and caused histological changes, includ-ing diffuse and severe apoptosis and diffuse eosinophilic leukocyte and lymphocyte infiltration, and pretreatment with vitamin C and vitamin E reduced oxidative dam-age as well as improved the histological profile.[120]

INFRARED RADIATION (IR)

The International Commission on Illumination (CIE) recommends the division of infrared radiation into the following bands with specific wavelengths: (Henderson, Roy, Wave length considerations, http://web.archive.org/web/20071028072)

IR-A: 0.7 µm–1.4 µm
IR-B: 1.4 µm–3 µm
IR-C: 3 µm–1000 µm

Astronomers divide the infrared spectrum into the following three groups with specific wave lengths: near-, mid-, and far-infrared (see http://www.ipac.caltech.edu/Outreach/Edu/Regions/irregions.html).

Near-infrared (NIR): (0.7–1 µm) –3 µm
Mid-infrared (MIR): 5 µm–(25–40 µm)
Far-infrared (FIR): (25–40 µm) to (200–350) µm

Infrared imaging is widely used for military and civilian purposes. Military applications include target acquisition, surveillance, night vision glasses, homing, and tracking. Civilian uses include thermal efficiency analysis, remote temperature sensing, short-range wireless communication, spectroscopy, and weather forecasting. In addition, astronomers use infrared sensor–equipped telescopes to penetrate dusty regions of space such as molecular clouds, detect objects such as planets, and to view highly red-shifted objects from the early days of the universe (see IR Astronomy: Overview, http://www.ipac.caltech.edu/Outreach/Edu/importance.html).

BIOLOGICAL EFFECTS OF INFRARED RADIATION

Very little information is available on the biological effects of infrared on humans or on animals. A review on the effects of infrared radiation and heat on the skin has shown that exposure to infrared radiation and heat individually induces cutaneous angiogenesis and inflammatory reactions, disrupts the dermal extracellular matrix by inducing MMPs, and alters dermal structural proteins. These changes can lead to premature aging of the skin.[121] Infrared radiation produces free radicals in the skin.[122] It has been proposed that supplementation with antioxidants may reduce infrared radiation–induced damage to the skin.[123] It was reported that even a lower 50 MHz of infrared irradiation increased markers of oxidative damage in the liver of guinea pigs, which was prevented by prior administration of NAC and epigallocatechin (EGCG).[124] The extent of increase in oxidative stress was dependent upon the duration of infrared radiation exposure. NAC and EGCG treatment reduced infrared radiation–induced oxidative damage in the liver of guinea pigs.[124]

EFFECT OF MAGNETIC FIELD

The strength of magnetic fields is expressed in teslas (1 tesla = 10,000 gauss). The strength of magnetic resonance imaging (MRI) may vary from 0.35 to 1.5 tesla. Most MRIs have a strength of 1.5 tesla. The MRI fields can be divided into three categories:

Low-field MRI: Under 0.2 tesla
Mid-field MRI: 0.2 to 0.6 tesla
High-field MRI: 1.0 to 1.5 tesla

In an epidemiologic study performed in Germany, it was observed that parental exposure to low-field MRI above 0.2 microtesla had no impact on cancer incidence among offspring.[125] Exposure of pregnant mice (genetically predisposed to develop eye abnormalities) to a 1.5-tesla magnetic field produced 15–37% eye malformation in comparison to control animals, which produced 2–19% abnormal eyes.[126] It has been reported that exposure to a 14.1-tesla magnetic field strength caused impairment in the vestibular apparatus of rats.[127] Short-term exposure to static magnetic fields (SMF) produced the following acute effects in humans.[128]

1. Vertigo, nausea, and a metallic taste, which may interfere in performance at work.
2. Changes in blood pressure and heart rate within the range of physiological variability after an exposure up to 8 tesla.
3. Induction of ectopic heartbeats and increased likelihood of reversible arrhythmia (possibly leading to ventricular fibrillation) in susceptible workers.
4. A decrease in working memory and eye–hand coordination in a dose-dependent manner after exposure to 1.5 to 3 tesla SMF.
5. Ability to perform intricate procedures may be impaired.

However, the authors concluded that no firm conclusions can be drawn on the effects of SMF on the above clinical end points because of the limitations of the available studies. Most studies suggest that there are no significant effects of 1.5-tesla SMFs on human health.

SUMMARY

All radiation is a form of energy with a wide electromagnetic spectrum. The electromagnetic field is characterized by its frequency or its corresponding wavelength. Electromagnetic radiation can be grouped in two major classes: ionizing radiation and nonionizing radiation. Ionizing radiation has a higher frequency and shorter wavelength and enough energy to knock out one electron from one of the orbits of the atom. Nonionizing radiation has a lower frequency and higher wavelength and does not have enough energy to remove an electron from one of the orbits of the atom. Nonionizing radiation includes ultraviolet (UV) radiation, infrared, visible light, sound waves, and radio waves, each of which is associated with a defined wavelength. The biological effects of nonionizing radiation have been extensively investigated with respect to UV radiation and electromagnetic radiation emitted from cell phones; however, some studies have been performed with infrared radiation and magnetic fields. UV radiation has been classified into three groups UVA, UVB, and UVC. UVA radiation has a wavelength in the range of 320–400 nm, and is not absorbed by the ozone layer. UVB has a wavelength in the range of 290–320 nm and is mostly absorbed by the ozone layer. UVC has a wavelength in the range of 100–290 nm and is completely absorbed by the ozone layer and atmosphere. Exposure to UVA and UVB radiation increases the risk of photoaging (premature aging), skin cancers (cutaneous melanoma, basal cell carcinoma, and squamous cell carcinoma), multiple sclerosis, mutations, and immune suppression. The combination of appropriately prepared sunscreen and antioxidants can provide protection against UV radiation–induced skin damage more than that produced by the individual agents.

Cell or mobile phone technology and its use throughout the world has exploded during last decade. The cell phone converts voices to impulses that are transmitted over radio waves at frequencies ranging from about 800 to over 1900 MHz. The effect of radio-frequency emitting cell phones on brain cancer has been the subject of intensive debate. Most epidemiologic studies showed that excessive use of cell phones over a long period of time was associated with the increased risk of brain cancers, such as glioma and acoustic neuroma; other studies failed to observe such an association. Increased oxidative damage can enhance the risk of cancer. The fact that animal studies showed that the electromagnetic radiation emitted from cell phones increased oxidative damage in the brain, which can be protected by pretreatment with antioxidants such as vitamin C, vitamin E, and N-acetylcysteine (NAC), suggests that there may be a causal relationship between overuse of cell phones and increased risk of brain cancers. Therefore, caution should be maintained regarding the overuse of this communication technology, especially by younger individuals because of their high sensitivity to electromagnetic radiation.

Very few studies have been performed on the effects of infrared and magnetic fields on human health. Exposure to infrared radiation can induce premature aging of the skin by generating free radicals and causing inflammation; however, supplementation with antioxidants reduced the adverse effects of infrared radiation on the skin. Magnetic field strength commonly used in MRIs has no significant long-term adverse health effects on humans.

Based on studies published, I propose that oral supplementation with a micronutrient preparation containing multiple dietary and endogenous antioxidants together with a topical application of sunscreen before exposure to UV radiation or infrared radiation may protect against the risk of developing premature aging, skin cancer, and other adverse health effects more than that produced by the individual agents. In addition, an oral supplementation with a similar micronutrient preparation may reduce the potential risk of excessive use of cell phones on brain cancers.

REFERENCES

1. Pavlou, P., Rallis, M., Deliconstantinos, G., Papaioannou, G., and Grando, S. A., In-vivo data on the influence of tobacco smoke and UV light on murine skin, *Toxicol Ind Health* 25 (4–5), 231–239, 2009.
2. Naylor, E. C., Watson, R. E., and Sherratt, M. J., Molecular aspects of skin ageing, *Maturitas* 69 (3), 249–256, 2011.
3. Kossodo, S., Wong, W. R., Simon, G., and Kochevar, I. E., Effects of UVR and UVR-induced cytokines on production of extracellular matrix proteins and proteases by dermal fibroblasts cultured in collagen gels, *Photochem Photobiol* 79 (1), 86–93, 2004.
4. Rijken, F. and Bruijnzeel, P. L., The pathogenesis of photoaging: The role of neutrophils and neutrophil-derived enzymes, *J Investig Dermatol Symp Proc* 14 (1), 67–72, 2009.
5. Krutmann, J., and Schroeder, P., Role of mitochondria in photoaging of human skin: The defective powerhouse model, *J Investig Dermatol Symp Proc* 14 (1), 44–49, 2009.
6. Haylett, A. K., Nie, Z., Brownrigg, M., Taylor, R., and Rhodes, L. E., Systemic photoprotection in solar urticaria with alpha-melanocyte-stimulating hormone analogue [Nle4-D-Phe7]-alpha-MSH, *Br J Dermatol* 164 (2), 407–414, 2011.
7. Schmitz, S., Garbe, C., Tebbe, B., and Orfanos, C. E., Long-wave ultraviolet radiation (UVA) and skin cancer [in German], *Hautarzt* 45 (8), 517–525, 1994.
8. De Fabo, E. C., Noonan, F. P., Fears, T., and Merlino, G., Ultraviolet B but not ultraviolet A radiation initiates melanoma, *Cancer Res* 64 (18), 6372–6376, 2004.
9. Balk, S. J., Ultraviolet radiation: A hazard to children and adolescents, *Pediatrics* 127 (3), e791–817, 2011.
10. Wolnicka-Glubisz, A., and Noonan, F. P., Neonatal susceptibility to UV induced cutaneous malignant melanoma in a mouse model, *Photochem Photobiol Sci* 5 (2), 254–260, 2006.
11. Kennedy, C., Bajdik, C. D., Willemze, R., De Gruijl, F. R., and Bouwes Bavinck, J. N., The influence of painful sunburns and lifetime sun exposure on the risk of actinic keratoses, seborrheic warts, melanocytic nevi, atypical nevi, and skin cancer, *J Invest Dermatol* 120 (6), 1087–1093, 2003.
12. Dulon, M., Weichenthal, M., Blettner, M., Breitbart, M., Hetzer, M., Greinert, R., Baumgardt-Elms, C., and Breitbart, E. W., Sun exposure and number of nevi in 5- to 6-year-old European children, *J Clin Epidemiol* 55 (11), 1075–1081, 2002.

13. Bauer, A., Diepgen, T. L., and Schmitt, J., Is occupational solar UV-irradiation a relevant risk factor for basal cell carcinoma? A systematic review and meta-analysis of the epidemiologic literature, *Br J Dermatol* 165 (3), 612–625, 2011.

14. Schmitt, J., Diepgen, T., and Bauer, A., Occupational exposure to non-artificial UV-light and non-melanocytic skin cancer: A systematic review concerning a new occupational disease, *J Dtsch Dermatol Ges* 8 (4), 250–263, 250–264, 2010.

15. Barysch, M. J., Hofbauer, G. F., and Dummer, R., Vitamin D, ultraviolet exposure, and skin cancer in the elderly, *Gerontology* 56 (4), 410–413, 2010.

16. Mohr, S. B., Garland, C. F., Gorham, E. D., Grant, W. B., and Garland, F. C., Ultraviolet B irradiance and vitamin D status are inversely associated with incidence rates of pancreatic cancer worldwide, *Pancreas* 39 (5), 669–674, 2010.

17. Mohr, S. B., Garland, C. F., Gorham, E. D., Grant, W. B., and Garland, F. C., Ultraviolet B irradiance and incidence rates of bladder cancer in 174 countries, *Am J Prev Med* 38 (3), 296–302, 2010.

18. Mohr, S. B., Gorham, E. D., Garland, C. F., Grant, W. B., and Garland, F. C., Low ultraviolet B and increased risk of brain cancer: An ecological study of 175 countries, *Neuroepidemiology* 35 (4), 281–290, 2010.

19. International Agency for Research on Cancer Working Group on Artificial Ultraviolet (UV) Light and Skin Cancer, The association of use of sunbeds with cutaneous malignant melanoma and other skin cancers: A systematic review, *Int J Cancer* 120, 1116–1122, 2007.

20. Mogensen, M., and Jemec, G. B., The potential carcinogenic risk of tanning beds: Clinical guidelines and patient safety advice, *Cancer Manag Res* 2, 277–282, 2010.

21. Choi, K., Lazovich, D., Southwell, B., Forster, J., Rolnick, S. J., and Jackson, J., Prevalence and characteristics of indoor tanning use among men and women in the United States, *Arch Dermatol* 146 (12), 1356–1361, 2010.

22. Orton, S. M., Wald, L., Confavreux, C., Vukusic, S., Krohn, J. P., Ramagopalan, S. V., Herrera, B. M., Sadovnick, A. D., and Ebers, G. C., Association of UV radiation with multiple sclerosis prevalence and sex ratio in France, *Neurology* 76 (5), 425–431, 2011.

23. Staples, J., Ponsonby, A. L., and Lim, L., Low maternal exposure to ultraviolet radiation in pregnancy, month of birth, and risk of multiple sclerosis in offspring: Longitudinal analysis, *BMJ* 340, c1640, 2010.

24. Pierrot-Deseilligny, C., and Souberbielle, J. C., Is hypovitaminosis D one of the environmental risk factors for multiple sclerosis?, *Brain* 133 (Pt 7), 1869–1888, 2010.

25. Morison, W. L., Effects of ultraviolet radiation on the immune system in humans, *Photochem Photobiol* 50 (4), 515–524, 1989.

26. Nishigori, C., Yarosh, D. B., Donawho, C., and Kripke, M. L., The immune system in ultraviolet carcinogenesis, *J Investig Dermatol Symp Proc* 1 (2), 143–146, 1996.

27. Ng, R. L., Bisley, J. L., Gorman, S., Norval, M., and Hart, P. H., Ultraviolet irradiation of mice reduces the competency of bone marrow-derived CD11c+ cells via an indomethacin-inhibited pathway, *J Immunol* 185 (12), 7207–7215, 2010.

28. Pfeifer, G. P., You, Y. H., and Besaratinia, A., Mutations induced by ultraviolet light, *Mutat Res* 571 (1–2), 19–31, 2005.

29. Ougolkov, A., Zhang, B., Yamashita, K., Bilim, V., Mai, M., Fuchs, S. Y., and Minamoto, T., Associations among beta-TrCP, an E3 ubiquitin ligase receptor, beta-catenin, and NF-kappaB in colorectal cancer, *J Natl Cancer Inst* 96 (15), 1161–1170, 2004.

30. Bhatia, N., Demmer, T. A., Sharma, A. K., Elcheva, I., and Spiegelman, V. S., Role of beta-TrCP ubiquitin ligase receptor in UVB mediated responses in skin, *Arch Biochem Biophys* 508 (2), 178–184, 2011.

31. Herrling, T., Fuchs, J., Rehberg, J., and Groth, N., UV-induced free radicals in the skin detected by ESR spectroscopy and imaging using nitroxides, *Free Radic Biol Med* 35 (1), 59–67, 2003.

32. Emri, G., Horkay, I., and Remenyik, E., The role of free radicals in the UV-induced skin damage: Photo-aging [in Hungarian], *Orv Hetil* 147 (16), 731–735, 2006.

33. Zaidi, M. R., Davis, S., Noonan, F. P., Graff-Cherry, C., Hawley, T. S., Walker, R. L., Feigenbaum, L., Fuchs, E., Lyakh, L., Young, H. A., Hornyak, T. J., Arnheiter, H., Trinchieri, G., Meltzer, P. S., De Fabo, E. C., and Merlino, G., Interferon-gamma links ultraviolet radiation to melanomagenesis in mice, *Nature* 469 (7331), 548–553, 2011.

34. Ivanchuk, S. M., Mondal, S., Dirks, P. B., and Rutka, J. T., The INK4A/ARF locus: Role in cell cycle control and apoptosis and implications for glioma growth, *J Neurooncol* 51 (3), 219–229, 2001.

35. Serrano, M., The INK4a/ARF locus in murine tumorigenesis, *Carcinogenesis* 21 (5), 865–869, 2000.

36. Recio, J. A., Noonan, F. P., Takayama, H., Anver, M. R., Duray, P., Rush, W. L., Lindner, G., De Fabo, E. C., DePinho, R. A., and Merlino, G., Ink4a/arf deficiency promotes ultraviolet radiation-induced melanomagenesis, *Cancer Res* 62 (22), 6724–6730, 2002.

37. Yang, G., Curley, D., Bosenberg, M. W., and Tsao, H., Loss of xeroderma pigmentosum C (Xpc) enhances melanoma photocarcinogenesis in Ink4a-Arf-deficient mice, *Cancer Res* 67 (12), 5649–5657, 2007.

38. Hirota, A., Kawachi, Y., Yamamoto, M., Koga, T., Hamada, K., and Otsuka, F., Acceleration of UVB-induced photoageing in nrf2 gene-deficient mice, *Exp Dermatol* 20 (8), 664–668, 2011.

39. Kawachi, Y., Xu, X., Taguchi, S., Sakurai, H., Nakamura, Y., Ishii, Y., Fujisawa, Y., Furuta, J., Takahashi, T., Itoh, K., Yamamoto, M., Yamazaki, F., and Otsuka, F., Attenuation of UVB-induced sunburn reaction and oxidative DNA damage with no alterations in UVB-induced skin carcinogenesis in Nrf2 gene-deficient mice, *J Invest Dermatol* 128 (7), 1773–1779, 2008.

40. Hirota, A., Kawachi, Y., Itoh, K., Nakamura, Y., Xu, X., Banno, T., Takahashi, T., Yamamoto, M., and Otsuka, F., Ultraviolet A irradiation induces NF-E2-related factor 2 activation in dermal fibroblasts: Protective role in UVA-induced apoptosis, *J Invest Dermatol* 124 (4), 825–832, 2005.

41. Tomas, D., Apoptosis, UV-radiation, precancerous and skin tumors [in Croatian], *Acta Med Croatica* 63 (Suppl 2), 53–58, 2009.

42. Gasparro, F. P., Mitchnick, M., and Nash, J. F., A review of sunscreen safety and efficacy, *Photochem Photobiol* 68 (3), 243–256, 1998.

43. Burnett, M. E., and Wang, S. Q., Current sunscreen controversies: A critical review, *Photodermatol Photoimmunol Photomed* 27 (2), 58–67, 2011.

44. Wu, Y., Matsui, M. S., Chen, J. Z., Jin, X., Shu, C. M., Jin, G. Y., Dong, G. H., Wang, Y. K., Gao, X. H., Chen, H. D., and Li, Y. H., Antioxidants add protection to a broad-spectrum sunscreen, *Clin Exp Dermatol* 36 (2), 178–187, 2011.

45. Inui, M., Ooe, M., Fujii, K., Matsunaka, H., Yoshida, M., and Ichihashi, M., Mechanisms of inhibitory effects of CoQ10 on UVB-induced wrinkle formation in vitro and in vivo, *Biofactors* 32 (1–4), 237–243, 2008.

46. Yue, Y., Zhou, H., Liu, G., Li, Y., Yan, Z., and Duan, M., The advantages of a novel CoQ_{10} delivery system in skin photo-protection, *Int J Pharm* 392 (1–2), 57–63, 2010.

47. Fuller, B., Smith, D., Howerton, A., and Kern, D., Anti-inflammatory effects of CoQ_{10} and colorless carotenoids, *J Cosmet Dermatol* 5 (1), 30–38, 2006.

48. Cotter, M. A., Thomas, J., Cassidy, P., Robinette, K., Jenkins, N., Florell, S. R., Leachman, S., Samlowski, W. E., and Grossman, D., N-acetylcysteine protects melanocytes against oxidative stress/damage and delays onset of ultraviolet-induced melanoma in mice, *Clin Cancer Res* 13 (19), 5952–5958, 2007.

49. Goodson, A. G., Cotter, M. A., Cassidy, P., Wade, M., Florell, S. R., Liu, T., Boucher, K. M., and Grossman, D., Use of oral N-acetylcysteine for protection of melanocytic nevi against UV-induced oxidative stress: Towards a novel paradigm for melanoma chemoprevention, *Clin Cancer Res* 15 (23), 7434–7440, 2009.
50. Morley, N., Curnow, A., Salter, L., Campbell, S., and Gould, D., N-acetyl-L-cysteine prevents DNA damage induced by UVA, UVB and visible radiation in human fibroblasts, *J Photochem Photobiol B* 72 (1–3), 55–60, 2003.
51. Burke, K. E., Interaction of vitamins C and E as better cosmeceuticals, *Dermatol Ther* 20 (5), 314–321, 2007.
52. Krol, E. S., Kramer-Stickland, K. A., and Liebler, D. C., Photoprotective actions of topically applied vitamin E, *Drug Metab Rev* 32 (3–4), 413–420, 2000.
53. Maalouf, S., El-Sabban, M., Darwiche, N., and Gali-Muhtasib, H., Protective effect of vitamin E on ultraviolet B light-induced damage in keratinocytes, *Mol Carcinog* 34 (3), 121–130, 2002.
54. Dreher, F., Gabard, B., Schwindt, D. A., and Maibach, H. I., Topical melatonin in combination with vitamins E and C protects skin from ultraviolet-induced erythema: A human study in vivo, *Br J Dermatol* 139 (2), 332–339, 1998.
55. Rozman, B., Gasperlin, M., Tinois-Tessoneaud, E., Pirot, F., and Falson, F., Simultaneous absorption of vitamins C and E from topical microemulsions using reconstructed human epidermis as a skin model, *Eur J Pharm Biopharm* 72 (1), 69–75, 2009.
56. Konger, R. L., A new wrinkle on topical vitamin E and photo-inflammation: Mechanistic studies of a hydrophilic gamma-tocopherol derivative compared with alpha-tocopherol, *J Invest Dermatol* 126 (7), 1447–1449, 2006.
57. Hochberg, M., Kohen, R., and Enk, C. D., Role of antioxidants in prevention of pyrimidine dimer formation in UVB irradiated human HaCaT keratinocytes, *Biomed Pharmacother* 60 (5), 233–237, 2006.
58. Greul, A. K., Grundmann, J. U., Heinrich, F., Pfitzner, I., Bernhardt, J., Ambach, A., Biesalski, H. K., and Gollnick, H., Photoprotection of UV-irradiated human skin: An antioxidative combination of vitamins E and C, carotenoids, selenium and proanthocyanidins, *Skin Pharmacol Appl Skin Physiol* 15 (5), 307–315, 2002.
59. Overvad, K., Thorling, E. B., Bjerring, P., and Ebbesen, P., Selenium inhibits UV-light-induced skin carcinogenesis in hairless mice, *Cancer Lett* 27 (2), 163–170, 1985.
60. Burke, K. E., Burford, R. G., Combs, G. F., Jr., French, I. W., and Skeffington, D. R., The effect of topical L-selenomethionine on minimal erythema dose of ultraviolet irradiation in humans, *Photodermatol Photoimmunol Photomed* 9 (2), 52–57, 1992.
61. Rafferty, T. S., McKenzie, R. C., Hunter, J. A., Howie, A. F., Arthur, J. R., Nicol, F., and Beckett, G. J., Differential expression of selenoproteins by human skin cells and protection by selenium from UVB-radiation-induced cell death, *Biochem J* 332 (Pt 1), 231–236, 1998.
62. Burke, K. E., Clive, J., Combs, G. F. Jr., and Nakamura, R. M., Effects of topical L-selenomethionine with topical and oral vitamin E on pigmentation and skin cancer induced by ultraviolet irradiation in Skh:2 hairless mice, *J Am Acad Dermatol* 49 (3), 458–472, 2003.
63. Rizwan, M., Rodriguez-Blanco, I., Harbottle, A., Birch-Machin, M. A., Watson, R. E., and Rhodes, L. E., Tomato paste rich in lycopene protects against cutaneous photodamage in humans in vivo: A randomized controlled trial, *Br J Dermatol* 164 (1), 154–162, 2011.
64. Aust, O., Stahl, W., Sies, H., Tronnier, H., and Heinrich, U., Supplementation with tomato-based products increases lycopene, phytofluene, and phytoene levels in human serum and protects against UV-light-induced erythema, *Int J Vitam Nutr Res* 75 (1), 54–60, 2005.

65. Stahl, W., Heinrich, U., Aust, O., Tronnier, H., and Sies, H., Lycopene-rich products and dietary photoprotection, *Photochem Photobiol Sci* 5 (2), 238–242, 2006.
66. Bando, N., Hayashi, H., Wakamatsu, S., Inakuma, T., Miyoshi, M., Nagao, A., Yamauchi, R., and Terao, J., Participation of singlet oxygen in ultraviolet-A-induced lipid peroxidation in mouse skin and its inhibition by dietary beta-carotene: An ex vivo study, *Free Radic Biol Med* 37 (11), 1854–1863, 2004.
67. Wong, W. C., Wu, J. Y., and Benzie, I. F., Photoprotective potential of *Cordyceps* polysaccharides against ultraviolet B radiation-induced DNA damage to human skin cells, *Br J Dermatol* 164 (5), 980–986, 2011.
68. Katiyar, S. K., Green tea prevents non-melanoma skin cancer by enhancing DNA repair, *Arch Biochem Biophys* 508 (2), 152–158, 2011.
69. Nichols, J. A., and Katiyar, S. K., Skin photoprotection by natural polyphenols: Anti-inflammatory, antioxidant and DNA repair mechanisms, *Arch Dermatol Res* 302 (2), 71–83, 2010.
70. Chan, C. F., Lien, C. Y., Lai, Y. C., Huang, C. L., and Liao, W. C., Influence of purple sweet potato extracts on the UV absorption properties of a cosmetic cream, *J Cosmet Sci* 61 (5), 333–341, 2010.
71. Verschooten, L., Smaers, K., Van Kelst, S., Proby, C., Maes, D., Declercq, L., Agostinis, P., and Garmyn, M., The flavonoid luteolin increases the resistance of normal, but not malignant keratinocytes, against UVB-induced apoptosis, *J Invest Dermatol* 130 (9), 2277–2285, 2010.
72. Lin, J. Y., Tournas, J. A., Burch, J. A., Monteiro-Riviere, N. A., and Zielinski, J., Topical isoflavones provide effective photoprotection to skin, *Photodermatol Photoimmunol Photomed* 24 (2), 61–66, 2008.
73. Chiu, T. M., Huang, C. C., Lin, T. J., Fang, J. Y., Wu, N. L., and Hung, C. F., In vitro and in vivo anti-photoaging effects of an isoflavone extract from soybean cake, *J Ethnopharmacol* 126 (1), 108–113, 2009.
74. Widyarini, S., Spinks, N., Husband, A. J., and Reeve, V. E., Isoflavonoid compounds from red clover (*Trifolium pratense*) protect from inflammation and immune suppression induced by UV radiation, *Photochem Photobiol* 74 (3), 465–470, 2001.
75. Afaq, F., Zaid, M. A., Khan, N., Dreher, M., and Mukhtar, H., Protective effect of pomegranate-derived products on UVB-mediated damage in human reconstituted skin, *Exp Dermatol* 18 (6), 553–561, 2009.
76. Vaid, M., and Katiyar, S. K., Molecular mechanisms of inhibition of photocarcinogenesis by silymarin, a phytochemical from milk thistle (*Silybum marianum* L. Gaertn.) (Review), *Int J Oncol* 36 (5), 1053–1060, 2010.
77. Reeve, V. E., Allanson, M., Arun, S. J., Domanski, D., and Painter, N., Mice drinking goji berry juice (*Lycium barbarum*) are protected from UV radiation-induced skin damage via antioxidant pathways, *Photochem Photobiol Sci* 9 (4), 601–607, 2010.
78. Liu, Y., Chan, F., Sun, H., Yan, J., Fan, D., Zhao, D., An, J., and Zhou, D., Resveratrol protects human keratinocytes HaCaT cells from UVA-induced oxidative stress damage by downregulating Keap1 expression, *Eur J Pharmacol* 650 (1), 130–137, 2011.
79. Vostalova, J., Zdarilova, A., and Svobodova, A., *Prunella vulgaris* extract and rosmarinic acid prevent UVB-induced DNA damage and oxidative stress in HaCaT keratinocytes, *Arch Dermatol Res* 302 (3), 171–181, 2010.
80. Fahlman, B. M., and Krol, E. S., UVA and UVB radiation-induced oxidation products of quercetin, *J Photochem Photobiol B* 97 (3), 123–131, 2009.
81. Scalia, S., and Mezzena, M., Photostabilization effect of quercetin on the UV filter combination, butyl methoxydibenzoylmethane-octyl methoxycinnamate, *Photochem Photobiol* 86 (2), 273–278, 2010.
82. Fahlman, B. M., and Krol, E. S., Inhibition of UVA and UVB radiation-induced lipid oxidation by quercetin, *J Agric Food Chem* 57 (12), 5301–5305, 2009.

83. Kimura, S., Warabi, E., Yanagawa, T., Ma, D., Itoh, K., Ishii, Y., Kawachi, Y., and Ishii, T., Essential role of Nrf2 in keratinocyte protection from UVA by quercetin, *Biochem Biophys Res Commun* 387 (1), 109–114, 2009.
84. Sime, S., and Reeve, V. E., Protection from inflammation, immunosuppression and carcinogenesis induced by UV radiation in mice by topical Pycnogenol, *Photochem Photobiol* 79 (2), 193–198, 2004.
85. Oresajo, C., Stephens, T., Hino, P. D., Law, R. M., Yatskayer, M., Foltis, P., Pillai, S., and Pinnell, S. R., Protective effects of a topical antioxidant mixture containing vitamin C, ferulic acid, and phloretin against ultraviolet-induced photodamage in human skin, *J Cosmet Dermatol* 7 (4), 290–297, 2008.
86. Oresajo, C., Yatskayer, M., Galdi, A., Foltis, P., and Pillai, S., Complementary effects of antioxidants and sunscreens in reducing UV-induced skin damage as demonstrated by skin biomarker expression, *J Cosmet Laser Ther* 12 (3), 157–162, 2010.
87. Darr, D., Dunston, S., Faust, H., and Pinnell, S., Effectiveness of antioxidants (vitamin C and E) with and without sunscreens as topical photoprotectants, *Acta Derm Venereol* 76 (4), 264–268, 1996.
88. Dreher, F. and Maibach, H., Protective effects of topical antioxidants in humans, *Curr Probl Dermatol* 29, 157–164, 2001.
89. Han, Y. Y., Kano, H., Davis, D. L., Niranjan, A., and Lunsford, L. D., Cell phone use and acoustic neuroma: The need for standardized questionnaires and access to industry data, *Surg Neurol* 72 (3), 216–222, 2009.
90. Kan, P., Simonsen, S. E., Lyon, J. L., and Kestle, J. R., Cellular phone use and brain tumor: A meta-analysis, *J Neurooncol* 86 (1), 71–78, 2008.
91. Hardell, L., and Sage, C., Biological effects from electromagnetic field exposure and public exposure standards, *Biomed Pharmacother* 62 (2), 104–109, 2008.
92. Hardell, L., Carlberg, M., Soderqvist, F., Mild, K. H., and Morgan, L. L., Long-term use of cellular phones and brain tumours: Increased risk associated with use for > or =10 years, *Occup Environ Med* 64 (9), 626–632, 2007.
93. Hardell, L., Carlberg, M., Soderqvist, F., and Hansson Mild, K., Meta-analysis of long-term mobile phone use and the association with brain tumours, *Int J Oncol* 32 (5), 1097–1103, 2008.
94. Takebayashi, T., Akiba, S., Kikuchi, Y., Taki, M., Wake, K., Watanabe, S., and Yamaguchi, N., Mobile phone use and acoustic neuroma risk in Japan, *Occup Environ Med* 63 (12), 802–807, 2006.
95. Croft, R. J., McKenzie, R. J., Inyang, I., Benke, G. P., Anderson, V., and Abramson, M. J., Mobile phones and brain tumours: A review of epidemiological research, *Australas Phys Eng Sci Med* 31 (4), 255–267, 2008.
96. Sato, Y., Akiba, S., Kubo, O., and Yamaguchi, N., A case-case study of mobile phone use and acoustic neuroma risk in Japan, *Bioelectromagnetics* 32 (2), 85–93, 2011.
97. Sadetzki, S., Chetrit, A., Jarus-Hakak, A., Cardis, E., Deutch, Y., Duvdevani, S., Zultan, A., Novikov, I., Freedman, L., and Wolf, M., Cellular phone use and risk of benign and malignant parotid gland tumors: A nationwide case-control study, *Am J Epidemiol* 167 (4), 457–467, 2008.
98. de Vocht, F., Burstyn, I., and Cherrie, J. W., Time trends (1998–2007) in brain cancer incidence rates in relation to mobile phone use in England, *Bioelectromagnetics* 32 (5), 334–339, 2011.
99. Ahlbom, A., Feychting, M., Green, A., Kheifets, L., Savitz, D. A., and Swerdlow, A. J., Epidemiologic evidence on mobile phones and tumor risk: A review, *Epidemiology* 20 (5), 639–652, 2009.
100. Collaborators, I. S. G., Brain tumor risk in relation to mobile telephone use: Results of the Interphone international case-control study, *Int J Epidemiol* 39, 675–694, 2010.

101. Lehrer, S., Green, S., and Stock, R. G., Association between number of cell phone contracts and brain tumor incidence in nineteen U.S. States, *J Neurooncol* 101 (3), 505–507, 2011.
102. Sorahan, T., Hamilton, L., Gardiner, K., Hodgson, J. T., and Harrington, J. M., Maternal occupational exposure to electromagnetic fields before, during, and after pregnancy in relation to risks of childhood cancers: Findings from the Oxford Survey of Childhood Cancers, 1953–1981 deaths, *Am J Ind Med* 35 (4), 348–357, 1999.
103. Elliott, P., Toledano, M. B., Bennett, J., Beale, L., de Hoogh, K., Best, N., and Briggs, D. J., Mobile phone base stations and early childhood cancers: Case-control study, *BMJ* 340, c3077, 2010.
104. Johansen, C., Electromagnetic fields and health effects: Epidemiologic studies of cancer, diseases of the central nervous system and arrhythmia-related heart disease, *Scand J Work Environ Health* 30 Suppl 1, 1–30, 2004.
105. Abdel-Rassoul, G., El-Fateh, O. A., Salem, M. A., Michael, A., Farahat, F., El-Batanouny, M., and Salem, E., Neurobehavioral effects among inhabitants around mobile phone base stations, *Neurotoxicology* 28 (2), 434–440, 2007.
106. Thomee, S., Harenstam, A., and Hagberg, M., Mobile phone use and stress, sleep disturbances, and symptoms of depression among young adults: A prospective cohort study, *BMC Public Health* 11, 66, 2011.
107. Danker-Hopfe, H., Dorn, H., Bornkessel, C., and Sauter, C., Do mobile phone base stations affect sleep of residents? Results from an experimental double-blind sham-controlled field study, *Am J Hum Biol* 22 (5), 613–618, 2010.
108. Augner, C., and Hacker, G. W., Associations between problematic mobile phone use and psychological parameters in young adults, *Int J Public Health*, In press.
109. Berg-Beckhoff, G., Blettner, M., Kowall, B., Breckenkamp, J., Schlehofer, B., Schmiedel, S., Bornkessel, C., Reis, U., Potthoff, P., and Schuz, J., Mobile phone base stations and adverse health effects: Phase 2 of a cross-sectional study with measured radio frequency electromagnetic fields, *Occup Environ Med* 66 (2), 124–130, 2009.
110. Hutter, H. P., Moshammer, H., Wallner, P., Cartellieri, M., Denk-Linnert, D. M., Katzinger, M., Ehrenberger, K., and Kundi, M., Tinnitus and mobile phone use, *Occup Environ Med* 67 (12), 804–808, 2010.
111. Panda, N. K., Modi, R., Munjal, S., and Virk, R. S., Auditory changes in mobile users: Is evidence forthcoming?, *Otolaryngol Head Neck Surg* 144 (4), 581–585, 2011.
112. Sievert, U., Eggert, S., Goltz, S., and Pau, H. W., Effects of electromagnetic fields emitted by cellular phone on auditory and vestibular labyrinth [in German], *Laryngorhinootologie* 86 (4), 264–270, 2007.
113. Balbani, A. P., and Montovani, J. C., Mobile phones: Influence on auditory and vestibular systems, *Braz J Otorhinolaryngol* 74 (1), 125–131, 2008.
114. Tillmann, T., Ernst, H., Ebert, S., Kuster, N., Behnke, W., Rittinghausen, S., and Dasenbrock, C., Carcinogenicity study of GSM and DCS wireless communication signals in B6C3F1 mice, *Bioelectromagnetics* 28 (3), 173–187, 2007.
115. Kim, J. Y., Hong, S. Y., Lee, Y. M., Yu, S. A., Koh, W. S., Hong, J. R., Son, T., Chang, S. K., and Lee, M., In vitro assessment of clastogenicity of mobile-phone radiation (835 MHz) using the alkaline comet assay and chromosomal aberration test, *Environ Toxicol* 23 (3), 319–327, 2008.
116. Verschaeve, L., Juutilainen, J., Lagroye, I., Miyakoshi, J., Saunders, R., de Seze, R., Tenforde, T., van Rongen, E., Veyret, B., and Xu, Z., In vitro and in vivo genotoxicity of radiofrequency fields, *Mutat Res* 705 (3), 252–268, 2010.
117. Ozgur, E., Guler, G., and Seyhan, N., Mobile phone radiation-induced free radical damage in the liver is inhibited by the antioxidants N-acetyl cysteine and epigallocatechin-gallate, *Int J Radiat Biol* 86 (11), 935–45, 2010.

118. Ozguner, F., Oktem, F., Armagan, A., Yilmaz, R., Koyu, A., Demirel, R., Vural, H., and Uz, E., Comparative analysis of the protective effects of melatonin and caffeic acid phenethyl ester (CAPE) on mobile phone-induced renal impairment in rat, *Mol Cell Biochem* 276 (1–2), 31–37, 2005.
119. Sokolovic, D., Djindjic, B., Nikolic, J., Bjelakovic, G., Pavlovic, D., Kocic, G., Krstic, D., Cvetkovic, T., and Pavlovic, V., Melatonin reduces oxidative stress induced by chronic exposure of microwave radiation from mobile phones in rat brain, *J Radiat Res (Tokyo)* 49 (6), 579–586, 2008.
120. Guney, M., Ozguner, F., Oral, B., Karahan, N., and Mungan, T., 900 MHz radiofrequency-induced histopathologic changes and oxidative stress in rat endometrium: Protection by vitamins E and C, *Toxicol Ind Health* 23 (7), 411–420, 2007.
121. Cho, S., Shin, M. H., Kim, Y. K., Seo, J. E., Lee, Y. M., Park, C. H., and Chung, J. H., Effects of infrared radiation and heat on human skin aging in vivo, *J Investig Dermatol Symp Proc* 14 (1), 15–19, 2009.
122. Darvin, M. E., Haag, S., Meinke, M., Zastrow, L., Sterry, W., and Lademann, J., Radical production by infrared A irradiation in human tissue, *Skin Pharmacol Physiol* 23 (1), 40–46, 2010.
123. Darvin, M. E., Haag, S. F., Meinke, M. C., Sterry, W., and Lademann, J., Determination of the influence of IR radiation on the antioxidative network of the human skin, *J Biophotonics* 4 (1–2), 21–29, 2011.
124. Guler, G., Turkozer, Z., Tomruk, A., and Seyhan, N., The protective effects of N-acetyl-L-cysteine and epigallocatechin-3-gallate on electric field-induced hepatic oxidative stress, *Int J Radiat Biol* 84 (8), 669–680, 2008.
125. Hug, K., Grize, L., Seidler, A., Kaatsch, P., and Schuz, J., Parental occupational exposure to extremely low frequency magnetic fields and childhood cancer: A German case-control study, *Am J Epidemiol* 171 (1), 27–35, 2010.
126. Tyndall, D. A., and Sulik, K. K., Effects of magnetic resonance imaging on eye development in the C57BL/6J mouse, *Teratology* 43 (3), 263–275, 1991.
127. Houpt, T. A., Cassell, J. A., Riccardi, C., DenBleyker, M. D., Hood, A., and Smith, J. C., Rats avoid high magnetic fields: Dependence on an intact vestibular system, *Physiol Behav* 92 (4), 741–747, 2007.
128. Franco, G., Perduri, R., and Murolo, A., Health effects of occupational exposure to static magnetic fields used in magnetic resonance imaging: A review [in Italian], *Med Lav* 99 (1), 16–28, 2008.

Index

Printed and bound by CPI Group (UK) Ltd, Croydon, CR0 4YY

21/10/2024

01777089-0009